Advances in Industrial Control

For other titles published in this series, go to
www.springer.com/series/1412

Other titles published in this series:

Digital Controller Implementation and Fragility
Robert S.H. Istepanian and James F. Whidborne (Eds.)

Optimisation of Industrial Processes at Supervisory Level
Doris Sáez, Aldo Cipriano and Andrzej W. Ordys

Robust Control of Diesel Ship Propulsion
Nikolaos Xiros

Hydraulic Servo-systems
Mohieddine Mali and Andreas Kroll

Model-based Fault Diagnosis in Dynamic Systems Using Identification Techniques
Silvio Simani, Cesare Fantuzzi and Ron J. Patton

Strategies for Feedback Linearisation
Freddy Garces, Victor M. Becerra, Chandrasekhar Kambhampati and Kevin Warwick

Robust Autonomous Guidance
Alberto Isidori, Lorenzo Marconi and Andrea Serrani

Dynamic Modelling of Gas Turbines
Gennady G. Kulikov and Haydn A. Thompson (Eds.)

Control of Fuel Cell Power Systems
Jay T. Pukrushpan, Anna G. Stefanopoulou and Huei Peng

Fuzzy Logic, Identification and Predictive Control
Jairo Espinosa, Joos Vandewalle and Vincent Wertz

Optimal Real-time Control of Sewer Networks
Magdalene Marinaki and Markos Papageorgiou

Process Modelling for Control
Benoît Codrons

Computational Intelligence in Time Series Forecasting
Ajoy K. Palit and Dobrivoje Popovic

Modelling and Control of Mini-Flying Machines
Pedro Castillo, Rogelio Lozano and Alejandro Dzul

Ship Motion Control
Tristan Perez

Hard Disk Drive Servo Systems (2nd Ed.)
Ben M. Chen, Tong H. Lee, Kemao Peng and Venkatakrishnan Venkataramanan

Measurement, Control, and Communication Using IEEE 1588
John C. Eidson

Piezoelectric Transducers for Vibration Control and Damping
S.O. Reza Moheimani and Andrew J. Fleming

Manufacturing Systems Control Design
Stjepan Bogdan, Frank L. Lewis, Zdenko Kovačić and José Mireles Jr.

Windup in Control
Peter Hippe

Nonlinear H_2/H_∞ Constrained Feedback Control
Murad Abu-Khalaf, Jie Huang and Frank L. Lewis

Practical Grey-box Process Identification
Torsten Bohlin

Control of Traffic Systems in Buildings
Sandor Markon, Hajime Kita, Hiroshi Kise and Thomas Bartz-Beielstein

Wind Turbine Control Systems
Fernando D. Bianchi, Hernán De Battista and Ricardo J. Mantz

Advanced Fuzzy Logic Technologies in Industrial Applications
Ying Bai, Hanqi Zhuang and Dali Wang (Eds.)

Practical PID Control
Antonio Visioli

(*continued after Index*)

Guowei Cai · Ben M. Chen · Tong Heng Lee

Unmanned Rotorcraft Systems

Springer

Guowei Cai
Temasek Laboratories
National University of Singapore
T-Lab Building, 5A, Engineering Drive 1
Singapore 117411
Singapore
cai_guowei@nus.edu.sg

Tong Heng Lee
Dept. Electrical & Computer Engineering
National University of Singapore
4 Engineering Drive 3
Singapore 117576
Singapore
eleleeth@nus.edu.sg

Ben M. Chen
Dept. Electrical & Computer Engineering
National University of Singapore
4 Engineering Drive 3
Singapore 117576
Singapore
bmchen@nus.edu.sg

ISSN 1430-9491
ISBN 978-0-85729-634-4 e-ISBN 978-0-85729-635-1
DOI 10.1007/978-0-85729-635-1
Springer London Dordrecht Heidelberg New York

British Library Cataloguing in Publication Data
A catalogue record for this book is available from the British Library

Library of Congress Control Number: 2011930654

Cover design: VTeX UAB, Lithuania

Printed on acid-free paper

Springer is part of Springer Science+Business Media (www.springer.com)

To
Our UAV Research Team
and
Our Families

Series Editors' Foreword

The series *Advances in Industrial Control* aims to report and encourage technology transfer in control engineering. The rapid development of control technology has an impact on all areas of the control discipline. New theory, new controllers, actuators, sensors, new industrial processes, computer methods, new applications, new philosophies, ..., new challenges. Much of this development work resides in industrial reports, feasibility study papers and the reports of advanced collaborative projects. The series offers an opportunity for researchers to present an extended exposition of such new work in all aspects of industrial control for wider and rapid dissemination.

Autonomous control and guidance has had a long and evolving history. Long ocean voyages, large vessels and advanced propulsion systems soon led to the development of autopilots for sea-going vessels. Control developments for aerial vehicles and aircraft followed a similar pattern. Although some similar ideas prevailed for terrestrial vehicles, it took the emergence of electronic implementations for sophisticated control techniques to become more widespread in road vehicles. This field of control-technological development was designed to ease the task of the sailor, the pilot or the driver while they were still present on the bridge, in the cockpit or in the driver's seat; however, to dispatch a range of tasks in hazardous environments or to perform routine tasks over significant geographical distances, *unmanned* vehicular technology has developed and made significant strides in recent years. In space, autonomous control and guidance is a prerequisite for accomplishing many tasks involving satellites and unmanned planet "rover" vehicles. Aerial vehicles, either fixed-wing aircraft or rotorcraft are well developed for surveillance and other tasks and autonomous unmanned subsea vehicles are a critical enabler in the success of the oil industry in exploiting offshore oil resources.

In recent *Advances in Industrial Control* monographs, Guillaume J.J. Ducard considered some autonomous control aspects in *Fault-tolerant Flight Control and Guidance Systems* (ISBN 978-1-84882-560-4, 2009) for an unmanned fixed-wing aerial craft, while a little further back in time, Pedro Castillo, Rogelio Lozano, Alejandro E. Dzul reported on the *Modelling and Control of Mini-Flying Machines* (ISBN 978-1-85233-957-9, 2005). To add to this literature, the Series Editors are now pleased to introduce this volume of *Advances in Industrial Control* entitled *Unmanned Rotorcraft Systems* authored by Guowei Cai, Ben M. Chen and Tong H. Lee that describes in comprehensive depth an exemplary development project to build an unmanned rotorcraft aerial system that was undertaken at the National University of Singapore.

Many monographs in the industrial control field, concentrate on the modeling, control design and simulation testing for a particular application but few contributions are able to emulate the authors of this one in describing the system development process, and trying to elucidate the general principles for constructing a complete system of which the control system is but one part. *Unmanned Rotorcraft Systems* has the advantage of being able to illustrate the steps of the development process through the design, construction and testing of a working prototype, the SheLion rotorcraft. Consequently, the monograph progresses through the hardware and software selection stages (Chaps. 3 and 4) and further shows how one advanced control technique (extended Kalman filtering) can be used to enhance the accuracy and performance of some low-cost technology (Chap. 5). The control system development involves modeling (Chap. 6), control design (Chap. 7) and experimental testing (Chap. 8). At each step of the way, the authors demonstrate the need to find solutions that balance technical sophistication against practical specifications, computational constraints and good operational performance. The authors also seek to identify genericity in the procedure so that the techniques can find a wider applicability in the rotorcraft and control system engineering fields. The experimental work reported involves computer simulation, hardware-in-the-loop trials and finally actual flight tests. The authors even extend an invitation to readers to view videos of the flight testing at http://uav.ece.nus.edu.sg/. The monograph closes with two chapters of more innovative work. One chapter (Chap. 10) describes how to achieve formation flying and the final chapter (Chap. 11) reports on the use of vision to follow a moving target. This chapter involves image processing and reports on how to integrate visual information into the rotorcraft control system.

The monograph reports a fascinating project that resulted in real-world outcomes. The comprehensive scope of the activities presented in this monograph and the demonstration of a very practical application of control engineering makes this inspirational material for a wide range of engineering students and researchers. Clearly, these rotorcraft project developments at the National University of Singapore have important industrial relevance and the monograph should be of interest to readers from the aerospace, control, signal processing and electrical engineering disciplines.

Industrial Control Centre
Glasgow, Scotland, UK

M.J. Grimble
M.A. Johnson

Preface

In recent years, research and development of unmanned systems have gained much attention in the academic and military communities worldwide. Topics like unmanned aircraft, underwater explorers, satellites, and intelligent robotics are widely investigated as they have potential applications in the military and civilian domains. They are developed to be capable of working autonomously without the interference of a human pilot. The challenge is that they need to deal with various situations that arise in very complicated and uncertain environments, such as unexpected obstacles, enemies attacking and device failures. Besides, they are required to communicate with technical personnel in the ground station. Consideration of a wide range of factors needs to be taken. Control systems for the unmanned vehicles are required to integrate not only basic input-output control laws but also high-level functionalities for decision making and task scheduling. Software systems for unmanned vehicles are required to perform tasks from hardware driving to the management of device operations, and from traditional input-output control law implementation to task scheduling and event management.

In this monograph, the authors aim to explore the research and development of fully functional miniature unmanned-aerial-vehicle (UAV) rotorcraft, which consist of a small-scale basic rotorcraft with all necessary accessories onboard and a ground station. The unmanned system is an integration of advanced technologies developed in the communications, computing, and control areas. It is an excellent test bed for testing and implementing modern control techniques. It is, however, a highly challenging process. The flight dynamics of small-scale rotorcraft such as a hobby helicopter is similar to its full-scale counterpart but owns some unique characteristics such as the utilization of a stabilizer bar, higher rotor stiffness, and yaw rate feedback control. Besides these, the strict limitation on payload also increases the difficulty in upgrading a small-scale rotorcraft to a UAV with full capacities. Based on its various characteristics and limitations, a lightweight but effective onboard avionic system with corresponding onboard/ground software should be carefully designed to realize the system identification and automatic flight requirements. These issues will be addressed in detail in this monograph. Research on utilizing the vision-based system for accomplishing ground target tracking and following, cooperative control, and flight formation of multiple unmanned rotorcraft is also highlighted.

The intended audience of this monograph includes practicing engineers in rotorcraft industry and researchers in the areas related to the development of unmanned aerial systems. An appropriate background for this book would be some senior level and/or first-year graduate level courses in aerodynamic engineering, control engineering, electrical engineering, and/or mechanical engineering.

The authors of this monograph are thankful to the whole UAV Research Team at the National University of Singapore. We would like to thank Dr. Feng Lin, Dr. Biao Wang, Dr. Kemao Peng, Dr. Miaobo Dong, Dr. Ben Yun, Xiangxu Dong, Xiaolian Zheng, Fei Wang, Shiyu Zhao, Swee-King Phang, Kevin Ang, and Jinqiang Cui for their help and contributions. We are particularly thankful to Dr. Feng Lin for his contribution to the results presented in Chap. 11 and to Dr. Biao Wang for his help to the material given in Chap. 10 of this monograph. We would also like to extend our thanks and appreciations to Ms. Charlotte Cross, Editorial Assistant of Springer, for her kindly help and assistance, and to the Springer's copy editor and series editor for their careful reading of the entire manuscript and their invaluable comments.

We have had the benefit of the collaboration of several coworkers and discussions with international visitors, from whom we have learned a great deal. Among them are Dr. Kai-Yew Lum and Dr. Hai Lin of National University of Singapore, Dr. Chang Chen and Dr. Rodney Teo of the DSO National Laboratories of Singapore, Professor Da-Zhong Zheng of Tsinghua University, Professor Clarence de Silva of the University of British Columbia, Professor Frank Lewis of the University of Texas at Arlington, Professor Lihua Xie of Nanyang Technological University, Professor Delin Luo of Xiamen University, Professor Hai-Bin Duan of Beijing University of Aeronautics and Astronautics, Professor Wei Kang of the Naval Postgraduate School, USA, and Dr. Siva Banda of the Air Force Research Laboratory, USA. We are indebted to them for their valuable contributions and/or comments.

The second author would like to thank particularly the Defence Science and Technology Agency (DSTA), Singapore, for granting him the Temasek Young Investigator Award in 2003 to initiate his research on unmanned systems. We would also like to acknowledge Temasek Laboratories and the Temasek Defence Systems Institute, the National University of Singapore, for their financial support and research funding over the years. We are thankful to the Department of Electrical and Computer Engineering and Temasek Laboratories, the National University of Singapore, for providing us generous laboratory spaces for housing our unmanned aircraft and related research activities.

Last, but certainly not the least, we owe a debt of gratitude to our families for their sacrifice, understanding, and encouragement during the course of preparing this monograph. It is very natural that we dedicate this work to our families and to our whole UAV Research Team.

Singapore, Singapore

Guowei Cai
Ben M. Chen
Tong H. Lee

Contents

Abbreviations

Acronyms

A/D	Analog-to-Digital
AHRS	Attitude Heading Reference System
AOA	Angle Of Attack
CAM	Camera
CAMSHIFT	Continuously Adaptive Mean Shift
CCD	Charge-Coupled-Device
CEP	Circular Error Probable
CF	Compact Flash
CG	Center of Gravity
CIFER	Comprehensive Identification from FrEquency Responses
CMOS	Complementary Metal-Oxide-Semiconductor
CMM	Communication
CORBA	Common Object Request Broker Architecture
CPU	Central Processing Unit
CTL	Control
D/A	Digital-to-Analog
DAQ	Data Acquisition
DARPA	Defense Advanced Research Projects Agency
DC	Direct Current
DGPS	Differential Global Positioning System
DLG	Data Logging
DOF	Degree Of Freedom
DSP	Digital Signal Processor
ECEF	Earth-Centered Earth-Fixed
EKF	Extended Kalman Filter
EMI	Electromagnetic Interference
FPS	Frames Per Second
GPS	Global Positioning System
GUI	Graphical User Interface
HITL	Hardware-In-The-Loop

HSV	Hue, Saturation, Value
IMG	Image
IMU	Inertial Measurement Unit
INS	Inertial Navigation System
I/O	Input/Output
JPEG	Joint Photographic Experts Group
Li-Po	Lithium-Polymer
LMM	Lightweight Multi-role Missile
LQR	Linear Quadratic Regulator
MAV	Micro Aerial Vehicle
MEMS	Micro-Electronic-Mechanical-System
MFC	Microsoft Foundation Class
MIMO	Multi-Input/Multi-Output
MTE	Mission Task Element
NA	Not Applicable
NAV	Navigation
NED	North-East-Down
Ni-Cd	Nickel-Cadmium
Ni-Mh	Nickel-Metal Hydride
NUS	National University of Singapore
OCP	Open Control Platform
OpenGL	Open Graphical Library
PCI	Peripheral Component Interconnect
PCM	Pulse Code Modulation
PD	Proportional-Derivative
PID	Proportional-Integral-Derivative
PPM	Pulse Position Modulation
RC	Radio-Controlled
RFI	Radio Frequency Interference
RGB	Red, Green, Blue
RPM	Revolutions Per Minute
RPT	Robust and Perfect Tracking
RTK	Real-Time Kinematic
RTOS	Real-Time Operating System
SAV	Save
SBC	Single Board Computer
SISO	Single-Input/Single-Output
SVO	Servo
TPP	Tip-Path-Plane
UAV	Unmanned Aerial Vehicle
UKF	Unscented Kalman Filter
VDE	Virtual Design Environment
WGS	World Geodetic System

Symbols[1]

A_{b_s}	Coupling effect from b_s to a_s (s^{-1})
A_{lon}	Linkage gain ratio of θ_{cyc,a_s} to δ_{lon} (rad)
a_s	Longitudinal TPP flapping angle of bare main rotor (rad)
B_{a_s}	Coupling effect from a_s to b_s (s^{-1})
B_{lat}	Linkage gain ratio of θ_{cyc,b_s} to δ_{lat} (rad)
b_{mr}	Main rotor blade number
b_s	Lateral TPP flapping angle of bare main rotor (rad)
b_{tr}	Tail rotor blade number
C_{D0}	Drag coefficient of main rotor blade
C_{lon}	Linkage gain ratio of stabilizer bar cyclic change to δ_{lon} (rad)
$C_{l\alpha,hf}$	Lift curve slope of horizontal fin (rad^{-1})
$C_{l\alpha,mr}$	Lift curve slope of main rotor blade (rad^{-1})
$C_{l\alpha,sb}$	Lift curve slope of stabilizer bar paddle (rad^{-1})
$C_{l\alpha,tr}$	Lift curve slope of tail rotor blade (rad^{-1})
$C_{l\alpha,vf}$	Lift curve slope of vertical fin (rad^{-1})
c_{mr}	Main rotor blade chord length (m)
c_s	Longitudinal TPP flapping angle of stabilizer bar (rad)
c_{sb}	Stabilizer bar paddle chord length (m)
c_{tr}	Tail rotor blade chord length (m)
D_{hf}	Horizontal fin location behind the CG (m)
D_{lat}	Linkage gain ratio of stabilizer bar cyclic change to δ_{lat} (rad)
D_{tr}	Tail rotor hub location behind the CG (m)
D_{vf}	Vertical fin position behind the CG (m)
d_s	Lateral TPP flapping angle of stabilizer bar (rad)
e_{mr}	Effective hinge offset of main rotor (m)
F_b	Aerodynamic force vector (N)
$F_{b,g}$	Gravity force vector projected onto the body frame (N)
H_{mr}	Main rotor hub location above the CG (m)
H_{tr}	Tail rotor hub location above the CG (m)
H_{vf}	Vertical fin location above the CG (m)
g	Local acceleration of gravity (m/s^2)
$\mathbf{J}$	A diagonal matrix of moments of inertia in body frame with its main diagonal elements being J_{xx}, J_{yy}, J_{zz} ($kg{\cdot}m^2$)
K_I	Integral gain of the yaw rate feedback controller
K_P	Proportional gain of the yaw rate feedback controller
K_a	Ratio of yaw rate to normalized rudder input (rad/s)
K_{col}	Ratio of main rotor blade collective pitch angle to collective pitch servo deflection (rad)
K_{ped}	Ratio of tail rotor blade collective pitch angle to rudder servo deflection (rad)
K_{sb}	Ratio of main rotor blade cyclic pitch to stabilizer bar flapping

[1]Listed in this section are all the key symbols and parameters associated with flight dynamics modeling of rotorcraft together with their physical descriptions and units (if any).

K_β	Rotor spring constant (N·m)
$L_{\mathrm{mr}}, M_{\mathrm{mr}}, N_{\mathrm{mr}}$	Aerodynamic moments generated by main rotor (kg·m^2)
$L_{\mathrm{vf}}, N_{\mathrm{vf}}$	Aerodynamic moments generated by vertical fin (kg·m^2)
$L_{\mathrm{tr}}, N_{\mathrm{tr}}$	Aerodynamic moments generated by tail rotor (kg·m^2)
M_{b}	Aerodynamic moment vector (kg·m^2)
M_{hf}	Aerodynamic moment generated by horizontal fin (kg·m^2)
m	Helicopter mass (kg)
n_{tr}	Gear ratio of tail rotor to main rotor
p, q, r	Angular velocities (rad/s)
P_{c}	Climbing power of main rotor (W)
P_{i}	Induced power of main rotor (W)
P_{n}	Local NED position vector with its elements being $x_{\mathrm{n}}, y_{\mathrm{n}}, z_{\mathrm{n}}$ (m)
P_{pa}	Parasitic power of main rotor (W)
P_{pr}	Profile power of main rotor (W)
$\mathbf{R}_{\mathrm{n/b}}$	Rotation matrix from the body frame to the local NED frame
R_{e}	Reynolds number
R_{mr}	Main rotor blade radius (m)
$R_{\mathrm{sb,in}}$	Inner radius of the stabilizer bar disc (m)
$R_{\mathrm{sb,out}}$	Outer radius of the stabilizer bar disc (m)
R_{tr}	Tail rotor blade radius (m)
$\mathbf{S}$	Transformation matrix from Euler angles derivatives to $\boldsymbol{\omega}^{\mathrm{b}}_{\mathrm{b/n}}$
S_{fx}	Effective longitudinal fuselage drag area (m^2)
S_{fy}	Effective lateral fuselage drag area (m^2)
S_{fz}	Effective vertical fuselage drag area (m^2)
S_{hf}	Effective horizontal fin area (m^2)
S_{vf}	Effective vertical fin area (m^2)
T_{mr}	Main rotor thrust (N)
T_{tr}	Tail rotor thrust (N)
$\mathbf{V}_{\mathrm{a}}$	Velocity vector relative to the air projected onto body frame with its elements being $u_{\mathrm{a}}, v_{\mathrm{a}}, w_{\mathrm{a}}$ (m/s)
$\mathbf{V}_{\mathrm{b}}$	Velocity vector projected onto body frame with its elements being u, v, w (m/s)
$\mathbf{V}_{\mathrm{n}}$	Velocity vector projected onto local NED frame with its elements being $u_{\mathrm{n}}, v_{\mathrm{n}}, w_{\mathrm{n}}$ (m/s)
$\mathbf{V}_{\mathrm{wind}}$	Wind gust velocity vector projected onto body frame with its elements being $u_{\mathrm{wind}}, v_{\mathrm{wind}}, w_{\mathrm{wind}}$ (m/s)
$v_{\mathrm{i,mr}}$	Main rotor induced velocity (m/s)
$v_{\mathrm{i,tr}}$	Tail rotor induced velocity (m/s)
v_{vf}	Local lateral airspeed at the vertical fin (m/s)
$\hat{v}^2_{\mathrm{mr}}$	Intermediate variable in main rotor thrust calculation (m^2/s^2)
$\hat{v}^2_{\mathrm{tr}}$	Intermediate variable in tail rotor thrust calculation (m^2/s^2)
w_{hf}	Local vertical airspeed at the horizontal fin (m/s)
x, y, z	Position coordinates in local NED frame (m)
$x_{\mathrm{n}}, y_{\mathrm{n}}, z_{\mathrm{n}}$	Position coordinates in local NED frame (m)
$X_{\mathrm{mr}}, Y_{\mathrm{mr}}, Z_{\mathrm{mr}}$	Aerodynamic forces generated by main rotor (N)

$X_{\text{fus}}, Y_{\text{fus}}, Z_{\text{fus}}$	Aerodynamic forces generated by fuselage (N)
Y_{tr}	Aerodynamic force generated by tail rotor (N)
Y_{vf}	Aerodynamic force generated by vertical fin (N)
Z_{hf}	Aerodynamic force generated by horizontal fin (N)
α_{st}	Critical angle of attack in stall (rad)
γ_{mr}	Lock number of main rotor blade
γ_{sb}	Lock number of the stabilizer bar
δ_{col}	Normalized collective pitch servo input $[-1, 1]$
δ_{lat}	Normalized aileron servo input $[-1, 1]$
δ_{lon}	Normalized elevator servo input $[-1, 1]$
δ_{ped}	Normalized rudder servo input $[-1, 1]$
$\delta_{\text{ped,int}}$	Intermediate state in yaw rate feedback controller dynamics (rad)
$\bar{\delta}_{\text{ped}}$	Rudder servo actuator deflection (rad)
θ_{col}	Collective pitch angle of main rotor blade (rad)
$\theta_{\text{cyc,a}_\text{s}}$	Longitudinal cyclic pitch angle of main rotor blade (rad)
$\theta_{\text{cyc,b}_\text{s}}$	Lateral cyclic pitch angle of main rotor blade (rad)
θ_{ped}	Collective pitch angle of tail rotor blade (rad)
λ_{vf}	Indicator of the vertical fin exposed to tail rotor wake
ρ	Air density ($\text{kg}\cdot\text{m}^3$)
τ_{mr}	Time constant of bare main rotor (s)
τ_{sb}	Time constant of stabilizer bar (s)
ϕ, θ, ψ	Euler angles (rad)
Ω_{mr}	Main rotor rotating speed (rad/s)
Ω_{tr}	Tail rotor rotating speed (rad/s)
$\boldsymbol{\omega}^{\text{b}}_{\text{b/n}}$	Angular velocity vector with its elements being p, q, r (rad/s)

Chapter 1
Introduction

1.1 Introduction

An unmanned aerial vehicle (UAV) is an aircraft that is equipped with necessary data processing units, sensors, automatic control, and communications systems and is capable of performing autonomous flight missions without the interference of a human pilot. Miniature (or mini) UAVs refer to those ranging from micro aerial vehicles (MAVs) with less than a 15 cm wing span or rotor span to vehicles with a payload of tens of kilograms. Characterized by some unique features such as low cost, small size, and great maneuverability, miniature UAVs have gained strong interest worldwide during the last two to three decades. They have been utilized to serve for numerous applications in both military and civilian domains. Driven by many rapid advances in areas such as sensor, manufacturing, and communication technologies, miniature UAVs are becoming smarter than ever and gradually becoming an indispensable assistant to human beings.

Based on their shapes and geometric structures, miniature UAVs can be characterized into the following four categories: (i) fixed-wing UAVs (see, e.g., [3, 197]), (ii) rotorcraft UAVs (see, e.g., [53, 89]), (iii) flapping-wing UAVs (see, e.g., [41, 215]), and (iv) other unconventional UAVs (see, e.g., [95, 132]). Among them, the former two types are currently the most popular choices for practical missions and scientific research, whereas the third type (e.g., unmanned flapping-wing vehicles) has received much attention in academic circles during the last decade. Some preliminary progresses are also being made for the flapping-wing UAVs, although it is still too early to talk about their real applications. When it comes to the unconventional UAVs, they still remain in the initial or even conceptual development stage for the time being.

It is well known that the research and development of miniature UAVs are diversified (see, e.g., [19, 186] and references therein). In this monograph, we focus our attention on a particular class of miniature UAVs, i.e., the miniature rotorcraft UAVs, which is in line with our work at the National University of Singapore (NUS). Miniature rotorcraft UAVs are commonly upgraded from radio-controlled (RC) hobby helicopters by equipping appropriate avionic systems for automatic flight

G. Cai et al., *Unmanned Rotorcraft Systems*, Advances in Industrial Control,
DOI 10.1007/978-0-85729-635-1_1,

Fig. 1.1 HeLion—the first unmanned rotorcraft constructed at NUS

control. Besides the aforementioned features of general UAVs, rotorcraft UAVs have a unique hovering capability, which makes them the best choice for applications conducted in confined areas. Shown in Figs. 1.1, 1.2, 1.3 and 1.4 are unmanned rotorcraft systems constructed by our NUS UAV Research Team [17, 20, 21, 143, 196]. The purpose of this monograph is to systematically document our research results over the last seven years under a single cover.

1.2 Brief History of Rotorcraft

Although the ancestors of modern rotorcraft (such as the Chinese bamboo dragon and the rotorcraft prototype invented by Da Vinci) can be traced many hundreds or even thousands of years back in history [91], miniature rotorcraft UAVs have been popular and widely utilized only in the last few decades. The technology boost starting from the 1970s has accelerated the birth of the modern UAV and further brought huge progress in its development. In this section, we briefly present the history of the modern miniature rotorcraft UAVs.

The development of the miniature rotorcraft UAVs naturally starts with the achievements made for hobby helicopters. Since the first full-scale helicopter was constructed by Igor Sikorsky in 1941, many attempts have been made to reduce its size and realize autonomous flight. The first hobby-based helicopter with sufficient

Fig. 1.2 SheLion (a twin of HeLion) in action

controllability was built by Dieter Schluter of West Germany in 1968. Its flight performance was further enhanced by Kavan Inc., via (i) applying the Bell and Hiller concepts [183] (i.e., adding a stabilizer bar) to the main rotor design in 1974 and (ii) installing a yaw rate feedback in 1978, respectively. Such a configuration was quickly adopted as a standard by the hobby manufacturers for mass production. Since the early 1980s, mature model helicopters have been available in hobby shops all over the world.

In the 1980s, the rapid development of the embedded system technology and micro electronic mechanical system (MEMS) technology formed the second catalyst. Their wide usage has greatly reduced the size and weight of the data processing units, navigation sensors, and communication devices, without losing computational power, measurement accuracy, and communication range. Such a development made it possible to build a light yet powerful avionic system and eventually led to the birth of modern rotorcraft UAVs.

Since the 1990s, the development of sophisticated and reliable rotorcraft UAVs has become an attractive research topic in academic communities worldwide. In 1991, the first international aerial robotics competition was held at the Georgia Institute of Technology. This competition started as a university-based event but soon grew into a major international arena for unmanned aerial systems. The successive years of the competition have seen the aerial robots growing in their capabilities from vehicles that can barely maintain themselves in the air to those that

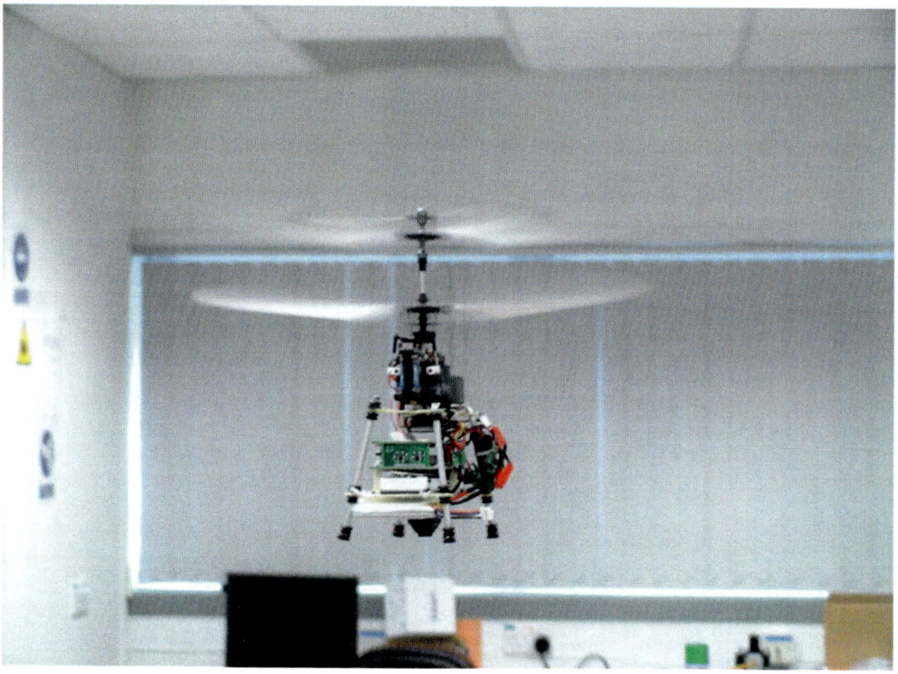

Fig. 1.3 BabyLion—an indoor coaxial rotorcraft UAV

Fig. 1.4 HeLion and SheLion in flight formation

are capable of self-stabilizing, self-navigating, and interacting with their surroundings, especially with objects on the ground [206]. Many research studies in this area have been carried out to develop more advanced hardware platforms, software systems, aerodynamic models, and automatic flight control systems. In 1992, the idea of using very small microdrones in military operations was discussed in a workshop conducted by the Defense Advanced Research Projects Agency (DARPA) of the United States [207]. A multi-year, US$35 million development program was eventually launched by DARPA in 1997, directly aiming to develop MAVs with less than 15 cm wing spans or rotor spans. The initial study of this DARPA project ended in 2001. Unfortunately, the results were somewhat negative, showing that a 15 cm UAV is simply too small to be useful or even workable, at least over the short term [207]. More attention has then been shifted to the development of the larger scale, that is, miniature UAVs as well as their practical applications.

The current popularity of the miniature rotorcraft UAVs is mainly contributed by the numerous research groups and companies who are actively conducting research in this area. We end the brief history overview by listing some representative groups along with their miniature unmanned rotorcraft products. Interested readers are referred to the corresponding references, if available, for more information.

- Baykar Machine Inc.—Malazgirt Mini Unmanned Helicopters [114]
- Beijing University of Aeronautics and Astronautics—FH Series UAVs [54]
- Carnegie Mellon University—Yamaha-R50-based UAV Helicopters [218]
- Chiba University—Sky Surveyor [168]
- Codarra Advanced Systems—AVATAR
- Draganfly Innovations Inc.—Draganflyer Series Multiple Rotor UAVs [48]
- Epson Tokyo R&D—Micro Flying Robot [125]
- ETH Zurich—AkroHeli [51], PIXHAWK [144], and muFly [127]
- Georgia Institute of Technology—GTMax [89]
- HighEye Aerial Service—HEF Series UAV Helicopters [81]
- Honeywell—Duct-fan-based UAV iSTAR [136]
- Israel Aerospace Industries—Naval Rotary UAV [131]
- Linkoping University—WITAS UAV System [214]
- Massachusetts Institute of Technology—MIT Quad-rotor MAV [182]
- Nanjing University of Aeronautics and Astronautics—YJL Quadrotor UAV
- NASA Ames Research Center—Yamaha-Rmax-based UAV Helicopters [30]
- National University of Singapore—Lion UAV Family [135]
- Rotomotion Inc.—SR Series VTOL UAVs [170]
- SAAB Aerosystems—Skeldar V-150 VTOL UAV [167]
- Shanghai Jiaotong University—Sky-Explorer
- Schiebel—Camcopter S-100 UAV System [22]
- Sikorsky Aircraft Inc.—Cypher and Cypher II [166]
- Technische University of Berlin—MARVIN Mark Series UAV [116]
- University of California at Berkeley—Ursa Major and Ursa Maxima [7]
- University of New South Wales—MAVstar [118]
- University of Southern California—AVATAR [185]
- University of Waterloo—Duct-fan UAV
- US Naval Research Laboratories—Dragon Warrior [134]
- Yamaha Inc.—R50 and Rmax UAV Helicopters [217]

We should note that the above list is far from completed.

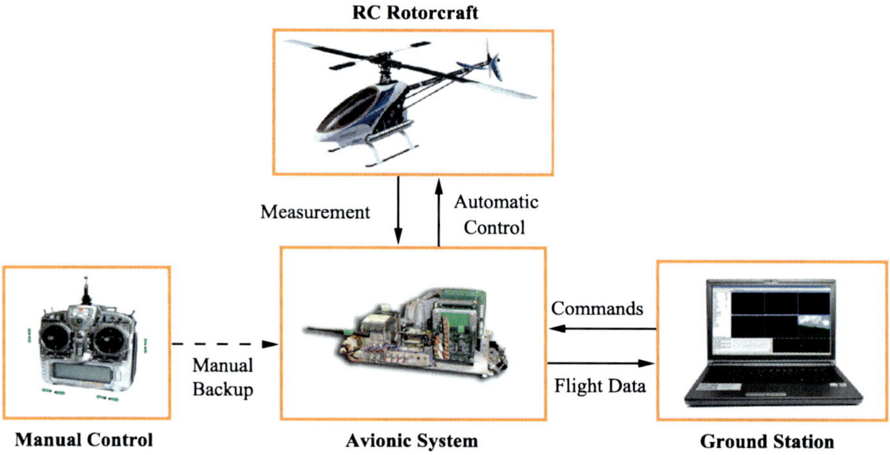

Fig. 1.5 A complete miniature rotorcraft UAV system

1.3 Essential Hardware Components

Hardware platform construction is the core in building a UAV system. Although the miniature rotorcraft UAVs are diversified in size, shape, payload, and application purpose, all of them share a similar system configuration. More specifically, a complete mini rotorcraft UAV system, as depicted in Fig. 1.5, consists of the following four parts:

1. A radio-controlled (RC) rotorcraft
2. An avionic system for collecting inflight data, performing automatic control laws, executing mission-oriented tasks, and communicating with the ground station
3. A manual control system consisting of a pilot and a wireless joystick
4. A ground station system for monitoring the flight states of the UAV and communicating with the avionic system

1.3.1 RC Rotorcraft

RC rotorcraft have experienced rapid development during the last three decades. Its maturity clearly appears in the existence of hundreds of professional hobby-purpose helicopter products and millions of aeromodeling fans. Despite the increasing popularity of multiple-rotor RC flying vehicles, single-rotor RC helicopters are still dominating the market and the research circle. The single-rotor RC helicopters commonly have higher thrust-to-weight ratio, reduced drag, stiffer rotors, and more aggressive head mixing. As such, they can generally achieve greater agility and are

Table 1.1 Key specifications of Type I, II, and III RC rotorcraft

Specification	Type I	Type II	Type III
Fuselage length	2–4 m	1.2–1.6 m	< 0.6 m
Main rotor span	2.5–4 m	1.3–1.7 m	< 0.7 m
Blade number	2 or 3	2	2
Flight endurance	60–360 minutes	15–40 minutes	< 7 minutes
No-load weight	15–60 kg	3.5–5 kg	< 1 kg
Maximum takeoff weight	30–100 kg	5–12 kg	< 0.5 kg
Fuselage material	aluminum or stainless steel	carbon fiber or plastic	carbon fiber or plastic
Power source	diesel or gasoline	nitro fuel or gasoline	battery

able to complete many acrobatic flight motions such as inverted hovering and flapping motion. Interested readers are referred to more acrobatic flight demonstrations on the RC-hobby websites (see, e.g., [55–57]).

From the viewpoint of their size and the way they are implemented, miniature RC rotorcraft can be generally categorized into three types based primarily on size. Their key specifications are summarized and compared in Table 1.1.

1. Type I RC rotorcraft generally have larger size, wider rotor span, longer flight endurance, and heavier payload. Examples for Type I RC rotorcraft include (i) Rmax [217] (from Yamaha Inc.), (ii) ZALA 421-02 [221] (from ZALA Aero Inc.), (iii) Schiebel S-100 [156] (from Schiebel Inc.), and (iv) AF25B [2] (from Copterworks Inc.). Type I RC rotorcraft have the capability of serving for many practical missions such as crop dusting and victim surveillance.
2. Type II covers the mainstream of the RC rotorcraft, which are designed for hobby purposes. There are hundreds of brands of Type II rotorcraft available in the market. Some representative examples are (i) Raptor 90 SE [150] (from Thunder Tiger Inc.), (ii) Observer Twin [137] (from Bergen RC Inc.), and (iii) Turbulence D3 [181] (from Hirobo Inc.). From Table 1.1, we can clearly observe that they are much less powerful than Type I helicopters. However, due to their suitable size, great maneuverability, and lower cost, they are the most popular choice in hobby circles. They are also prevalent in the academic community for research-based UAV projects (see, e.g., [21, 37, 68, 128]). They have recently been adopted by some commercial companies to carry out some short-endurance missions such as aerial photography (see, e.g., [5, 58]).
3. Type III is the smallest and youngest group of the RC rotorcraft and has appeared only during the last decade. Besides single-rotor helicopters, multiple-rotor helicopters are commonly seen in this group. Examples for this category are (i) single-rotor TREX 450 [179] (from Align Inc.), (ii) coaxial Lama V4 [105] (from ESky Inc.), and (iii) quadrotor Draganflyer X4 [48] (from Draganfly Innovations Inc.). They are ultra small in size and light in weight, which makes them the most suitable for indoor purposes.

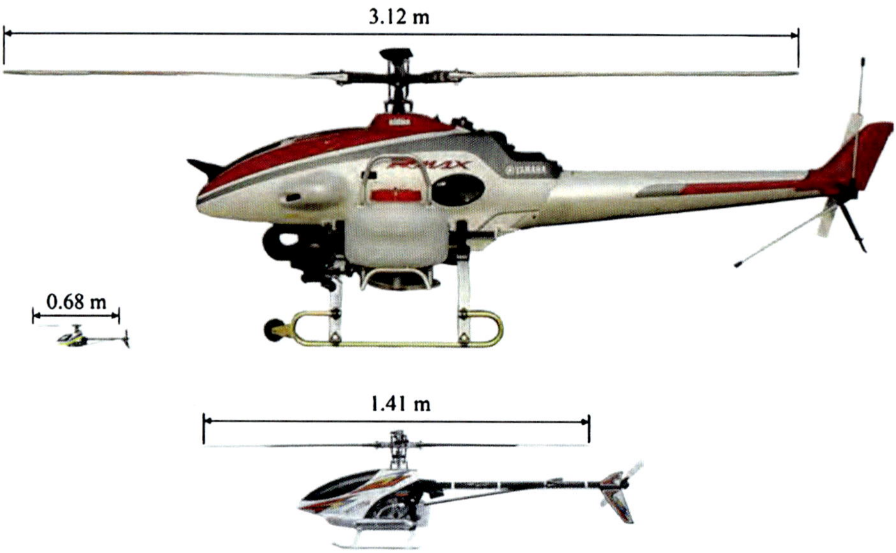

Fig. 1.6 Examples of Type I (Rmax), II (Raptor 90 SE), and III (TREX 450) rotorcraft

Figure 1.6 shows the representative examples of Type I to III single-rotor helicopters, i.e., Rmax, Raptor 90 SE, and TREX 450, respectively, in accordance with their actual scaling ratios.

1.3.2 Avionic System

The avionic system is an essential part in a miniature UAV rotorcraft. Its components selection, system design and integration are one of our primary research focuses, which are to be presented in detail later in Chap. 3. We highlight here the functions and features of the essential components adopted in a typical avionic system.

1.3.2.1 Avionic Processing Stack

The avionic processing stack consists of at least one computer (e.g., flight control computer) and associated extension boards to realize the fundamental tasks, including: (i) analyzing inflight information, (ii) executing automatic control laws, (iii) communicating with the ground station, and (iv) logging necessary inflight data. If extra missions (see, e.g., vision-based target following) are required, we commonly include more computer board(s) to handle the corresponding work. For the miniature UAVs, the avionic processing stack is dominated by PC/104(-plus)-based embedded single board computers (SBCs). PC/104(-plus) is an embedded computer standard, which is defined by the PC/104 Consortium [140] for embedded applications in industry. The standard size of PC/104(-plus) modules is about 90 × 96 mm with a height of about 10 to 15 mm and with an average weight of 100 g. The PC/104(-plus) standard adopts a pin-socket connection, which is designed for some specialized embedded computing environments where applications depend on reliable data acquisition in the face of strong disturbances and vibrations [208]. This feature is particularly suitable for rotorcraft UAVs that operate airborne with persistent vibrations. The pin-socket connection of the PC/104(-plus) SBCs also provides greater compatibility for integrating additional modules, such as serial extension boards, analog-to-digital (A/D) conversion boards, power regulator boards and frame grabber boards (for image and video converting), into the system without reconfiguration. It is worth noting that some strong interest has recently been shown in implementing palm-size or even finger-size SBCs, which are ultra small in size with sufficient processing speed (40∼600 MHz) and low power consumption (commonly < 2 W). However, such a new trend is still at a very initial stage (see, e.g., [13, 190, 196]).

1.3.2.2 Navigation Sensors

Navigation sensors provide all the necessary inflight measurement data for automatic flight control. The integrated navigation scheme, in which multiple navigation sensors are combined together to obtain the best-achievable navigation accuracy and reliability, is widely adopted by miniature UAVs in outdoor applications. Generally, a complete navigation solution consists of part of or all the following four components, including (i) an inertial measurement unit (IMU), (ii) a global positioning system (GPS), (iii) a magnetometer, and (iv) a sophisticated estimation algorithm.

IMU [205] refers to a sensor cluster box containing three accelerometers and three gyroscopes. The accelerometers, which are mounted along strictly orthogonal axes, are utilized to measure inertial accelerations. The gyroscopes are placed along the same orthogonal pattern, providing measurement on angular rates. For miniature UAV applications, MEMS-based (micro-electronic-mechanical-system-based) accelerometers and gyroscopes are dominantly used due to several distinguishing advantages such as low cost, ultra small size, and sufficient resolution and accuracy. However, the MEMS-based IMU commonly suffers from integration error, noise, and large measurement drift.

GPS [204] is widely used nowadays in numerous civil and military applications. The majority of GPS receivers that are commercially available work at L1-band (1.57542 GHz). Information provided by the GPS receivers commonly includes (i) position in the geodetic coordinate system with about 3 m CEP (3 m circular error probable [32]), (ii) velocity in the vehicle-carried NED frame (with the accuracy about 0.5 m/s), and (iii) time when the information package is sent. The accuracy of position and velocity can be further improved to the meter or even centimeter level by employing advanced positioning methodologies such as differential GPS (DGPS) and real-time kinematic (RTK) navigation methods. However, a stationary

GPS base station is required in order to achieve high accuracy measurement. GPS measurement is drift-free and thus commonly blended into the estimation algorithm addressed below as the periodical reference signal. Two major deficiencies of GPS receivers are the vulnerability in a poor-visibility environment and low update rate (commonly 1 to 4 Hz).

A magnetometer is a navigational instrument for determining direction relative to the magnetic poles of the earth [200]. In the miniature rotorcraft UAV systems, the magnetometer can effectively provide initial reference and periodical correction for the heading angle via measuring the strength of the magnetic field. Most of the commercial magnetometers are MEMS-based and with a sufficient resolution (milli-gauss level) and sampling rate (up to 10 Hz). For the specific implementation in the miniature UAVs, special attention should be paid to electromagnetic interference (EMI) shielding and hard- and soft-iron calibrations.

Finally, we note that almost all the commercial navigation sensing systems have integrated some estimation algorithms to enhance their measurement accuracy and to overcome problems such as the immeasurability of Euler angles, insufficient sampling rate of GPS-based positions, and measurement drifting. The well-known extended Kalman filter (EKF) technique is the most popular choice adopted in the commercial products. The estimation algorithm is executed either on an independent digital signal processor (DSP) or directly on the flight control processing unit.

1.3.2.3 Fail-Safe Servo Controller and Wireless Links

The fail-safe servo controller is to drive the servo actuators and to realize smooth switching between the manual control and automatic control modes in real time. It is equipped in most of the miniature UAV rotorcraft to substantially enhance the airborne safety. The wireless links provide communications and data exchanges between the avionic system and the ground control station. The configuration for basic wireless communications (such as inflight data downlink and command/trajectory uplink) is to use a pair of wireless modules equipped on the avionic system and the ground station, respectively. In some projects such as those reported in [120, 162], extra set(s) of wireless modules are utilized for special requirements (e.g., transmitting GPS-based calibration signals).

1.3.3 Manual Backup

In principle, a manual backup is not necessary for a fully functional UAV. However, for safety, a great majority of existing unmanned systems still retain such a backup, which in fact is commonly assigned with higher control authority than the automatic flight control system. The manual control is realized either through an RC joystick or directly through ground station manipulation. For most of the mini rotorcraft UAV systems, the manual control signal is generally modulated onto 29 to 72 MHz based on pulse position modulation (PPM) or more robust pulse code modulation (PCM) techniques. We should also note that the 2.4 GHz frequency hopping modulation technology is a new trend developed within the last five years in the RC flying circle.

1.3.4 Ground Control Station

The last essential part of the overall unmanned system is the ground control station. Its main responsibility is to realize effective communications between the avionic system and the ground users and pilots. To fulfill this aim, the ground station is generally required to have the following fundamental capabilities: (i) displaying and monitoring the inflight status, (ii) displaying images captured by the onboard system, (iii) generating and updating flight trajectories, (iv) sending control commands to the avionic system, (v) facilitating the ground piloted control or automatic control, especially in unexpected situations such as emergency landing and cruise, and (vi) logging inflight data. Other features such as displaying the reconstruction of the actual flight status in a 3D virtual environment can be very helpful to the ground users when the UAV is flying out of sight (see, e.g., [42]).

1.4 Software Design and Integration

A sophisticated software system is required to ensure that all of the hardware components for a UAV system work properly and effectively and to ensure good communications and coordinations between the onboard system and the ground station. It can naturally be divided into two parts, one for the avionic system onboard and one for the ground station.

1.4.1 Avionic Real-Time Software System

The avionic software system coordinates all the hardware components onboard in an appropriate sequence. For most of the UAV avionic systems, the essential tasks include (i) navigation data collection, (ii) flight control algorithm execution, (iii) servo actuator or motor driving, (iv) communication with the ground station system, and (v) inflight data logging, which are required to be executed strictly and precisely in every execution cycle. As such, the development of the avionic software system is dominantly carried out in a real-time operating system (RTOS) environment, which can effectively guarantee that the final system performs in a deterministic way, based on certain scheduling, intertask communications, resource sharing, interrupt handling, and memory allocation algorithms [211]. Currently, the three most

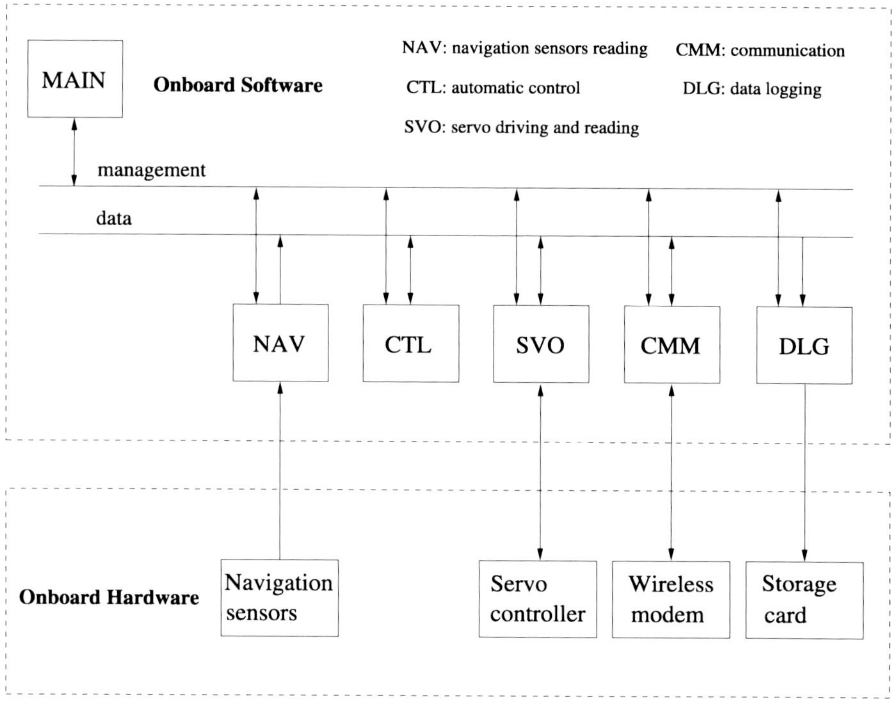

Fig. 1.7 Typical framework of an avionic software system

popular real-time operating systems adopted in the UAV development are the QNX Neutrino [149], VxWorks [193], and RTLinux [152]. Shown in Fig. 1.7 is a typical framework of the avionic software system, in which a multi-thread structure is adopted and each block is designed for a specific device and task. More specifically,

1. NAV is a block interacting with the navigation sensors and collecting necessary measurement data.
2. CTL is for implementing the automatic flight control laws.
3. SVO is for driving the servo actuators or motors.
4. CMM is for communicating between the avionic system and the ground station through the wireless links.
5. DLG is for data logging, which is usually designed as a background task to save the inflight data.
6. Finally, the MAIN block is for managing all tasks.

Note that the avionic system might have more functions for additional applications.

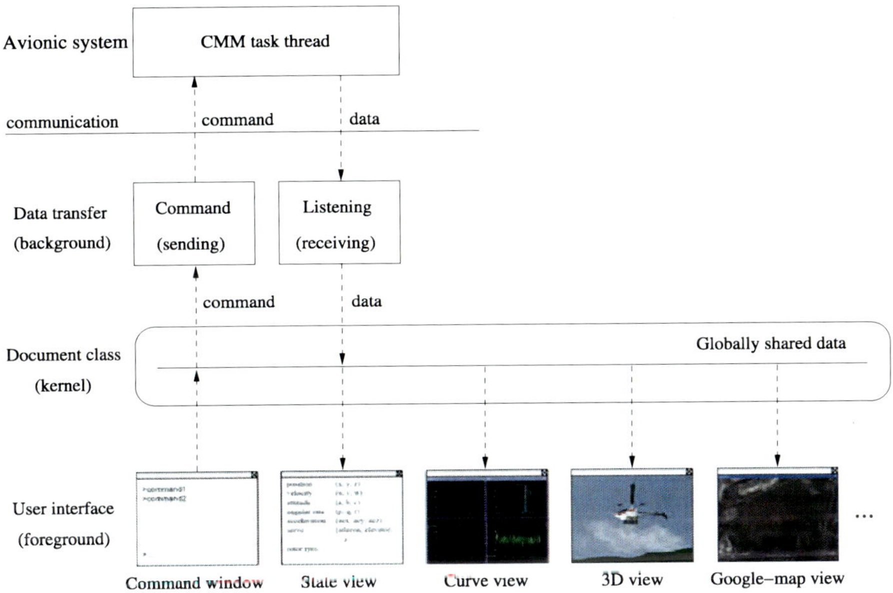

Fig. 1.8 Framework of the ground station software system

1.4.2 Ground Control Station Software Structure

Compared with its avionic counterpart, the real-time feature for the ground station software system is preferable but not strictly compulsory. As such, many ground station software systems, particularly for scientific research and commercial purposes, are not developed under an RTOS environment. Instead, other powerful programming environments with rich interface capacities, such as Windows-based Visual C++ [192], are commonly adopted. Shown in Fig. 1.8 is the framework of the ground station software system adopted in our UAV systems (see also [43, 45]). Generally, the ground station software system consists of three layers:

1. The first is a background layer for data transferring. To be more specific, it communicates with the avionic system through the wireless channel to realize the data receiving and command/trajectory uploading.
2. The second is a foreground layer to display inflight data and to issue flight commands.
3. The third is a kernel layer to allow the background and foreground layers to cooperate harmoniously and to dynamically realize data exchanges via a globally shared database.

1.5 Flight Dynamics Modeling

It is crucial to obtain a fairly comprehensive model of a UAV if one wishes to design an advanced automatic flight control system by incorporating multivariable control techniques such as the linear quadratic regulator (LQR) and H_∞ control, and nonlinear control. Flight dynamics modeling of a miniature rotorcraft UAV, especially for its wide flight envelope dynamics, is an extremely challenging task. Due to the nature and physical structure of the rotorcraft UAV, dynamic modeling with inflight data and with parameter identification approach has been proven to be an ideal choice to derive a fairly accurate model of mini UAVs.

1.5.1 First-Principles Approach

The first-principles modeling approach is well developed for obtaining dynamical models for full-scale manned and unmanned rotorcraft. As reported in [171, 177], such a modeling approach is generally labor intensive and requires the estimation or measurement of the aerodynamic, inertial, and structural properties of the rotorcraft. The models adopted are commonly of high order with complicated structures (see, e.g., [11, 139]) and need to be iteratively tuned based on flight-test data and existing databases. For these reasons, such an approach is only recommended to those who are interested in deriving a nonlinear dynamics model of a miniature UAV system in its wide flight envelope.

Flight dynamics modeling associated with the wide flight envelope for the mini rotorcraft UAVs has become one of the key research focuses in the last five to ten years. A comprehensive nonlinear model covering the wide flight envelope is greatly essential both for the flight control system design and for implementing a sophisticated and meaningful hardware-in-the-loop simulation system (see, e.g., [15]). It is particularly useful for certain flight conditions, in which flight motions are so aggressive and dangerous that conducting actual flight tests for model identification is extremely difficult or even impractical. Some initial success, in which the first-principles modeling technique is applied to small-scale UAV helicopters, has been documented in the literature. For example, in [67], a 17th-order nonlinear model is derived for an X-Cell 60 mini rotorcraft UAV. In another example [33], a novel first-principles modeling approach, named MOSCA, is proposed for off-flight simulation and nonlinear/linear model generation. However, it is our belief that there are many issues associated with the first-principles modeling of miniature rotorcraft UAV systems, including proper structure determination, parameter identification, and model validation, which are yet to be fully resolved.

1.5.2 System and Parameter Identification

System and parameter identification is an effective approach commonly adopted for modeling of flight systems. It is suitable for obtaining a linearized model and can be conducted in either the time domain or the frequency domain or both.

For the time-domain identification, the dynamic model is identified by matching predicted time histories against measured ones [177]. For the inherently unstable platform like single-rotor UAVs, the time-domain identification approach is not the best choice because (i) the equations of motion must be numerically integrated in time for each iterative update in the parameters [177], which causes great difficulty in identifying the parameters related to unstable and weakly stable modes, and (ii) the large amount of historical data involved in the iteration generates a heavy computational burden. As such, the time-domain approach generally does not guarantee the production of an accurate model. For example, the prediction error method is used in [126] to identify a 6-degree-of-freedom (6-DOF) model of a mini rotorcraft UAV at the hovering flight condition. The bandwidth limitation, caused by the inability to process long data records, decreases the accuracy of the identified model.

The frequency-domain identification method is based on frequency responses generated from flight test data, in which the dynamic model is identified by minimizing the error between the predicted frequency histories and measured frequency responses. Compared with its time-domain counterpart, the frequency-domain approach is more suitable for identifying systems with inherent instability and has unique features such as efficient noise elimination, direct and accurate time-delay identification, and less computational cost. A comprehensive comparison between the frequency-domain and time-domain methods for rotorcraft systems can be found in [177]. Some persuasive examples of the utilization of the frequency-domain approach for rotorcraft have been reported in the literature. For instance, in [123], a frequency-response-based identification software package, named Comprehensive Identification from FrEquency Responses (or CIFER), is implemented on a Yamaha R50-based UAV helicopter, and a reliable 11th-order state-space model for the hovering flight condition is successfully identified. This software package also has been implemented in [30] and [122] to identify extended higher-order dynamic models with clearer physical meanings, for hovering and forward flight conditions.

Lastly, we would like to conclude this section by noting that a fairly comprehensive and accurate nonlinear model has been obtained in [16] for SheLion, a Raptor 90 SE-based UAV helicopter [150], through the combination of both the first-principles approach and the system and parameter identification method. The result will be given in detail later in Chap. 6.

1.6 Flight Control Systems

The automatic flight control system is essential for a UAV to carry out flight missions with minimal, or even without, interference from human pilots. The classical

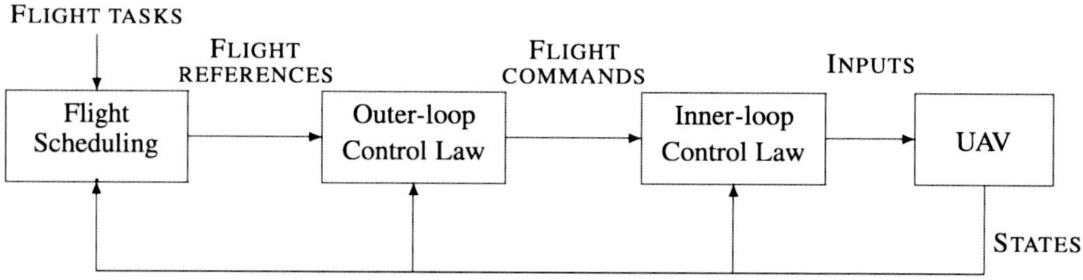

Fig. 1.9 Structure of the hierarchical flight control system

single-input/single-output (SISO) feedback control method (i.e., PD or PID control) is one of the most common choices because of its simplicity in structure with less requirement on the accuracy of the dynamical model of the UAV. Examples include the CMU-R50 UAV helicopter [121], in which an SISO PD control law is adopted and further optimized using CONDUIT for both hovering and forward flight, and the Ursa Major 3 UAV helicopter [162], in which an SISO PID control is implemented for automatic hovering. To improve flight control performance, much research is devoted to the study of implementing more advanced control techniques on the miniature rotorcraft UAVs. For example, a flight control system using an MIMO (multi-input/multi-output) H_∞ control approach has been designed and implemented for the mini rotorcraft UAVs in [199]. It is reported that the resulting system has clearly outperformed the classical method. Other cases reported in the literature include systems designed by using (i) a decentralized decoupled model predictive approach [163], (ii) a neural network method [5, 49, 194], (iii) an adaptive control technique [38], (iv) a fuzzy logic approach [93], (v) μ-synthesis [198], (vi) an approximate linearization method [100], (vii) a nonlinear feed-forward method [9], (viii) a differential geometry technique [85], (ix) H_∞ static output-feedback control [66], (x) a learning control technique [50], and (xi) intelligent control methods [173], to name a few. Although there is a large number of works that have been performed along these lines, many of them are still in the simulation stage. They are far from being ready for actual implementation onto the real platform.

Recently, Cai et al. [18] and Peng et al. [142] have proposed a flight control scheme consisting of three parts, namely, the inner-loop control law, outer-loop and flight scheduling (see Fig. 1.9). In [18], the function of the inner-loop control law, designed using the H_∞ control approach, is to guarantee the asymptotic stability of the aircraft motion and to have good disturbance rejection with respect to wind gusts. The role of the outer-loop is to produce flight commands or references to the inner-loop control layer, and finally the task of the flight scheduling part is to generate the flight references for pre-scheduled flight missions.

1.7 Application Examples

Although it is still a long way for the unmanned systems to undertake many missions that are currently carried out by manned systems, some current technologies developed in the unmanned systems are mature enough to be integrated into some real-life applications in both the military and civilian domains. We list in the following some examples, which show a good illustration of the applications of the technologies developed in the unmanned systems.

Battlefield To some extent, warfare applications can be regarded as a crucial reason for the birth of the unmanned systems including unmanned aerial vehicles and unmanned ground vehicles. The unmanned rotorcraft systems are mostly utilized for combat gunfire support, and surveillance and reconnaissance, on the battlefield.

1. COMBAT GUNFIRE SUPPORT. The current miniature unmanned combat aerial vehicles are dominated by fixed-wing UAVs due to their superior agility and longer endurance. The miniature rotorcraft UAVs, however, are capable of providing unique combat support in battles in confined areas such as streets. Two representative application examples of the rotorcraft UAV for combat gunfire support are the Autocopter Gunship [5] equipped with up to 2 AA-12 guns for air shooting and the Schiebel S-100 equipped with lightweight multi-role missiles (LMM).
2. SURVEILLANCE AND RECONNAISSANCE. The unique features of the rotorcraft UAV for being able to hover and to operate at a low altitude make it an ideal platform for surveillance and reconnaissance in some crucial tasks such as range safety monitoring, arms transfer monitoring, and mine detecting. A famous battlefield application is the successful deployment of a Honeywell duct-fan UAV in Iraq for explosive detection.

The rotorcraft UAVs can be of great assistance to civilian defense and law enforcement as well (see, e.g., the case reported in [124]).

Scientific Exploration Equipped with advanced sensors, a miniature rotorcraft UAV can serve as an excellent platform for scientific exploration. An impressive example is the implementation of an R50 UAV helicopter at Carnegie Mellon University. In a project called Haughton Crater Mission, the R50 UAV helicopter has demonstrated its validity in geological surveys, helping the researchers to gain a deeper understanding of the environment on Mars by studying similarities between Haughton Crater and Mars [24].

Agriculture and Forestry One successful application of the rotorcraft UAVs in the civil domain is the pesticide spraying in agriculture and forestry. The most representative examples are the R50 and Rmax helicopters developed by Yamaha. Their origin can be traced back to a request in 1983 from the Japanese Ministry of Agriculture, Forestry and Fisheries for unmanned helicopters for crop dusting that could help reduce labor and costs in the labor-intensive rice farming industry. As a result, the R50 and Rmax series eventually emerged as the dominant UAV technology for pesticide spraying service nowadays.

Courtesy of Flying-Cam

Fig. 1.10 Sports and entertainment broadcasting

Courtesy of Control and Robotics Laboratory, Chiba University

Fig. 1.11 Sky Surveyor inspecting a power transmission line

Sports Broadcasting and Movie Making The audience is filled with wonder at the sight of the bird's eye view of a splendid stadium or special effects in movies. However, they seldom realize that many of the spectacular sights are aerially photographed by mini rotorcraft UAVs (see Fig. 1.10 for illustrations). Interested readers are referred to [58] for more information on aerial photography.

Engineering and Construction With the help of high resolution cameras or video recorders, rotorcraft UAVs can be utilized to assist workers and engineers to solve problems in engineering and construction, which are difficult to resolve through conventional approaches. For instance, the Automatic Lab of Chiba University has successfully developed a rotorcraft UAV called Sky Surveyor [168], equipped with a vibration-free pan/tilt/zoom camera (see Fig. 1.11) to help Chugoku Electric Power conduct inspections on power transmission lines between conjuncted high-voltage towers.

1.8 Preview of Each Chapter

The preview of each chapter is given next. In Chap. 2, we summarize several coordinate systems used in our work. More specifically, we are to present the concepts of the geodetic coordinate system, the earth-centered earth-fixed coordinate system, the local north-east-down (NED) coordinate system, the vehicle-carried NED coordinate system, the body coordinate system of an unmanned flying vehicle, and the coordinate transformations among them, which are heavily used in navigation, guidance, and control of aircraft.

Chapter 3 presents a systematic methodology for constructing a fully functional unmanned rotorcraft system, which involves key steps such as virtual design environment selection, selection of hardware components, design and integration of the avionic system, and performance and reliability evaluation. A Raptor 90 SE-based UAV, named SheLion, a twin of HeLion, but with an additional onboard vision system, is used to illustrate the design procedure throughout the entire chapter.

We aim in Chap. 4 to develop a comprehensive software system that can be universally adopted in unmanned aerial systems. The software system developed consists of two main parts, the onboard software subsystem and the ground control station software subsystem. The onboard subsystem provides reliable support for high precision timer and synchronization operations; processes vision images captured; operates hardware components such as navigation sensors, data acquisition boards, and servo systems; logs data in flying processes; communicates with the ground control station, and implements automatic control algorithms. The ground control station subsystem is programmed for data transferring with the onboard system and for monitoring the flight status of the unmanned system.

Chapter 5 aims to introduce an integration of a low-cost inertial attitude and position reference system for mini UAV helicopters by utilizing the well-known extended EKF technique. More specifically, we propose a systematic signal enhancement procedure for measurement sensors adopted in miniature unmanned aerial systems. The procedure yields more accurate measurement of the Euler angles, angular rates, positions, and velocities of the UAV with a self-integrated navigation unit.

In Chap. 6, we present a comprehensive modeling process to obtain a highly accurate nonlinear dynamical model for our unmanned systems, SheLion (also applicable to HeLion). We first derive a minimum-complexity model structure, which covers all the important dynamic features necessary for flight control law design. Based on this structured model, we develop a five-step procedure, a systematic combination of the first-principles and system identification approaches, to determine all the associated model parameters. We then carry out a thorough validation process to verify the fidelity of the flight dynamics model in the wide flight envelope. Finally, we proceed to determine the flight envelope of the obtained flight dynamics model, which is essential before proceeding to conduct flight control law design and flight experiments.

We propose a three-layer automatic flight control system for our unmanned vehicles based on the time scales of the state variables of the helicopter, which consists of the inner loop, the outer loop, and the flight scheduling layers. The inner loop stabilizes the dynamics of the helicopter associated with its angular velocities and Euler angles. The outer loop controls the position of the unmanned system. Lastly, the outmost layer, i.e., the flight scheduling layer, generates the necessary trajectories for predefined flight missions. Chapter 7 presents the design of the inner-loop control law using an H_∞ control technique based on the linearized model obtained in Chap. 6. More specifically, we focus on issues related to design specification selection, problem formulation, flight control law design, and overall performance evaluation. Design specifications for military rotorcraft set for US army aviation [1] are adopted throughout the whole process to guarantee a top level performance defined in [1].

We utilize a so-called robust and perfect tracking control technique in Chap. 8 to design the outer-loop control law to control the position of the unmanned system, which is capable of achieving much better performance for situations when complicated maneuvers are required. The robust and perfect tracking control technique is to design a controller such that the resulting closed-loop system is asymptotically stable and the controlled output almost perfectly tracks a given reference signal in the presence of any initial conditions and external disturbances. It makes use of all possible information including the system measurement output and the command reference signal together with all its derivatives, if available, for control. Such a unique feature is particularly useful for the outer-loop layer, in which the position reference and its velocity as well as acceleration all can be measured by the onboard avionic system.

In Chap. 9, we present a fairly comprehensive evaluation of the overall flight control system designed in Chaps. 7 and 8 through hardware-in-the-loop simulations and actual flight tests. We aim to evaluate its performance and robustness by a careful selection of mission-task-elements (MTEs) adopted from ADS-33D-PRF [1], which is set for military rotorcraft by US army aviation. The selected mission-task-elements for test include depart/abort (forward flight), hover, depart/abort (backward flight), hovering turn, vertical maneuver, lateral reposition, turn-to-target, slalom, and pirouette. The results obtained clearly indicate that our design is very successful. The unmanned rotorcraft system is capable of achieving the desired performance in accordance with the military standard under examination.

To conclude the whole monograph, we feature respectively in Chaps. 10 and 11 the applications of the unmanned rotorcraft systems constructed. More specifically, in Chap. 10, we present some basic results on flight formation and collision avoidance of multiple unmanned systems. We adopt the leader-follower pattern to maintain a fixed geometrical formation while navigating the unmanned rotorcraft following certain trajectories. In order to avoid possible collisions in the actual formation flight test, a collision avoidance scheme based on some predefined alert zones and protected zones is employed. Simulations and experimental results are presented to verify our design. We should note that with the utilization of the RPT control technique presented in Chap. 8, the flight formation control is rather straightforward.

Finally, in Chap. 11, we document the design and implementation of a comprehensive vision system for an unmanned rotorcraft to realize missions such as ground target detection and following. To realize the autonomous ground target seeking and following, a sophisticated vision algorithm is proposed to detect the target and estimate relative distance to the target using an onboard color camera together with necessary navigation sensors. The vision feedback is then integrated with the automatic flight control system to guide the unmanned helicopter to follow the ground target inflight. The overall vision system is tested in actual flight missions, and the results obtained show that it is robust and efficient.

Chapter 2
Coordinate Systems and Transformations

2.1 Introduction

In navigation, guidance, and control of an aircraft or rotorcraft, there are several coordinate systems (or frames) intensively used in design and analysis (see, e.g., [171]). For ease of references, we summarize in this chapter the coordinate systems adopted in our work, which include

1. the geodetic coordinate system,
2. the earth-centered earth-fixed (ECEF) coordinate system,
3. the local north-east-down (NED) coordinate system,
4. the vehicle-carried NED coordinate system, and
5. the body coordinate system.

The relationships among these coordinate systems, i.e., the coordinate transformations, are also introduced.

We need to point out that miniature UAV rotorcraft are normally utilized at low speeds in small regions, due to their inherent mechanical design and power limitation. This is crucial to some simplifications made in the coordinate transformation (e.g., omitting unimportant items in the transformation between the local NED frame and the body frame). For the same reason, partial transformation relationships provided in this chapter are not suitable for describing flight situations on the oblate rotating earth.

2.2 Coordinate Systems

Shown in Figs. 2.1 and 2.2 are graphical interpretations of the coordinate systems mentioned above, which are to be used in sensor fusion, flight dynamics modeling, flight navigation, and control. The detailed description and definition of each of these coordinate systems are given next.

G. Cai et al., *Unmanned Rotorcraft Systems*, Advances in Industrial Control,
DOI 10.1007/978-0-85729-635-1_2,

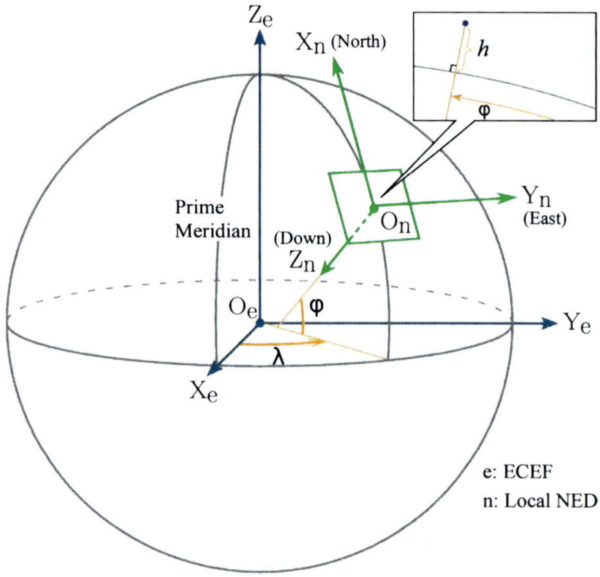

Fig. 2.1 Geodetic, ECEF, and local NED coordinate systems

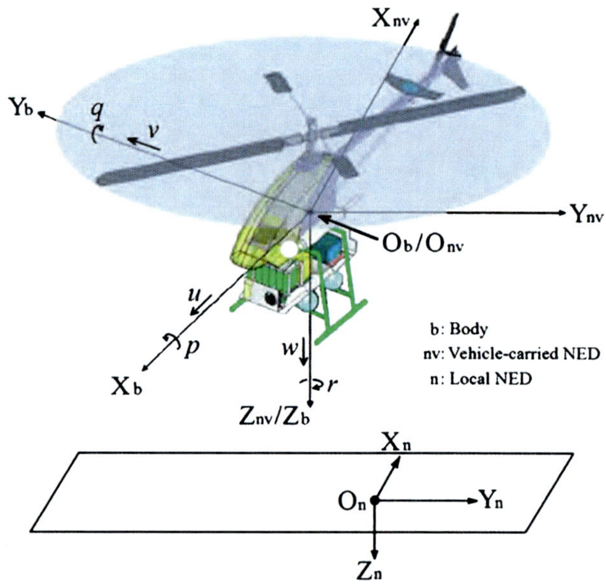

Fig. 2.2 Local NED, vehicle-carried NED, and body coordinate systems

2.2.1 Geodetic Coordinate System

The geodetic coordinate system (see Fig. 2.1) is widely used in GPS-based navigation. We note that it is not a usual Cartesian coordinate system but a system that characterizes a coordinate point near the earth's surface in terms of longitude, latitude, and height (or altitude), which are respectively denoted by λ, φ, and h. The longitude measures the rotational angle (ranging from $-180°$ to $180°$) between the Prime Meridian and the measured point. The latitude measures the angle (ranging from $-90°$ to $90°$) between the equatorial plane and the normal of the reference ellipsoid that passes through the measured point. The height (or altitude) is the local vertical distance between the measured point and the reference ellipsoid. It should be noted that the adopted geodetic latitude differs from the usual geocentric latitude (φ'), which is the angle between the equatorial plane and a line from the mass center of the earth. Lastly, we note that the geocentric latitude is not used in our work. Coordinate vectors expressed in terms of the geodetic frame are denoted with a subscript g, i.e., the position vector in the geodetic coordinate system is denoted by

$$P_{\rm g} = \begin{pmatrix} \lambda \\ \varphi \\ h \end{pmatrix}. \tag{2.1}$$

Important parameters associated with the geodetic frame include

1. the semi-major axis $R_{\rm Ea}$,
2. the flattening factor $\mathbf{f}$,
3. the semi-minor axis $R_{\rm Eb}$,
4. the first eccentricity $\mathbf{e}$,
5. the meridian radius of curvature $M_{\rm E}$, and
6. the prime vertical radius of curvature $N_{\rm E}$.

These parameters are either defined (items 1 and 2) or derived (items 3 to 6) based on the WGS 84 (world geodetic system 84, which was originally proposed in 1984 and lastly updated in 2004 [212]) ellipsoid model. More specifically, we have

$$R_{\rm Ea} = 6{,}378{,}137.0 \text{ m}, \tag{2.2}$$

$$\mathbf{f} = 1/298.257223563, \tag{2.3}$$

$$R_{\rm Eb} = R_{\rm Ea}(1-\mathbf{f}) = 6{,}356{,}752.0 \text{ m}, \tag{2.4}$$

$$\mathbf{e} = \frac{\sqrt{R_{\rm Ea}^2 - R_{\rm Eb}^2}}{R_{\rm Ea}} = 0.08181919, \tag{2.5}$$

$$M_{\rm E} = \frac{R_{\rm Ea}(1-\mathbf{e}^2)}{(1-\mathbf{e}^2\sin^2\varphi)^{3/2}}, \tag{2.6}$$

$$N_{\rm E} = \frac{R_{\rm Ea}}{\sqrt{1-\mathbf{e}^2\sin^2\varphi}}. \tag{2.7}$$

2.2.2 Earth-Centered Earth-Fixed Coordinate System

The ECEF coordinate system rotates with the earth around its spin axis. As such, a fixed point on the earth surface has a fixed set of coordinates (see, e.g., [202]). The origin and axes of the ECEF coordinate system (see Fig. 2.1) are defined as follows:

1. The origin (denoted by O_e) is located at the center of the earth.
2. The Z-axis (denoted by Z_e) is along the spin axis of the earth, pointing to the north pole.
3. The X-axis (denoted by X_e) intersects the sphere of the earth at 0° latitude and 0° longitude.
4. The Y-axis (denoted by Y_e) is orthogonal to the Z- and X-axes with the usual right-hand rule.

Coordinate vectors expressed in the ECEF frame are denoted with a subscript e. Similar to the geodetic system, the position vector in the ECEF frame is denoted by

$$P_e = \begin{pmatrix} x_e \\ y_e \\ z_e \end{pmatrix}. \tag{2.8}$$

2.2.3 Local North-East-Down Coordinate System

The local NED coordinate system is also known as a navigation or ground coordinate system. It is a coordinate frame fixed to the earth's surface. Based on the WGS 84 ellipsoid model, its origin and axes are defined as the following (see also Figs. 2.1 and 2.2):

1. The origin (denoted by O_n) is arbitrarily fixed to a point on the earth's surface.
2. The X-axis (denoted by X_n) points toward the ellipsoid north (geodetic north).
3. The Y-axis (denoted by Y_n) points toward the ellipsoid east (geodetic east).
4. The Z-axis (denoted by Z_n) points downward along the ellipsoid normal.

The local NED frame plays a very important role in flight control and navigation. Navigation of small-scale UAV rotorcraft is normally carried out within this frame. Coordinate vectors expressed in the local NED coordinate system are denoted with a subscript n. More specifically, the position vector, $\mathbf{P}_n$, the velocity vector, $\mathbf{V}_n$, and the acceleration vector, $\mathbf{a}_n$, of the NED coordinate system are adopted and are, respectively, defined as

$$\mathbf{P}_n = \begin{pmatrix} x_n \\ y_n \\ z_n \end{pmatrix}, \quad \mathbf{V}_n = \begin{pmatrix} u_n \\ v_n \\ w_n \end{pmatrix}, \quad \mathbf{a}_n = \begin{pmatrix} a_{x,n} \\ a_{y,n} \\ a_{z,n} \end{pmatrix}. \tag{2.9}$$

We also note that in our work, we normally select the takeoff point, which is also the sensor initialization point, in each flight test as the origin of the local NED frame. When it is clear in the context, we also use the following definition throughout the monograph for the position vector in the local NED frame,

$$\mathbf{P}_n = \begin{pmatrix} x \\ y \\ z \end{pmatrix}. \tag{2.10}$$

Furthermore, $h = -z$ is used to denote the actual height of the unmanned system.

2.2.4 Vehicle-Carried North-East-Down Coordinate System

The vehicle-carried NED system is associated with the flying vehicle. Its origin and axes (see Fig. 2.2) are given by the following:

1. The origin (denoted by O_{nv}) is located at the center of gravity (CG) of the flying vehicle.
2. The X-axis (denoted by X_{nv}) points toward the ellipsoid north (geodetic north).
3. The Y-axis (denoted by Y_{nv}) points toward the ellipsoid east (geodetic east).
4. The Z-axis (denoted by Z_{nv}) points downward along the ellipsoid normal.

Strictly speaking, the axis directions of the vehicle-carried NED frame vary with respect to the flying-vehicle movement and are thus not aligned with those of the local NED frame. However, as mentioned earlier, the miniature rotorcraft UAVs fly only in a small region with low speed, which results in the directional difference being completely neglectable. As such, it is reasonable to assume that the directions of the vehicle-carried and local NED coordinate systems constantly coincide with each other.

Coordinate vectors expressed in the vehicle-carried NED frame are denoted with a subscript nv. More specifically, the velocity vector, $\mathbf{V}_{nv}$, and the acceleration vector, $\mathbf{a}_{nv}$, of the vehicle-carried NED coordinate system are adopted and are, respectively, defined as

$$\mathbf{V}_{nv} = \begin{pmatrix} u_{nv} \\ v_{nv} \\ w_{nv} \end{pmatrix}, \qquad \mathbf{a}_{nv} = \begin{pmatrix} a_{x,nv} \\ a_{y,nv} \\ a_{z,nv} \end{pmatrix}. \tag{2.11}$$

2.2.5 Body Coordinate System

The body coordinate system is vehicle-carried and is directly defined on the body of the flying vehicle. Its origin and axes (see Fig. 2.2) are given by the following:

1. The origin (denoted by O_b) is located at the center of gravity (CG) of the flying vehicle.
2. The X-axis (denoted by X_b) points forward, lying in the symmetric plane of the flying vehicle.
3. The Y-axis (denoted by Y_b) is starboard (the right side of the flying vehicle).
4. The Z-axis (denoted by Z_b) points downward to comply with the right-hand rule.

Coordinate vectors expressed in the body frame are appended with a subscript b. Next, we define

$$\mathbf{V}_b = \begin{pmatrix} u \\ v \\ w \end{pmatrix} \tag{2.12}$$

to be the vehicle-carried NED velocity, i.e., $\mathbf{V}_{\mathrm{nv}}$, projected onto the body frame, and

$$\mathbf{a}_{\mathrm{b}} = \begin{pmatrix} a_{\mathrm{x}} \\ a_{\mathrm{y}} \\ a_{\mathrm{z}} \end{pmatrix} \tag{2.13}$$

to be the vehicle-carried NED acceleration, i.e., $\mathbf{a}_{\mathrm{nv}}$, projected onto the body frame. These two vectors are intensively used in capturing the 6-DOF rigid-body dynamics of unmanned systems.

2.3 Coordinate Transformations

The transformation relationships among the adopted coordinate frames are introduced in this section. We first briefly introduce some fundamental knowledge related to Cartesian-frame transformations before giving the detailed coordinate transformations.

2.3.1 Fundamental Knowledge

We summarize in this subsection the basic concepts of the Euler rotation and rotation matrix, Euler angles, and angular velocity vector used in flight modeling, control and navigation.

2.3.1.1 Euler Rotations

The orientation of one Cartesian coordinate system with respect to another can always be described by three successive Euler rotations [171]. For aerospace application, the Euler rotations perform about each of the three Cartesian axes consequently, following the right-hand rule. Shown in Fig. 2.3 is a simple example, in which Frames C1 and C2 are two Cartesian systems with the aligned Z-axes pointing toward us. We take Frame C2 as the reference and can obtain Frame C1 through a Euler rotation (by rotating Frame C2 counter-clockwise with an angle of ξ). Then, it is straightforward to verify that the position vectors of any given point expressed in Frame C1, say $\mathbf{P}_{\mathrm{C1}}$, and in Frame C2, say $\mathbf{P}_{\mathrm{C2}}$, are related by

$$\mathbf{P}_{\mathrm{C1}} = \mathbf{R}_{\mathrm{C1/C2}} \mathbf{P}_{\mathrm{C2}}, \tag{2.14}$$

where $\mathbf{R}_{\mathrm{C1/C2}}$ is defined as a rotation matrix that transforms the vector $\mathbf{P}$ from Frame C2 to Frame C1 and is given as

$$\mathbf{R}_{\mathrm{C1/C2}} = \begin{bmatrix} \cos\xi & \sin\xi & 0 \\ -\sin\xi & \cos\xi & 0 \\ 0 & 0 & 1 \end{bmatrix}. \tag{2.15}$$

It is simple to show that

$$\mathbf{R}_{\mathrm{C2/C1}} = \mathbf{R}_{\mathrm{C1/C2}}^{-1} = \mathbf{R}_{\mathrm{C1/C2}}^{\mathrm{T}}. \tag{2.16}$$

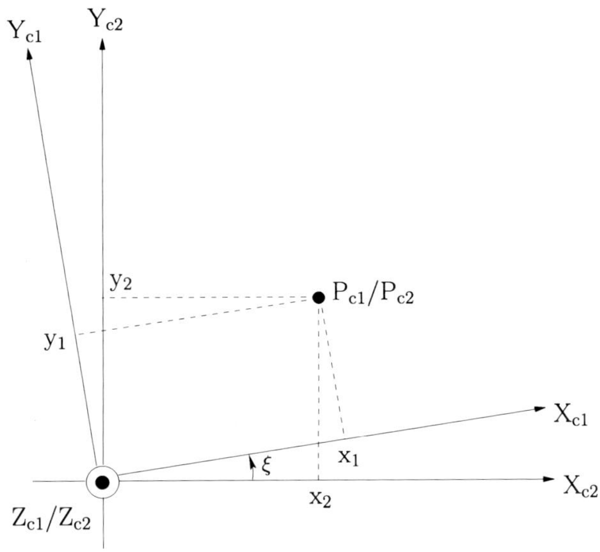

Fig. 2.3 Illustration of a Euler rotation

2.3.1.2 Euler Angles

The Euler angles are three angles introduced by Euler to describe the orientation of a rigid body. Although the relative orientation between any two Cartesian frames can be described by Euler angles, we focus in this monograph merely on the transformation between the vehicle-carried (or the local) NED and the body frames, following a particular rotation sequence. More specifically, the adopted Euler angles move the reference frame to the referred frame, following a Z-Y-X (or the so-called 3–2–1) rotation sequence. These three Euler angles are also known as the yaw (or heading), pitch, and roll angles, which are defined as the following (see Fig. 2.4 for graphical illustration):

1. YAW ANGLE, denoted by ψ, is the angle from the vehicle-carried NED X-axis to the projected vector of the body X-axis on the X-Y plane of the vehicle-carried NED frame. The right-handed rotation is about the vehicle-carried NED Z-axis. After this rotation (denoted by $\mathbf{R}_{\mathrm{int1/nv}}$), the vehicle-carried NED frame transfers to a once-rotated intermediate frame.
2. PITCH ANGLE, denoted by θ, is the angle from the X-axis of the once-rotated intermediate frame to the body frame X-axis. The right-handed rotation is about the Y-axis of the once-rotated intermediate frame. After this rotation (denoted by $\mathbf{R}_{\mathrm{int2/int1}}$), we have a twice-rotated intermediate frame whose X-axis coincides with the X-axis of the body frame.
3. ROLL ANGLE, denoted by ϕ, is the angle from the Y-axis (or Z-axis) of the twice-rotated intermediate frame to that of the body frame. This right-handed rotation (denoted by $\mathbf{R}_{\mathrm{b/int2}}$) is about the X-axis of the twice-rotated intermediate frame (or the body frame).

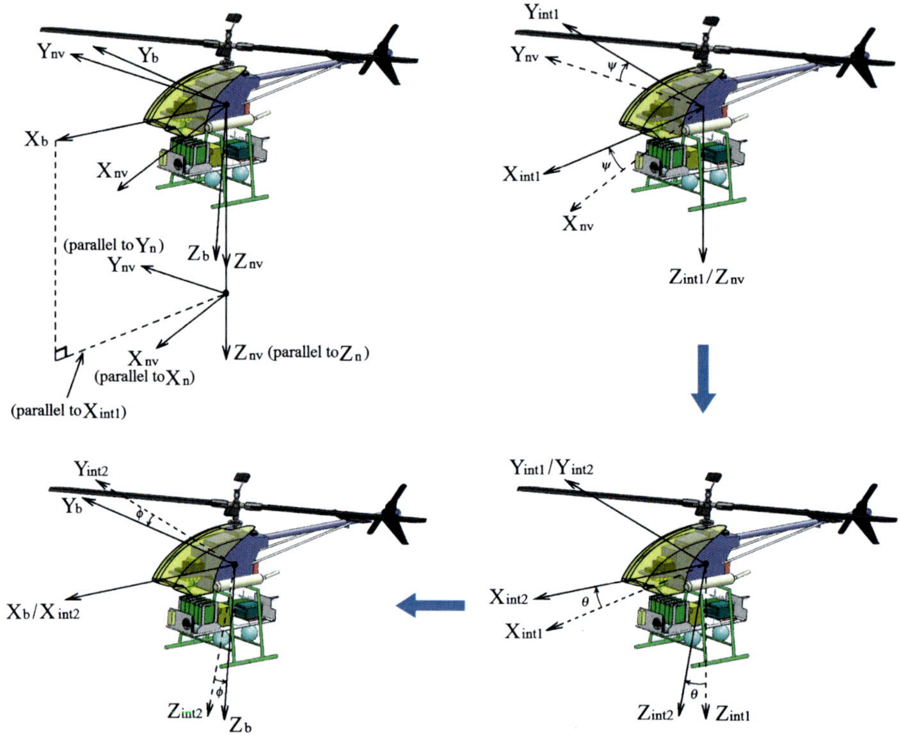

Fig. 2.4 Euler angles and yaw-pitch-roll rotation sequence

The three relative rotation matrices are respectively given by

$$\mathbf{R}_{\mathrm{int1/nv}} = \begin{bmatrix} \cos\psi & \sin\psi & 0 \\ -\sin\psi & \cos\psi & 0 \\ 0 & 0 & 1 \end{bmatrix}, \tag{2.17}$$

$$\mathbf{R}_{\mathrm{int2/int1}} = \begin{bmatrix} \cos\theta & 0 & -\sin\theta \\ 0 & 1 & 0 \\ \sin\theta & 0 & \cos\theta \end{bmatrix}, \tag{2.18}$$

and

$$\mathbf{R}_{\mathrm{b/int2}} = \begin{bmatrix} 1 & 0 & 0 \\ 0 & \cos\phi & \sin\phi \\ 0 & -\sin\phi & \cos\phi \end{bmatrix}. \tag{2.19}$$

2.3.1.3 Angular Velocities

The angular velocities (or angular rates) are associated with the relative motion between two coordinate systems. Considering that Frame C1 is rotating with respect to Frame C2, the angular velocity is denoted by

$$\boldsymbol{\omega}^{*}_{\mathrm{C1/C2}} = \begin{pmatrix} \omega_{\mathrm{x}} \\ \omega_{\mathrm{y}} \\ \omega_{\mathrm{z}} \end{pmatrix}, \tag{2.20}$$

where $*$ is a coordinate frame on which the angular velocity vector is projected. We note that the coordinate frame $*$ can be C1 or C2 or any another frame. It is simple to verify that the angular velocity vector of Frame C2 rotating with respect to Frame C1 is given by

$$\boldsymbol{\omega}^{*}_{\mathrm{C2/C1}} = -\boldsymbol{\omega}^{*}_{\mathrm{C1/C2}}. \tag{2.21}$$

2.3.2 Coordinate Transformations

We proceed to present the necessary coordinate transformations among the coordinate systems adopted, of which the first three transformations are mainly employed for rotorcraft spatial navigation, the fourth one is commonly adopted for flight control purposes, and finally, the last one focuses on an approximation particularly suitable for the miniature rotorcraft.

2.3.2.1 Geodetic and ECEF Coordinate Systems

The position vector transformation from the geodetic system to the ECEF coordinate system is an intermediate step in converting the GPS position measurement to the local NED coordinate system. Given a point in the geodetic system, say

$$P_{\mathrm{g}} = \begin{pmatrix} \boldsymbol{\lambda} \\ \boldsymbol{\varphi} \\ h \end{pmatrix},$$

its coordinate in the ECEF frame is given by

$$\mathbf{P}_{\mathrm{e}} = \begin{pmatrix} x_{\mathrm{e}} \\ y_{\mathrm{e}} \\ z_{\mathrm{e}} \end{pmatrix} = \begin{pmatrix} (N_{\mathrm{E}} + h)\cos\boldsymbol{\varphi}\cos\boldsymbol{\lambda} \\ (N_{\mathrm{E}} + h)\cos\boldsymbol{\varphi}\sin\boldsymbol{\lambda} \\ [N_{\mathrm{E}}(1-\mathbf{e}^2) + h]\sin\boldsymbol{\varphi} \end{pmatrix}, \tag{2.22}$$

where $\mathbf{e}$ and N_{E} are as given in (2.5) and (2.7), respectively.

2.3.2.2 ECEF and Local NED Coordinate Systems

The position transformation from the ECEF frame to the local NED frame is required together with the transformation from the geodetic system to the ECEF frame to form a complete position conversion from the geodetic to local NED frames. More specifically, we have

$$\mathbf{P}_{\mathrm{n}} = \mathbf{R}_{\mathrm{n/e}}(\mathbf{P}_{\mathrm{e}} - \mathbf{P}_{\mathrm{e,ref}}), \tag{2.23}$$

where $\mathbf{P}_{\mathrm{e,ref}}$ is the position of the origin of the local NED frame (i.e., O_{n}, normally the takeoff point in UAV applications) in the ECEF coordinate system, and $\mathbf{R}_{\mathrm{n/e}}$ is the rotation matrix from the ECEF frame to the local NED frame, which is given by

$$\mathbf{R}_{\mathrm{n/e}} = \begin{bmatrix} -\sin\boldsymbol{\varphi}_{\mathrm{ref}}\cos\boldsymbol{\lambda}_{\mathrm{ref}} & -\sin\boldsymbol{\varphi}_{\mathrm{ref}}\sin\boldsymbol{\lambda}_{\mathrm{ref}} & \cos\boldsymbol{\varphi}_{\mathrm{ref}} \\ -\sin\boldsymbol{\lambda}_{\mathrm{ref}} & \cos\boldsymbol{\lambda}_{\mathrm{ref}} & 0 \\ -\cos\boldsymbol{\varphi}_{\mathrm{ref}}\cos\boldsymbol{\lambda}_{\mathrm{ref}} & -\cos\boldsymbol{\varphi}_{\mathrm{ref}}\sin\boldsymbol{\lambda}_{\mathrm{ref}} & -\sin\boldsymbol{\varphi}_{\mathrm{ref}} \end{bmatrix}, \tag{2.24}$$

and where $\boldsymbol{\lambda}_{\mathrm{ref}}$ and $\boldsymbol{\varphi}_{\mathrm{ref}}$ are the geodetic longitude and latitude corresponding to $\mathbf{P}_{\mathrm{e,ref}}$.

2.3.2.3 Geodetic and Vehicle-Carried NED Coordinate Systems

In aerospace navigation, a kinematical relationship between geodetic position and vehicle-carried NED velocity is of great importance. The derivative of the geodetic position can be expressed in terms of the vehicle-carried NED velocity as the following:

$$\dot{\boldsymbol{\lambda}} = \frac{v_{\mathrm{nv}}}{(N_{\mathrm{E}} + h)\cos\boldsymbol{\varphi}}, \tag{2.25}$$

$$\dot{\boldsymbol{\varphi}} = \frac{u_{\mathrm{nv}}}{M_{\mathrm{E}} + h}, \tag{2.26}$$

and

$$\dot{h} = -w_{\mathrm{nv}}. \tag{2.27}$$

We note that the first two equations are derived based on spherical triangles, whereas the third one can be easily obtained from the definitions of h and w_{nv}.

The derivatives of the vehicle-carried NED velocities are respectively given by

$$\dot{u}_{\mathrm{nv}} = -\frac{v_{\mathrm{nv}}^2\sin\boldsymbol{\varphi}}{(N_{\mathrm{E}} + h)\cos\boldsymbol{\varphi}} + \frac{u_{\mathrm{nv}}w_{\mathrm{nv}}}{M_{\mathrm{E}} + h} + a_{\mathrm{mx,nv}}, \tag{2.28}$$

$$\dot{v}_{\mathrm{nv}} = \frac{u_{\mathrm{nv}}v_{\mathrm{nv}}\sin\boldsymbol{\varphi}}{(N_{\mathrm{E}} + h)\cos\boldsymbol{\varphi}} + \frac{v_{\mathrm{nv}}w_{\mathrm{nv}}}{N_{\mathrm{E}} + h} + a_{\mathrm{my,nv}}, \tag{2.29}$$

and

$$\dot{w}_{\mathrm{nv}} = -\frac{v_{\mathrm{nv}}^2}{N_{\mathrm{E}} + h} - \frac{u_{\mathrm{nv}}^2}{M_{\mathrm{E}} + h} + g + a_{\mathrm{mz,nv}}, \tag{2.30}$$

where g is the gravitational acceleration, and

$$\mathbf{a}_{\mathrm{mea,nv}} = \begin{pmatrix} a_{\mathrm{mx,nv}} \\ a_{\mathrm{my,nv}} \\ a_{\mathrm{mz,nv}} \end{pmatrix} \tag{2.31}$$

is the projection of $\mathbf{a}_{\mathrm{mea,b}}$, the proper acceleration measured on the body frame, onto the vehicle-carried NED frame. The proper acceleration is an acceleration relative to a free-fall observer who is momentarily at rest relative to the object being measured [209]. In the above equations, we omit terms related to the earth's self-rotation, which is reasonable for small-scale UAV rotorcraft working in a small confined area.

2.3.2.4 Vehicle-Carried NED and Body Coordinate Systems

Kinematical relationships between the vehicle-carried NED and the body frames are important to flight dynamics modeling and automatic flight control. For translational kinematics, we have

$$\mathbf{V}_{\mathrm{b}} = \mathbf{R}_{\mathrm{b/nv}} \mathbf{V}_{\mathrm{nv}}, \tag{2.32}$$

$$\mathbf{a}_{\mathrm{b}} = \mathbf{R}_{\mathrm{b/nv}} \mathbf{a}_{\mathrm{nv}}, \tag{2.33}$$

and

$$\mathbf{a}_{\mathrm{mea,b}} = \mathbf{R}_{\mathrm{b/nv}} \mathbf{a}_{\mathrm{mea,nv}}, \tag{2.34}$$

where $\mathbf{R}_{\mathrm{b/nv}}$ is the rotation matrix from the vehicle-carried NED frame to the body frame and is given by

$$\mathbf{R}_{\mathrm{b/nv}} = \begin{bmatrix} \mathbf{c}_{\theta}\mathbf{c}_{\psi} & \mathbf{c}_{\theta}\mathbf{s}_{\psi} & -\mathbf{s}_{\theta} \\ \mathbf{s}_{\phi}\mathbf{s}_{\theta}\mathbf{c}_{\psi} - \mathbf{c}_{\phi}\mathbf{s}_{\psi} & \mathbf{s}_{\phi}\mathbf{s}_{\theta}\mathbf{s}_{\psi} + \mathbf{c}_{\phi}\mathbf{c}_{\psi} & \mathbf{s}_{\phi}\mathbf{c}_{\theta} \\ \mathbf{c}_{\phi}\mathbf{s}_{\theta}\mathbf{c}_{\psi} + \mathbf{s}_{\phi}\mathbf{s}_{\psi} & \mathbf{c}_{\phi}\mathbf{s}_{\theta}\mathbf{s}_{\psi} - \mathbf{s}_{\phi}\mathbf{c}_{\psi} & \mathbf{c}_{\phi}\mathbf{c}_{\theta} \end{bmatrix}, \tag{2.35}$$

and where $\mathbf{s}_{*}$ and $\mathbf{c}_{*}$ denote $\sin(*)$ and $\cos(*)$, respectively.

For rotational kinematics, we focus on the angular velocity vector $\boldsymbol{\omega}^{\mathrm{b}}_{\mathrm{b/nv}}$, which describes the rotation of the vehicle-carried NED frame with respect to the body frame projected onto the body frame. Following the definition and sequence of the Euler angles, it can be expressed as

$$\begin{aligned} \boldsymbol{\omega}^{\mathrm{b}}_{\mathrm{b/nv}} := \begin{pmatrix} p \\ q \\ r \end{pmatrix} &= \begin{pmatrix} \dot{\phi} \\ 0 \\ 0 \end{pmatrix} + \mathbf{R}_{\mathrm{b/int2}} \left[\begin{pmatrix} 0 \\ \dot{\theta} \\ 0 \end{pmatrix} + \mathbf{R}_{\mathrm{int2/int1}} \begin{pmatrix} 0 \\ 0 \\ \dot{\psi} \end{pmatrix} \right] \\ &= \mathbf{S} \begin{pmatrix} \dot{\phi} \\ \dot{\theta} \\ \dot{\psi} \end{pmatrix}, \end{aligned} \tag{2.36}$$

where p, q, and r are the standard symbols adopted in the aerospace community for the components of $\boldsymbol{\omega}^{\mathrm{b}}_{\mathrm{b/nv}}$, $\mathbf{R}_{\mathrm{int2/int1}}$ and $\mathbf{R}_{\mathrm{b/int2}}$ are respectively given as in (2.18) and (2.19), and lastly, $\mathbf{S}$ is the lumped transformation matrix given by

$$\mathbf{S} = \begin{bmatrix} 1 & 0 & -\sin\theta \\ 0 & \cos\phi & \sin\phi\cos\theta \\ 0 & -\sin\phi & \cos\phi\cos\theta \end{bmatrix}. \tag{2.37}$$

It is simple to verify that

$$\mathbf{S}^{-1} = \begin{bmatrix} 1 & \sin\phi\tan\theta & \cos\phi\tan\theta \\ 0 & \cos\phi & -\sin\phi \\ 0 & \sin\phi/\cos\theta & \cos\phi/\cos\theta \end{bmatrix}. \tag{2.38}$$

We note that (2.36) is known as the Euler kinematical equation and that $\theta = \pm 90^\circ$ causes singularity in (2.37), which can be avoided by using quaternion expressions.

2.3.2.5 Local and Vehicle-Carried NED Coordinate Frames

As mentioned in Sect. 2.2.4, under the assumption that there is no directional difference between the local and vehicle-carried NED frames, we have

$$\mathbf{V}_{\mathrm{n}} = \mathbf{V}_{\mathrm{nv}}, \qquad \boldsymbol{\omega}^{\mathrm{b}}_{\mathrm{b/n}} = \boldsymbol{\omega}^{\mathrm{b}}_{\mathrm{b/nv}}, \qquad \mathbf{a}_{\mathrm{n}} = \mathbf{a}_{\mathrm{nv}}, \qquad \mathbf{a}_{\mathrm{mea,n}} = \mathbf{a}_{\mathrm{mea,nv}}, \tag{2.39}$$

where $\mathbf{a}_{\mathrm{mea,n}}$ is the projection of the proper acceleration measured on the body frame, i.e., $\mathbf{a}_{\mathrm{mea,b}}$, onto the local NED frame. These properties will be used throughout the entire monograph.

Chapter 3
Platform Design and Construction

3.1 Introduction

Constructing a miniature rotorcraft UAV is a challenging task. One has to carefully consider issues such as hardware components selection, layout design, weight balance, and anti-vibration. Although some miniature rotary UAV platforms have been built and used in practical missions, it is difficult to find a uniform, time-saving, and effective platform design methodology in the literature.

In this chapter, we present a systematic design methodology for constructing an unmanned rotorcraft system. For ease of understanding, we use the construction procedure of SheLion, a twin of HeLion but with an additional onboard vision system, as an example. It consists of the following four key steps:

1. Virtual design environment selection
2. Hardware component selection
3. Avionic system design and integration
4. Performance and reliability evaluation

The outline of this chapter is as follows. In Sect. 3.2, we introduce a virtual design environment (VDE), SolidWorks, which is intensively used for the layout design of the unmanned system. Section 3.3 focuses on the hardware components selection for SheLion. Next, we provide in Sect. 3.4 the design and integration procedure, based on the selected hardware components. Finally, the evaluation of the performance and reliability of the overall design is presented in Sect. 3.5.

3.2 Virtual Design Environment Selection

Our experiences have shown that it is important to choose a suitable VDE for constructing a reliable UAV, which can be a great help in the layout design and components integration. We have selected a powerful virtual design environment, called SolidWorks, a 3D-design environment [172], in building SheLion and all other unmanned systems in our group. SolidWorks has the following useful features:

G. Cai et al., *Unmanned Rotorcraft Systems*, Advances in Industrial Control,
DOI 10.1007/978-0-85729-635-1_3, © Springer-Verlag London Limited 2011

Fig. 3.1 SheLion and its virtual counterpart

1. Easy to use: Users can be familiar with functions in SolidWorks via a short self-training, based on several key tutorials and examples.
2. Powerful 3D and 2D design: For any selected component, its 3D virtual counterpart can be precisely molded in terms of shape, size, and color. When a 3D virtual component is created, its associated 2D views and blueprints, which are essential for mechanical manufacturing or modification, are generated at the same time.
3. Precise physical description: Each virtual component can be parameterized with necessary physical parameters such as size, dimension, and weight. Furthermore, SolidWorks integrates a professional function to generate various parameters that cannot be directly measured (e.g., the center gravity location and area of a rugged surface).
4. Animation function: For some components that can move or rotate (such as the pan/tilt servo mechanism addressed later in this chapter), SolidWorks can virtually emulate the motions via the integrated animation function and thus provide straightforward visualization to designers during the component integration procedure.

Such a VDE-based design concept is a remarkable feature of our proposed design methodology. Shown in Fig. 3.1 are the real and virtual SheLion. It is the carefully built virtual SheLion that leads to the quick and smooth construction of its counterpart in the real world.

3.3 Hardware Components Selection

The hardware configuration for SheLion is illustrated in Fig. 3.2, in which each solid block represents a specific device. As mentioned earlier in Chap. 1, the essential hardware components involved in a fully functional rotorcraft unmanned system are

1. an RC rotorcraft;
2. an avionic system for collecting inflight data, executing automatic control laws, and communicating with the ground station;

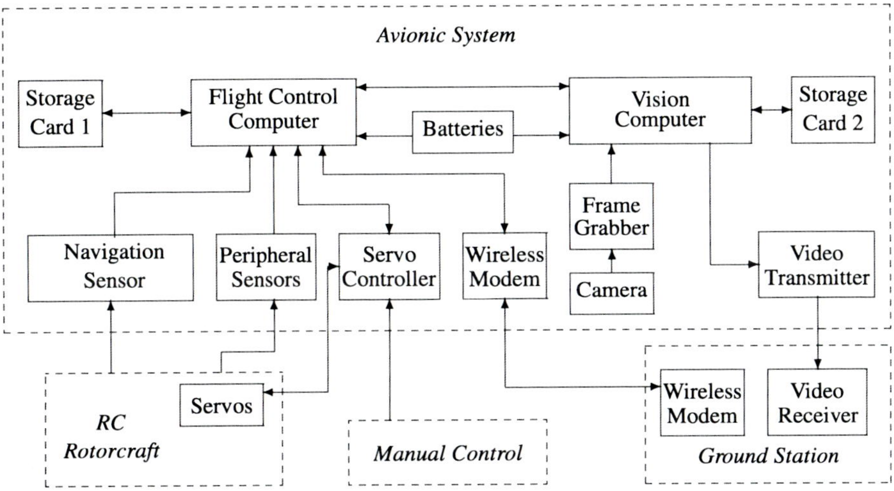

Fig. 3.2 Hardware configuration of SheLion UAV system

Fig. 3.3 Raptor 90 SE helicopter and its virtual counterpart

3. a manual control system consisting of a pilot and a wireless joystick; and
4. a ground station system for monitoring the flight states of the UAV and communicating with the avionic system.

3.3.1 RC Helicopter

The first component we need to select is a high-performance RC rotorcraft, which is also known as a hobby-based or model rotorcraft. For SheLion, we choose a Raptor 90 SE, which is illustrated in Fig. 3.3 along with its virtual counterpart. Some key specifications of the Raptor 90 SE are listed in Table 3.1. The primary reasons for selecting this helicopter are the following:

Table 3.1 Key specifications of Raptor 90 SE helicopter

Specification	Raptor 90 SE helicopter
Fuselage length	1.39 m
Main rotor span	1.41 m
Tail rotor span	0.25 m
Flight endurance	15 minutes
No-load weight	5.95 kg
Maximum takeoff weight	11 kg
Gear ratio	1:8.45:4.65 (main rotor:engine:tail rotor)
Fuselage material	Carbon fiber
Power source	Nitro fuel

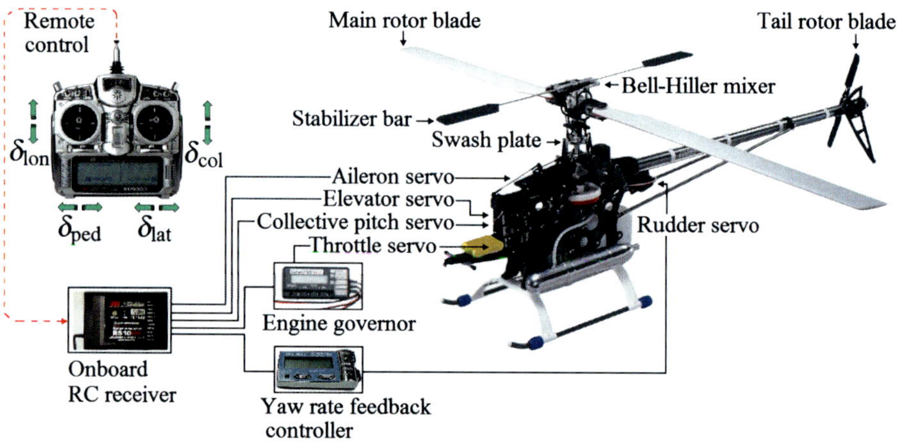

Fig. 3.4 Operating principle of Raptor 90 SE helicopter

1. Low cost and high performance: Compared with most Type II RC rotorcraft such as Turbulence D3 and Observer Twin, the Raptor 90 SE provides equivalent flight performance and reliable structural design, at around half the price.
2. Great maneuverability: The Raptor 90 SE was originally designed for acrobatic flight. Its agility and maneuverability are well known in the RC hobby flight circle.
3. Sufficient payload: From Table 3.1, we note that the maximum takeoff weight for the Raptor 90 SE is 11 kg. By deducting its no-load weight, the effective payload of the Raptor 90 SE is more than 5 kg, which is beyond our budget weight, 4 kg in total for the avionic system, landing skid, and vibration isolators.

The operating principle of the Raptor 90 SE (see Fig. 3.4) is standard and widely adopted by the hobby community. Five digital servo actuators, which are manufactured by well-known vendors Futaba [65] and JR-Propo [92], are employed to receive the control signal (from either the human pilot or the avionic system) and drive various control surfaces of the Raptor 90 SE:

1. The aileron servo, which produces a driving signal δ_{lat}, is in charge of the leftward and rightward tilting motion of the swash plate. Such a movement changes the cyclic pitch angle of the main rotor blades and results in both a rolling motion and lateral translation.
2. The elevator servo with a driving signal δ_{lon} is responsible for the forward and backward tilting motion of the swash plate. This tilting also changes the cyclic pitch angle of the main rotor blades but results in a pitching motion and longitudinal translation.
3. The collective pitch servo through its driving signal δ_{col} changes the vertical position of the swash plate. As a result, the collective pitch angle of the main rotor blades is changed to generate heave motion. It should be noted that such a feature might not be existent for some low-end fixed-pitch RC helicopters.
4. The rudder servo, which generates a driving signal δ_{ped}, cooperates with a factory-installed yaw rate feedback controller to realize the yaw rate and heading control via controlling the collective pitch angle of the tail rotor blade. We note that in the RC rotorcraft, the yaw rate controller is added to allow a human pilot to control with the over sensitive dynamics of a bare yaw channel.
5. Finally, the throttle servo works with an RC-purpose engine governor to control a constant rotation speed of the main rotor. We note that in all the works presented in this monograph, the throttle servo is pre-set to generate a constant rotation speed. This servo is not used in the automatic control loop.

We note that a Bell-Hiller stabilizer bar [183], the most distinguished feature of the RC helicopter (see also in Fig. 3.4), is very often employed to cooperate with the swash plate for achieving desired rotor flapping responses.

3.3.2 Flight Control Computer

The flight control computer is the brain of the avionic system. Its primary functions include

1. analyzing various flight data delivered by onboard sensors and the vision computer;
2. executing flight control law;
3. communicating with the ground station; and
4. logging flight data to a compact flash (CF) card (Storage Card 1 in Fig. 3.2) for post-flight analysis.

Because of the unique characteristics of the UAV systems, in selecting a flight computer, special attention shall be paid to its size, weight, input/output (I/O) ports configuration, expendability, anti-vibration property, and power consumption.

The final choice for SheLion is an embedded computer board PC/104 ATHENA, as shown in Fig. 3.5 along with its virtual counterpart. It is manufactured based on

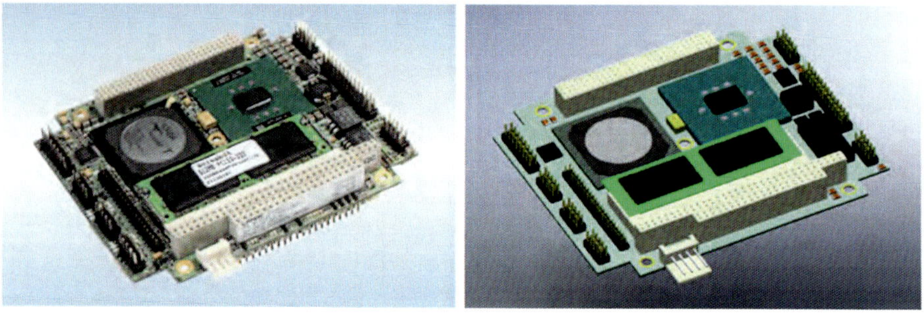

Fig. 3.5 ATHENA computer board and its virtual counterpart

the PC/104(-plus) standard, which is defined by the PC/104 Consortium [140] and commonly used in embedded system applications. ATHENA is well suited to our UAV application because of the following features:

1. Small size: PC/104 ATHENA is a handheld computer board with the dimensions of 96 mm × 90 mm × 10 mm.
2. Lightweight: With all of the wires and cables connected, the total weight of ATHENA is 140 g.
3. Sufficient processing speed: The processing speed of ATHENA is 500 MHz, which is far beyond the requirement of the onboard software system.
4. Rich I/O ports: ATHENA provides rich I/O ports for communicating with external devices, including four RS-232 serial ports, four USB ports, two counter/timer ports, one 24-way digital I/O port, one 16-way analog I/O port, and one 100 Mb Ethernet port. This feature provides sufficient freedom for the sensor selection and facilitates the online program debugging.
5. Anti-vibration capacity: The PC/104(-plus) standard adopts a unique multiple pin-hole connection method, which results in good connection reliability and anti-vibration capacity.
6. Expendability: The PC/104(-plus)-based computer boards are unified in connection and size. As such, other PC/104(-plus) boards, which are designed for special purposes (such as vision data conversion and I/O extension), can be easily attached to ATHENA to form a new computer stack.
7. Low power consumption: The power consumption of ATHENA for full working load is about 12.5 W per hour.

3.3.3 Navigation Sensors

Navigation sensors provide reliable measurement for the flight status of the flying vehicle. Many commercial navigation sensors are available on the market, varying in manufacturing technology, material, estimation algorithm, measuring range, size, and weight. In accordance with the working principle, a complete navigation solution generally falls into one of the following three types: (i) INS (inertial navigation

Fig. 3.6 MNAV and its virtual counterpart

system), (ii) INS/GPS (INS calibrated by GPS), or (iii) GPS-aided AHRS (attitude heading reference system aided by GPS). For SheLion, we aim to customize a low-cost and small-size GPS-aided AHRS with sufficient accuracy.

The GPS-aided AHRS consists of four key components: (i) an inertial measurement unit (IMU), (ii) a GPS receiver, (iii) a magnetometer, and (iv) a data fusion and estimation algorithm. Differing from the traditional AHRS that can only deliver a solution to attitude angles, the GPS-aided AHRS is capable of providing measurement data required for navigation and control, which include position, velocity, attitude, angular rate, and acceleration. For SheLion, a small-size navigation sensor suite, named MNAV, is adopted as the baseline to form the GPS-aided AHRS. It is shown in Fig. 3.6 along with its virtual counterpart. The main features of MNAV are as follows:

1. Complete integration: MNAV is a proper integration of a MEMS-based IMU, a MEMS-based three-axis magnetometer, and a finger-size GPS receiver. All of the raw inflight data can be captured by the MNAV.
2. Compact layout: As shown in Fig. 3.6, MNAV comes with a precise orthogonal pattern, with the accelerometers, gyroscopes, and magnetometers mounted properly.
3. Sufficient range and resolution: The key specifications of the MNAV are given in Table 3.2. With the listed measuring ranges and resolutions, the MNAV can easily handle any normal helicopter maneuver, even for high speed flight conditions.
4. Small size and lightweight: The weight and dimensions of MNAV are 33 g and 57 mm × 45 mm × 11 mm, respectively. It is one of the lightest and smallest navigation sensors commercially available.
5. Serial-based output: The MNAV integrates a 16 MHz single board computer (SBC), which is mainly responsible for conducting the A/D conversion of the raw analog data and outputting the digital signal based on the standard serial protocol. Such a function facilitates the data I/O programming of the onboard software.

Table 3.2 Key specifications of MNAV

Specification	MNAV
Acceleration range	± 2 g[†]
Angular velocity range	± 200 deg/s
Magnetometer range	± 0.75 G
GPS accuracy in CEP	3 m
Update rate	1–100 Hz programmable
Size	$57 \times 45 \times 11$ mm
Weight	33 g
Power consumption	≤ 0.8 W

[†] g is the gravitational acceleration.

3.3.4 Peripheral Sensors

We have employed two peripheral sensors, i.e., an ultrasonic sonar and an RPM (revolutions per minute) sensor in SheLion. Although the GPS unit of MNAV is acceptable to the airborne position control, its accuracy is not good enough for realizing precise automatic landing. As such, we add an additional small-size (40 mm $\times$ 60 mm $\times$ 15 mm) and lightweight (45 g) ultrasonic sonar, namely, a UPK-2500 [184], for measuring the vertical height when SheLion is near the ground. Its effective range is 2 m with a resolution in the millimeter level. The output of the UPK-2500 is voltage signal (0 to 10 V corresponding to 0 to 2 m), which is fed to an analog input channel of the ATHENA computer board.

The RPM sensor is adopted to provide real-time RPM of the main rotor to the flight control computer. It is an essential component to airborne security and automatic landing. The RPM sensor equipped in SheLion is customized by our NUS UAV Research Team. It mainly consists of (i) a magnetic sensor to sense the rotation of the helicopter engine and (ii) a Schmidt trigger chip to conduct the A/D conversion and output the triggering signal to a counter/timer port residing in the ATHENA computer board.

3.3.5 Fail-Safe Servo Controller

The fail-safe servo controller (or servo controller in short) is another important device to guarantee the airborne security of the miniature UAV helicopters. It is mainly responsible for decoding both piloted and computer-generated servo control commands and selecting desired decoded signals to drive multiple servo actuators. In the case of accidents that might occur during an autonomous flight, with the assistance of the servo controller, the human pilot might have a chance to retrieve the UAV platform. For SheLion, a commercial servo controller board, namely an HBC-101, is selected and shown in Fig. 3.7. It has the following attractive features:

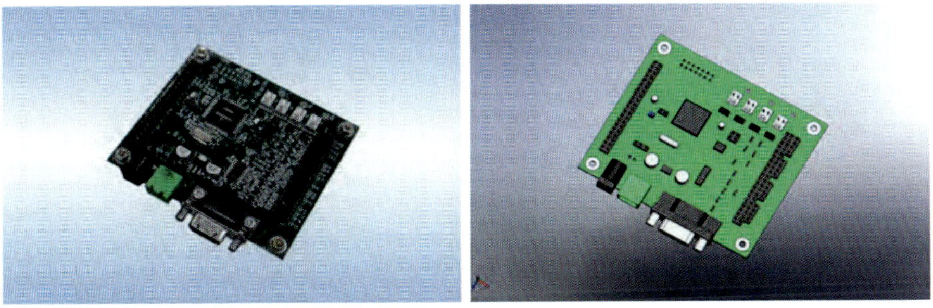

Fig. 3.7 HBC-101 and its virtual counterpart

1. Reliable switching function: At any time, the HBC-101 is capable of realizing smooth switching between automatic control and manual control signals. A special channel is pre-determined for receiving the switching signal that is generated by the manual control joystick. With such a design, the human pilot owns the highest authority to determine which side (automatic or manual input) is mapped to the HBC-101 servo output.
2. Sufficient input and output channels: As mentioned earlier, five servo actuators are used in the Raptor 90 SE helicopter. The HBC-101 provides eight input and eight output channels to fulfill our requirement.
3. Signal recording capacity: The HBC-101 has the capacity to sample both the input and output signals up to 100 Hz. The logged data can be sent to the ATHENA computer board through the serial communication protocol. This function is particularly important to flight dynamics modeling and automatic control performance analysis.
4. High resolution: The input-recording and servo-driving can be regarded as specified A/D and D/A conversion. A resolution up to 0.009 deg provided by the HBC-101 can substantially enhance the data quality and practical control performance.

3.3.6 Wireless Modem

A pair of wireless modems are used to establish communications between the UAV helicopter and the ground control station. Inflight status down link and command/trajectory uploading are both done through this wireless system. Communication range and reliability are the most important issues when we select the wireless modems for SheLion. Our final choice is an IM-500 (see Fig. 3.8) working at a 2.4 GHz bandwidth. This communication board has a handheld size and a weight of 75 g. Some of its distinguished features are listed as follows:

1. Extremely long range: With a clear line of sight, the maximum communication range of the IM-500 is 32 km.

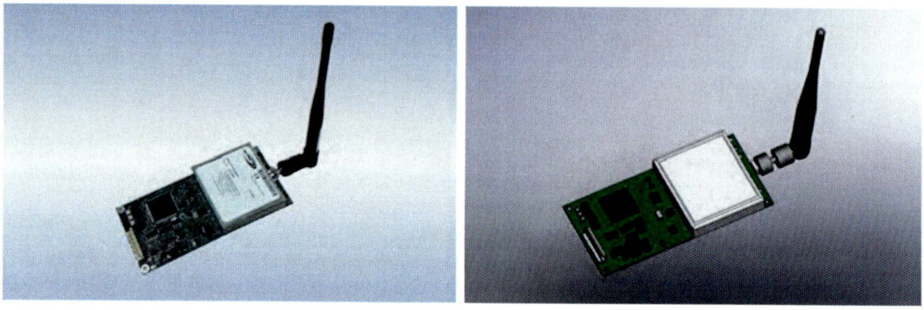

Fig. 3.8 IM-500 and its virtual counterpart

2. Serial protocol communication: The IM-500 adopts a serial communication protocol.
3. Sufficient throughput rate: The data throughput of the IM-500 is 115.2 Kbps (kilo bits per second), which is good enough for our applications.

3.3.7 Batteries

Three types of battery candidates are widely seen in civilian applications: lithium-polymer (Li-Po), nickel-metal hydride (Ni-Mh), and nickel-cadmium (Ni-Cd). Compared with the other two types, a Li-Po battery has a far superior performance in terms of (i) energy density, (ii) charge/discharge efficiency, (iii) self-discharge rate, (iv) usage durability, and (v) cycle durability. As such, we choose Li-Po batteries for SheLion. More specifically, four Li-Po batteries, produced by Thunder Power RC Inc., are used to provide power to the equipped electronic components. Two of them have an energy capacity at 7.4 V/1300 mAh (1 hour continuously working at 7.4 V with 1300 mA discharging current) with a weight of 70 g each, whereas the other two are at 11.1 V/2100 mAh with a weight of 190 g each. We need to highlight that the battery capacities are determined after a careful power supply design, which will be presented in detail later in Sect. 3.4.3.

3.3.8 Vision Computer

The vision computer coordinates the vision control part of the avionic system. Its main functions include (i) receiving raw vision information from a frame grabber, (ii) analyzing image data based on selected algorithms, (iii) communicating with the flight control computer to realize vision-based control, and (iv) transmitting vision data to the ground control station. The key reasons for adopting two independent computers for flight control and vision processing are as follows:

1. The computational load for vision processing is much heavier than for flight control. Integrating both in a single computer would greatly increase the possibility of software system jams due to data blocking, which might cause serious accidents when conducting actual flight tests.
2. The execution frequency of flight control is much faster than that of vision processing. Furthermore, in many implementations the vision signal may not even be utilized. A separated configuration makes the onboard software system development much easier in (i) execution time allocating, (ii) task thread scheduling, and (iii) data exchange.

For SheLion, another PC/104(-plus) computer board, named Cool-Road-Runner III, is used. It has a higher working speed (933 MHz) to account for the requirement on vision data processing, but less I/O port (not required for the vision processing part). An extra heat sink increases its total weight to 190 g. Except for these differences, Cool-Road-Runner III shares similar features with the ATHENA computer board detailed earlier.

3.3.9 Vision Sensor

The vision sensor, i.e., a camera, is responsible for obtaining inflight visual information on the surrounding environment. In our research, both static and dynamic features (see, e.g., the color and shape of landmarks and the ground target motion) of the collected image are required to obtain sufficient resolution. Our final choice is a compact-size CCD (charge-coupled-device) camera, an Edmund Optics color DSP board camera. It features 380 TV-line resolution, a 40-degree field of view, and an ultra light weight of 30 g. For low-level end (less than $1,000), the CCD camera, compared with its similar scale CMOS counterpart, generally provides a higher quality and sharper image, which is beneficial to the feature extraction (particularly, for color feature extraction) conducted in the vision processing algorithm. It is noted that the selected camera outputs an analog image signal in the PAL TV standard. The signal is first fed to the frame grabber (introduced below) for A/D conversion.

3.3.10 Frame Grabber

The primary function of a frame grabber is to perform the A/D conversion of the analog video signals and then output the digitalized data to the vision computer for further processing. Our selection is a PC/104(-plus)-standard frame grabber, a Colory 104, which has the following features:

1. High resolution: The Colory 104 is capable of providing a resolution up to 720 × 576 (pixels), which is sufficient for online processing.
2. Multiple video inputs: It is able to collect data from multiple cameras.

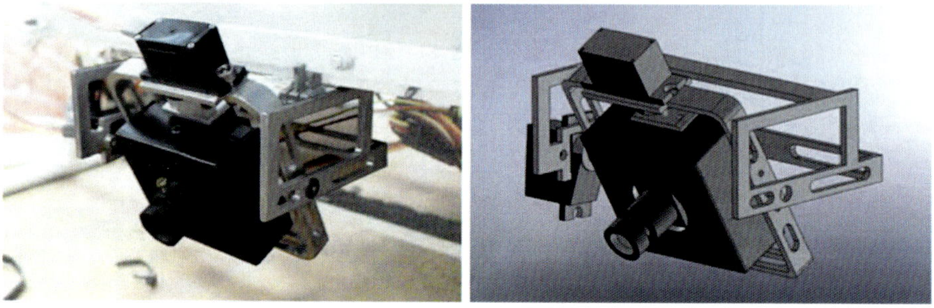

Fig. 3.9 Pan/tilt servo mechanism and its virtual counterpart

3. Sufficient processing rate: The maximum A/D conversion rate is 30 frames per seconds (FPS), which is higher than our onboard vision processing rate, i.e., 10 FPS.
4. Featured processing method: The Colory 104 can handle two image processing tasks in parallel to digitalize image signals into two different formats. One task is configured to convert the captured video signal to the RGB format and transfer it to the random-access memory (RAM) of the vision computer through a PCI bus for real-time image processing. The other is configured to compress the video signal to the JPEG format (with relatively smaller file size) and save it to a CF card (Storage Card 2 in Fig. 3.2) for post-flight analysis.

3.3.11 Servo Mechanism

In vision-based ground target following, it is common to maintain the target objects in the view field of the vision sensor. Attaching the vision sensor to a servo mechanism with pan/tilt freedom can make the realization much easier. Although some air photography gimbals are commercially available on the market, they are rather heavy (1 kg on average) for our miniature UAV with a limited payload. As such, we decide to carry out a custom design, which is shown in Fig. 3.9. It mainly consists of two digital servos and a compact mounting frame. The two servo actuators are controlled by another servo controller, named Pololu servo controller. It can only decode the computer-generated driving signal and without an input recording function. Although the Pololu servo controller is less functional compared with the HBC-101, it is suitable for servo mechanism driving and with a much lighter weight (10 g). The total weight of the custom-made pan/tilt servo mechanism is less than 40 g.

3.3.12 Video Transmitter and Receiver

In order to provide ground users with a real-time visualization in monitoring the onboard vision processing work, the video captured by the onboard camera is transmitted back to the ground control station, via the liaison between the video transmitter (the UAV side) and the video receiver (the ground control station side). To minimize interferences between the transmissions of the video data and the inflight data, we choose a set of video transmitter and receiver with a working bandwidth of 5.8 GHz, which is far away from the 2.4 GHz working frequency of the wireless modem for the inflight data transmission. The video data transmission rate is up to 11 mega bits per second with an effective range of 200 m. Such a range is sufficient for our experimental purpose.

3.3.13 Manual Control

For SheLion, manual control is realized via a high-quality radio controller, namely, a JR PCM-10X. Its robustness is well acknowledged in the RC community circle. The PCM-10X has 10 programmable signal channels with sufficient resolution. Among them, four channels are assigned to realize the servo-driving scheme introduced in Sect. 3.3.1 (note that the collective and throttle servos conventionally share one channel in manual control). To fulfill the requirement on real-time switching between the automatic and manual control modes, one channel of the PCM-10X is particularly programmed and allocated to send switching signals when necessary. The working frequency of the selected PCM-10X joystick is 72 MHz.

3.3.14 Ground Control Station

As mentioned earlier, the main responsibility of the ground control station is to realize effective communications between the avionic system and the ground users and pilots. It should include the following functions: (i) displaying and monitoring real-time inflight status, (ii) displaying images received from the video receiver, (iii) online generating flight trajectories, (iv) sending real-time commands to the avionic system, (v) facilitating the ground users and pilots in automatic control, especially in unexpected occasions such as emergency landing, and (vi) logging the inflight data as a backup of the onboard data recording. In our work, a rugged laptop with a special hardware and structural design for working in dusty and vibrational conditions is adopted as the ground control station. The detailed structure of the software system for the ground control station is presented in Chap. 4.

3.4 Avionic System Design and Integration

With all the hardware components selected in Sect. 3.3, we have successfully integrated them to form an effective avionic system, which has a total weight of 3.06 kg. Special consideration is given to (i) the overall layout, (ii) anti-vibration, (iii) the power supply, and (iv) interference shielding.

3.4.1 Layout Design

Layout design for the avionic system generally includes four steps as illustrated in Fig. 3.10, which are highlighted in the following:

1. DETERMINING GPS-AIDED AHRS LOCATION. Ideally, the optimal mounting position of the GPS-aided AHRS is the CG of the unmanned rotorcraft. However, it is generally impractical since the UAV CG is commonly located at the helicopter fuselage (near the engine). Any mounting offset of the GPS-aided AHRS can cause the so-called *lever arm effect*: The measured accelerations are biased due to the rotational motion of the rotorcraft, as shown in Fig. 3.11. It should be noted that such a bias is unavoidable and what we can do is to minimize it and simplify the data correction procedure. To realize these two aims, we have decided to strictly line up all four CGs of (i) the bare RC helicopter, (ii) the GPS-aided AHRS, (iii) the avionic system, and (iv) the overall rotorcraft UAV, along the Z-axis of the UAV body frame. Since the CG location of the RC helicopter can be precisely measured, the horizontal CG location of the GPS-aided AHRS can be easily determined. The remaining layout design affects only the vertical offset between the CGs of the GPS-aided AHRS and the UAV helicopter.
2. DETERMINING VISION SENSOR LOCATION. Since the vision sensor is employed for collecting visual information on the surrounding environment, a clear and wide eyesight is the most important issue when we determine its location. As a result, the selected CCD camera is suspended at the front of the avionic system via the custom built pan/tilt servo mechanism.
3. CG BALANCING. We need to determine the proper positions of (i) two PC/104(-plus) computers, (ii) the frame grabber, (iii) the servo controller board, (iv) the wireless modem, and (v) the batteries. Items 1 to 3 form a computer stack, which is to be located at the front side. The batteries and wireless modem are to be grouped and mounted at the rear side for longitudinal CG balancing. Furthermore, we have to ensure that the weight is laterally symmetrical and the overall size of the avionic system is as compact as possible. With the help of the SolidWorks VDE, we can precisely determine the slots between each of the two groups and the GPS-aided AHRS together with the consideration of extra space needed for wire/cable connection.
4. MOUNTING LIGHTWEIGHT COMPONENTS. The remaining lightweight (less than 50 g) components include (i) the ultrasonic sonar, (ii) the Schmidt trigger chip, (iii) the video transmitter, and (iv) a custom toggle panel. We have carefully selected their mounting positions to achieve a fine tuning of the CG of the avionic system.

3.4.2 Anti-vibration Design

There are three main vibration sources in SheLion: (i) the main rotor (30.8 Hz), (ii) the engine (260.5 Hz), and (iii) the tail rotor (143.4 Hz). These frequencies are

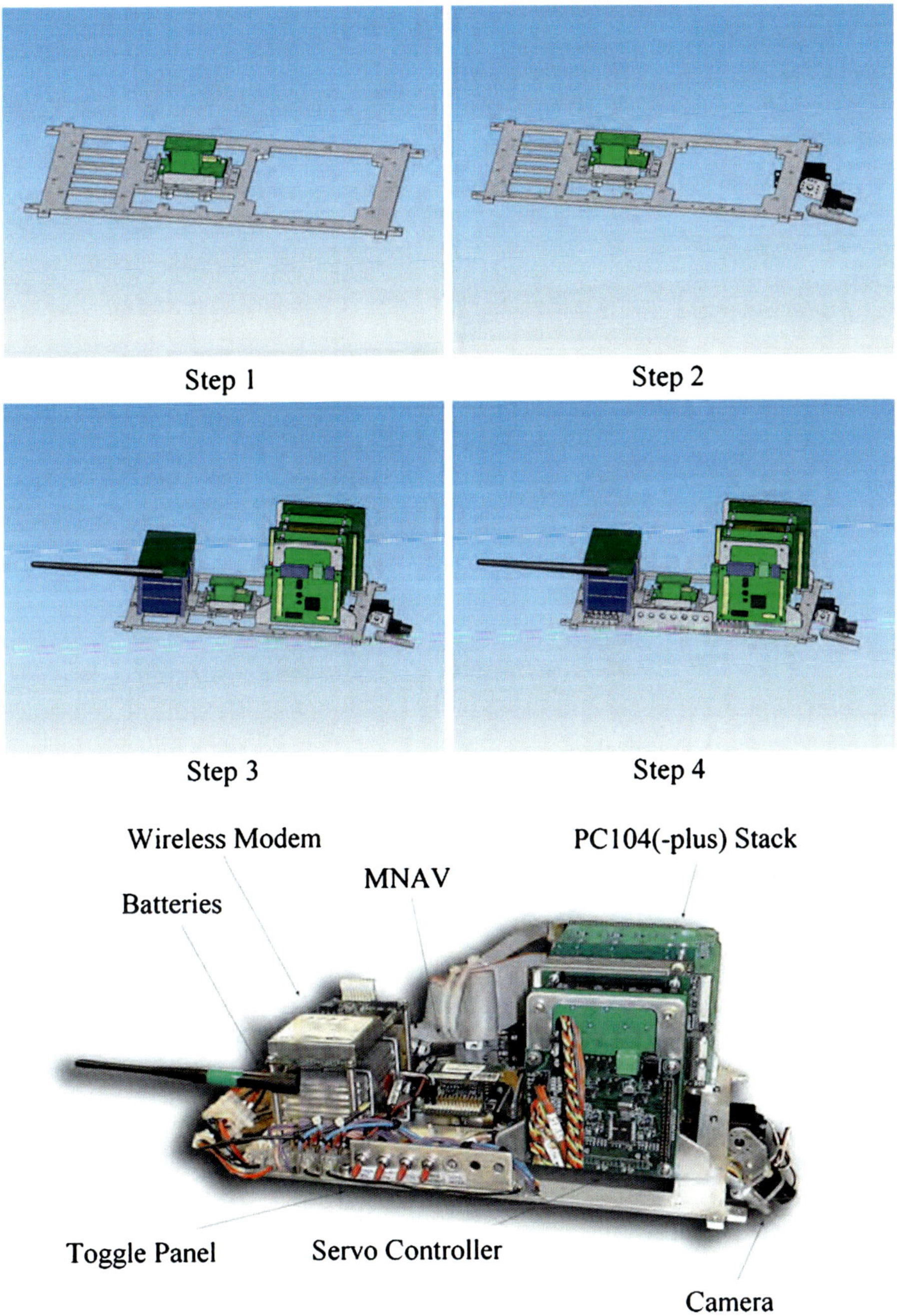

Fig. 3.10 Layout design procedure and the complete avionic system

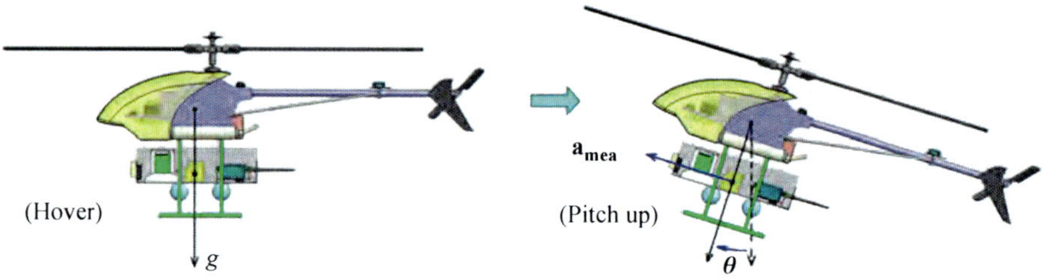

Fig. 3.11 Lever arm effect

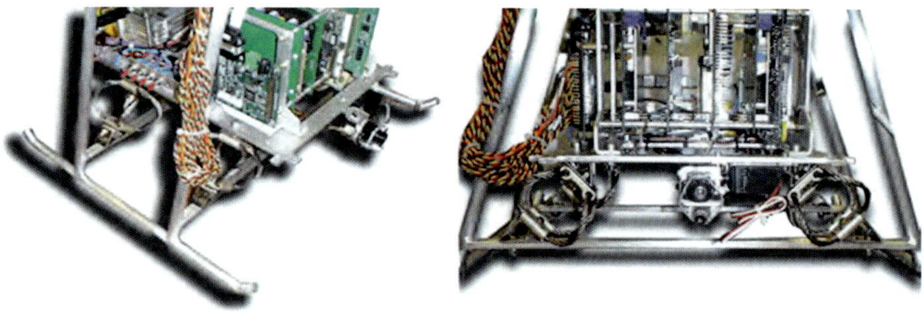

Fig. 3.12 Anti-vibration design for the onboard computer system (left: side view, right: front view)

calculated based on a governed main motor speed of 1,850 RPM. The combined vibration has an amplitude of about 0.8 g, i.e., 7.82 m/s^2, along all of the three axes in the body frame, which generates bias in the measurement data of the acceleration and angular velocity. Furthermore, it may cause a loose connection or malfunction of the hardware components. As such, an effective anti-vibration design is essential in constructing the avionic system to ensure its reliable functioning.

For SheLion, we use four wire-rope isolators for the anti-vibration purpose, which are mounted symmetrically around the CG of the avionic system (shown in Fig. 3.12) with the following particular features:

1. 45-degree compression mounting: As illustrated in Fig. 3.13, the selected wire rope isolators can be mounted in three different ways: (i) vertical compression, (ii) suspending shear/roll, and (iii) 45-degree compression/roll. We decide to use the third one since it can provide an effective vibration isolation in not only vertical but also horizontal directions. It is particularly suitable for the avionic system that is affected by vibration from multiple axes.
2. Good transmitting rate: The transmitting rate is defined as the ratio of the attenuated vibration level to the raw vibration level. According to the manufacturing data sheet given in Fig. 3.14 (see also [35]), the wire-rope isolators can achieve a less than 20% transmitting rate if the raw vibration frequency is three times larger than the system natural frequency (with the wire rope isolators attached). Considering the vibration source with the lowest frequency (30.8 Hz from the main rotor) of SheLion, we need to ensure that the system natural frequency is

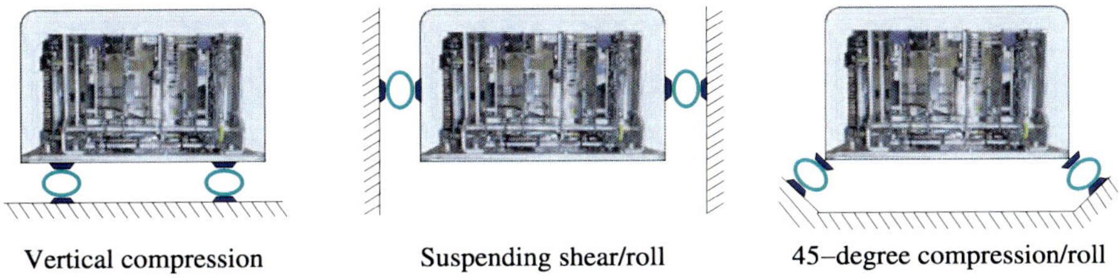

Fig. 3.13 Three mounting methods of the selected wire-rope isolators

Fig. 3.14 Working point of the selected wire-rope isolators

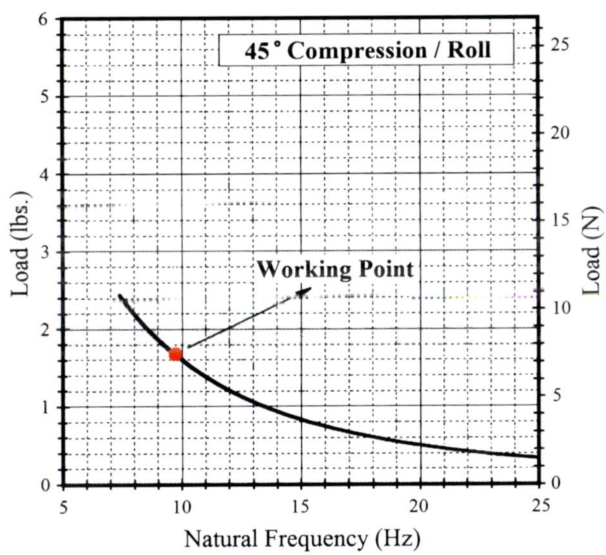

less than 10.26 Hz. We also have to take another important issue into account when selecting the system natural frequency—it should have a sufficient margin from the normal helicopter maneuver frequency to avoid any harmful resonance. The finally selected wire-rope isolators feature the working property depicted in Fig. 3.14. Since each isolator shares a load of 7.5 N (based on the total weight of the avionic system, 3.06 kg), the system natural frequency is about 10 Hz. It is sufficiently spaced from the working frequency of the helicopter maneuver, which is normally below 4 Hz.

Our anti-vibration design has demonstrated its efficiency in reducing raw vibration and increasing overall safety. Its actual performance will be further examined in Sect. 3.5.

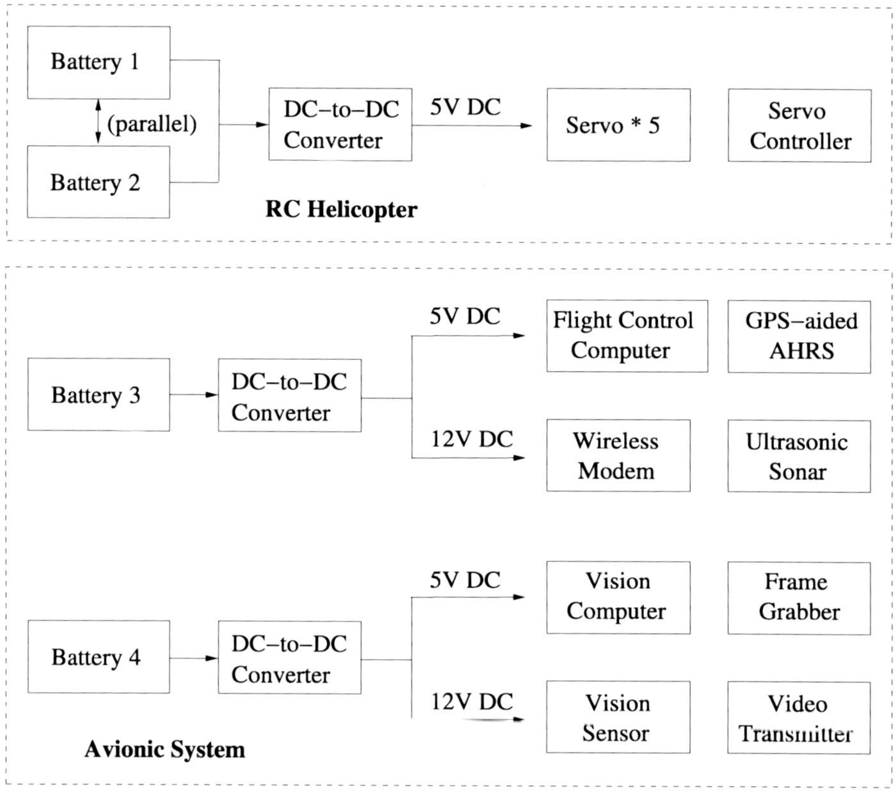

Fig. 3.15 Power supply design for SheLion

3.4.3 Power Supply Design

The primary aim of the power supply design is to achieve the best tradeoff between minimizing the battery weight and providing a sufficient power supply and safety margin. Figure 3.15 illustrates our solution, which is addressed in detail as follows.

Batteries 1 and 2, which have a capacity of 7.4 V/1300 mAh, are used to power the onboard servos and servo controllers through a high-efficiency (90% power transmission rate) DC-to-DC converter board. Although a single battery is sufficient to power these components, we have decided to use two batteries instead to enhance the safety margin. We need to highlight that in our design, the servo controller is also powered by Batteries 1 and 2. The main reason for such a configuration is again to maximumly ensure overall safety. With such a configuration, the pilot would still have a chance to retrieve the unmanned helicopter in certain unexpected situations such as onboard-software breakdown or out-of-power Batteries 3 and 4 (or even one of Batteries 1 and 2).

Table 3.3 Power consumption list for SheLion UAV helicopter

Hardware component	Power consumption (Wh)
Flight control computer	12.5 (at 5 V)
GPS-aided AHRS	0.5 (at 5 V)
Wireless modem	3.9 (at 12 V)
Ultrasonic sonar	0.9 (at 12 V)
Vision computer	16.5 (at 5 V)
Frame grabber	0.5 (at 5 V)
Vision sensor	0.6 (at 12 V)
Video transmitter	1.2 (at 12 V)

To avoid the potential conflict of the power supply between the flight control computer and the vision computer, we employ two additional batteries (Batteries 3 and 4) to separately provide power to them. Two more DC-to-DC converter boards are used to regulate the raw voltages of Batteries 3 and 4 to 5 V and 12 V, considering the different voltage levels of the involved hardware components. It can be observed from Table 3.3 that the total power consumption of these two groups is very similar (i.e., 17.8 Wh and 18.8 Wh, respectively). We thus select two identical batteries with a capacity of 11.1 V/2100 mAh (23.31 Wh). It is noted that the sufficient power supply margin and the 10% power loss in the DC-to-DC converter boards have been taken into account in our decision.

3.4.4 Shielding Design

The shielding design minimizes the harmful effects caused by both the electromagnetic interference (EMI) and the radio frequency interference (RFI). For the miniature UAVs, such interferences would result in side effects including (i) biased measurement of the magnetometer caused by the EMI, (ii) servo glitching caused by the RFI, (iii) malfunction of the GPS receiver caused by both the EMI and the RFI, and (iv) reduced range or malfunction of the RC control and the wireless (for both data and video signals) communications. In constructing SheLion, we have verified that the above side effects can be reduced or even eliminated by simply using aluminum-made boxes or foils to isolate the EMI/RFI sources.

3.5 Performance Evaluation

We have conducted a series of ground experiments to initially evaluate the performance and reliability of the SheLion UAV system. These experiments have proven that SheLion can yield very good performances in all categories. During the ground tests, SheLion is placed on level ground with its engine running at 85%

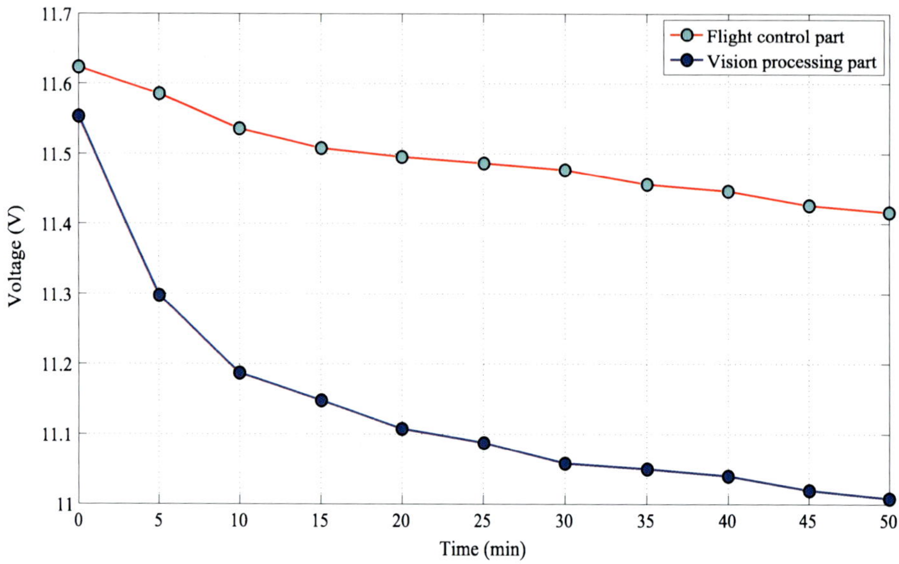

Fig. 3.16 Output voltages of lithium-polymer batteries

of 1,850 RPM (for the normal hover flight condition of SheLion). The ground control station is placed about 500 m away from SheLion. Each ground test lasts more than 12 minutes and the following items are thoroughly examined:

1. RANGE CHECK. The range check evaluates the communications range of the RC control. It is obviously essential for both manual and automatic control modes. It has been verified in intensive ground tests that our EMI/RFI shielding design is very successful and the original RC communications range (50 m without extending the antenna of the joystick) is well maintained.
2. WIRELESS COMMUNICATIONS RELIABILITY. The wireless communications reliability between SheLion and the ground station system is tested through continuously transmitting the inflight data packages in 50 Hz. Our ground tests have proven that the communications between the ground station and the SheLion UAV are perfectly reliable within 500 m.
3. POWER CONSUMPTION. To ensure the reliability of our power supply design, we perform a ground test that lasts 50 minutes. The input voltages for both the flight control and the vision processing groups are recorded periodically with a time interval of 5 minutes. The resulting output voltages of Batteries 3 and 4 are plotted in Fig. 3.16. As expected, the output voltages of both groups drop but with reasonable slopes. The final values stay respectively at 11.42 V (note that 11.1 V for Batteries 3 and 4 is the rated voltage) for the flight control part and 11.01 V for the vision processing part, which are within the safety level for the overall system. This result indicates that the selected batteries have sufficient power to continuously supply the overall avionic system during the whole experimental period.

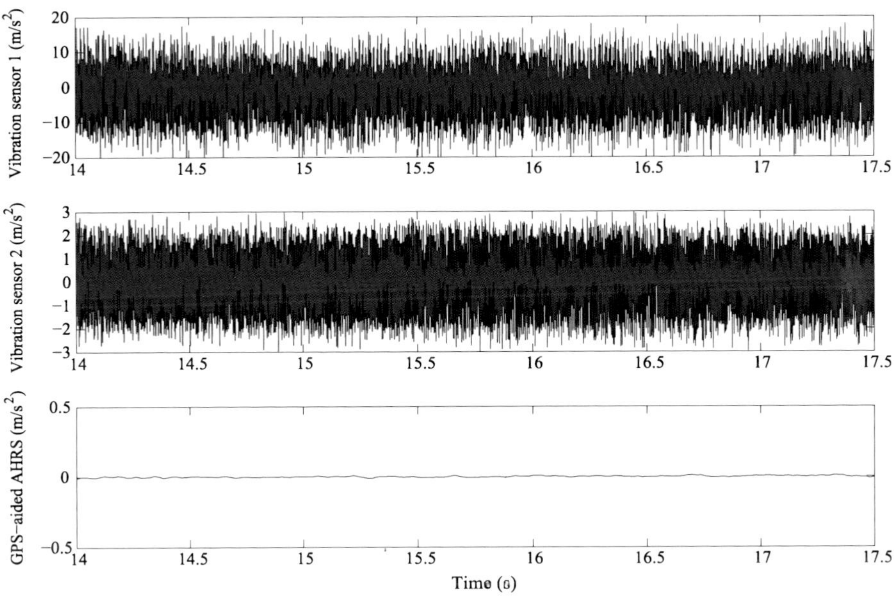

Fig. 3.17 Comparison of vibrational amplitudes along the body-frame Z-axis

4. ANTI-VIBRATION PERFORMANCE. To evaluate the effectiveness of the anti-vibration performance, two sets of three-axis vibration-detection sensors are used, of which one (vibration sensor 1) is attached to a lever of the landing skid and the other (vibration sensor 2) is attached underneath the aluminum plate of the avionic system. Figure 3.17 shows a test sample of the Z-axis acceleration measured by the sensors and the measured acceleration data of the GPS-aided AHRS. With the wire-rope isolators, the resulting vibration transmitting rate is in the range of 20%–25%, which indicates that our anti-vibration design is very effective. Similar results are also obtained for the other two axes. We note that the remaining 20%–25% vibration can be further eliminated through a digital filtering design developed in the onboard software system.
5. PERTURBATION TEST. The perturbation test is conducted in the air. In this experiment, we intend to evaluate the shaking motion of the UAV due to the frequency sweep (a technique commonly used in the system identification of aircraft or rotorcraft). It is actually a simulation of the system identification flight process, which will be addressed later in Chap. 6. More specifically, we first manually command SheLion to be stabilized at a hover flight condition and then inject a frequency-sweep signal to the aileron channel to rotate SheLion with a perturbation level up to $\pm 27^\circ$ in rolling. The main aim for this test is to evaluate the performance and feasibility of the UAV hardware components in drastic flight actions. Figures 3.18, 3.19, 3.20, 3.21 show the recorded data. It is proven that the UAV helicopter constructed, including its onboard hardware as well as software systems, can work properly in such a severe condition.

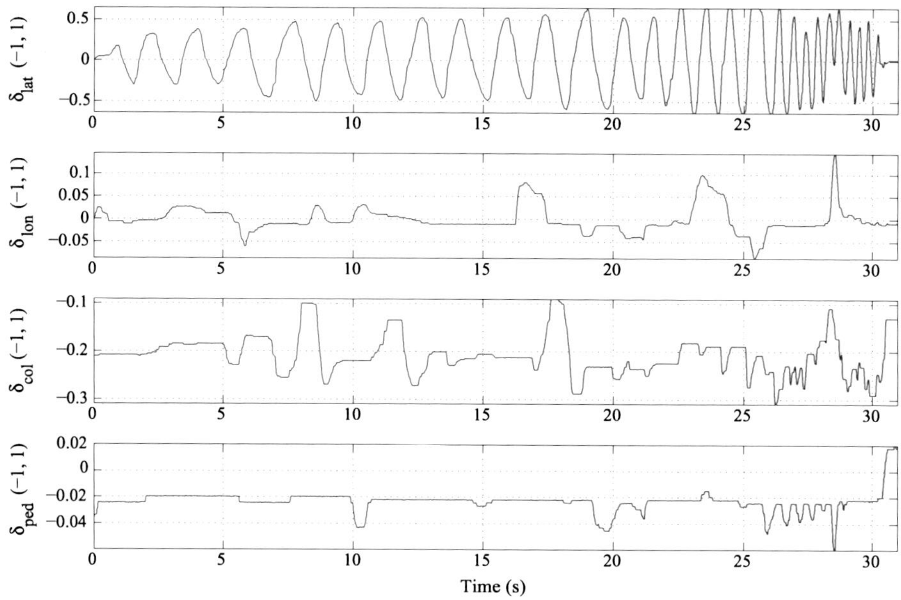

Fig. 3.18 Input signals recorded in the perturbation test

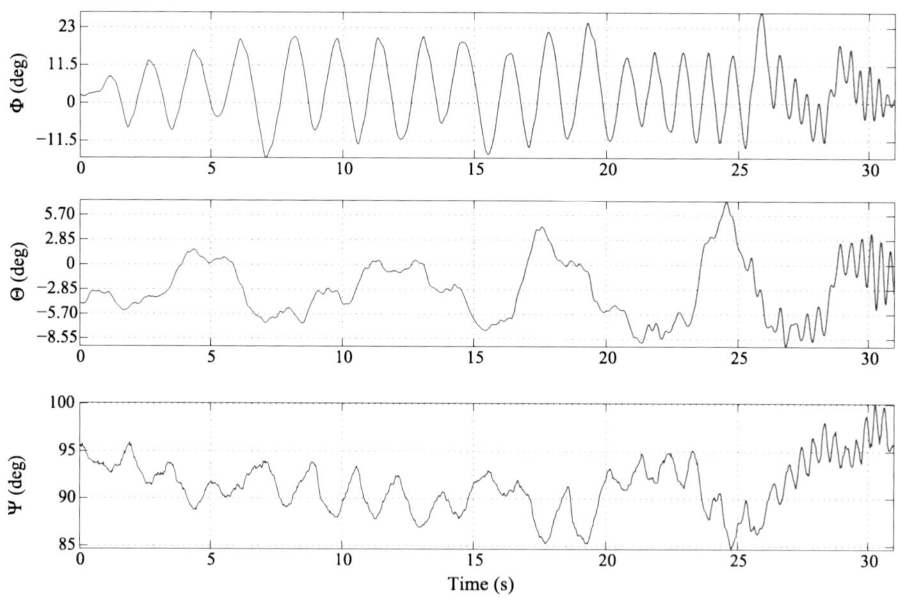

Fig. 3.19 Euler angles recorded in the perturbation test

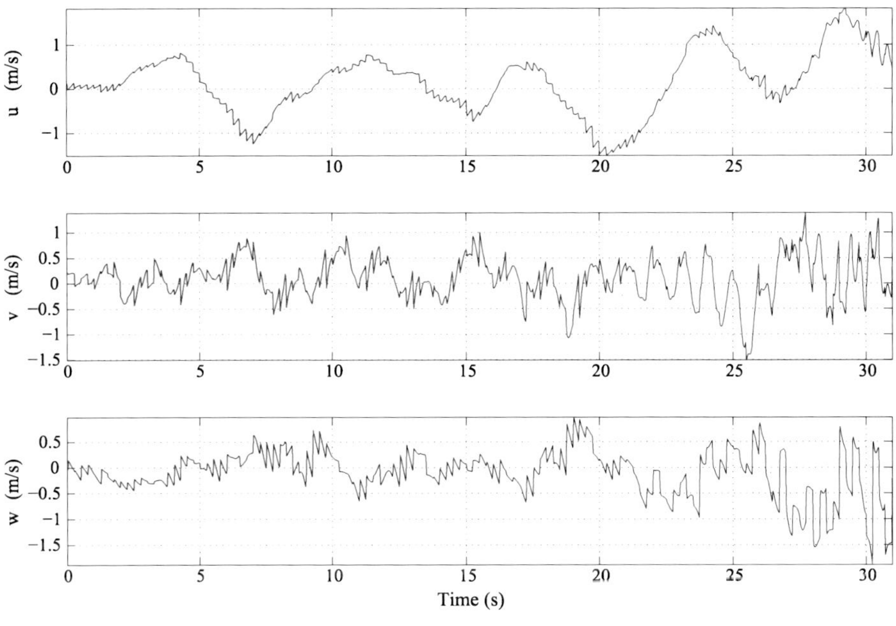

Fig. 3.20 Body-frame velocities recorded in the perturbation test

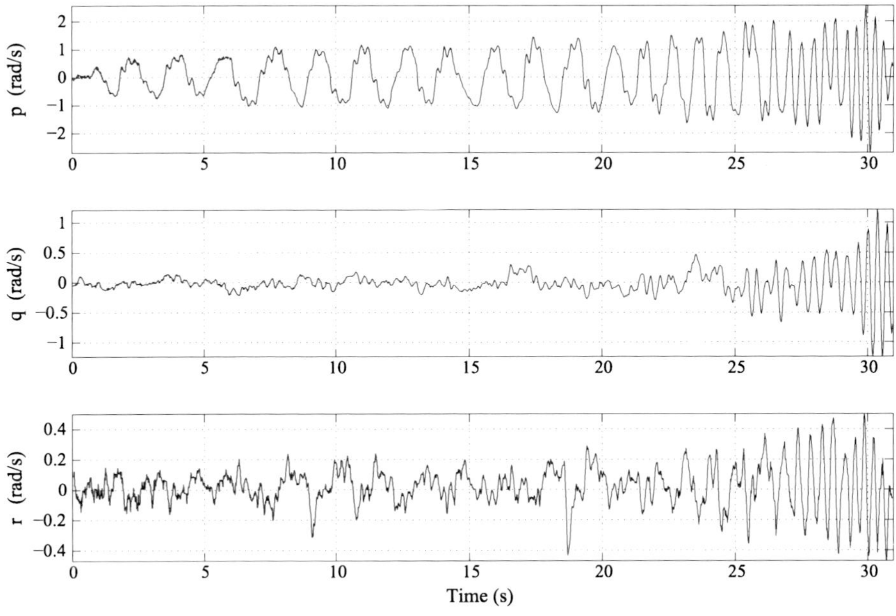

Fig. 3.21 Angular velocities recorded in the perturbation test

Chapter 4
Software Design and Integration

4.1 Introduction

Software systems for the unmanned aerial vehicles perform tasks such as (i) hardware driving, (ii) input-output control law implementation, (iii) device operation management, (iv) multiple-task scheduling, and (v) event management. A number of works have been published in the literature on software system architectures, control law realization, and software implementation for unmanned vehicles (see, e.g., [39, 43, 45, 75, 96, 117, 138, 188, 213]). An introduction of the onboard software implementation for a model helicopter is presented by Kottmann [101], in which onboard tasks such as (i) communications, (ii) data logging, and (iii) control, are briefly described. Wills et al. [213] developed a so-called open control platform (OCP) with reconfigurability and interoperability for complex dynamic systems and presented a demonstration prototype of a simulation platform for an unmanned helicopter based on the common object request broker architecture (CORBA). A practical solution for the software implementation for a fixed-wing UAV is given by Jang and Tomlin [88], which uses inter-process communications and synchronization to schedule the onboard tasks and adopts OCP3 for ground station software.

In this chapter, we follow the recent development reported in [43, 45] to present a comprehensive software system that offers good flexibility and extensibility for new modules and control functionalities and can be universally implemented in unmanned aerial systems. The system consists of two parts, i.e., the onboard software system and the ground station software system.

The onboard software system is developed using a real-time operating system (RTOS), namely QNX Neutrino, which provides reliable support for high precision timer and synchronization operations. It consists of two software modules, i.e., the flight control and vision processing modules. For the former, a multiple thread framework is employed to perform multiple tasks, such as (i) operating hardware components including the navigation sensors and servo actuators, (ii) logging the inflight data, (iii) communicating with the ground control station, and (iv) implementing the automatic control algorithms. To effectively realize automatic control, a behavior-based architecture [44] has been developed. In such an architecture, the

DOI 10.1007/978-0-85729-635-1_4,

operation of a UAV helicopter is organized in a variety of behaviors. A hierarchical and modulized structure is used to execute these behaviors and to integrate multiple control algorithms. A similar software structure is adopted by the onboard vision computer for real-time image processing.

The ground station software system runs on a commercial operating system, Windows XP Professional, which offers a powerful environment for developing the graphical user interfaces (GUIs). The ground control station is equipped with a wireless modem for communicating with the avionic system. The software structure for the ground control station has three layers, i.e., a GUI layer in the foreground, a data transferring layer in the background, and lastly a kernel layer acting for coordinating actions between the previous two layers.

4.2 Onboard Software System

We present in this section the development of the onboard software system for our UAV helicopter system. According to the hardware configuration of the avionic system (see Fig. 3.2), the onboard software system can be separated into two independent modules, i.e., the flight control software module and the vision processing software module. Both of them share a similar framework design and task management scheme. In what follows, we first describe the flight control software module, which includes issues such as (i) framework design, (ii) task management, (iii) automatic control implementation, and (iv) emergency handling, and then briefly introduce the features and differences of the vision processing software module.

4.2.1 Framework Design

The framework of the flight control software for SheLion is depicted in Fig. 4.1. The onboard software portion consists of several blocks, each of which corresponds to a device or task.

1. The NAV block interacts with the GPS-aided AHRS, retrieves the measured inflight data, and estimates the unmeasured flight states, which are essential for automatic control. As mentioned earlier in Chap. 3, the selected MNAV is able to provide measurements on (i) acceleration, (ii) angular velocity, (iii) magnetic field, (iv) geodetic position (in longitude, latitude, and altitude), and (v) NED-based velocity, with a sampling rate of 50 Hz. A comprehensive signal enhancement procedure is also conducted in the NAV block to realize the estimation of the Euler angles, the drift correction of the raw inflight data, and the synchronization of the complete data set. This signal enhancement procedure will be detailed in Chap. 5.
2. The DAQ block is for collecting additional information (e.g., the RPM of the main rotor), which may not be directly utilized for flight control but is very important for enhancing the safety of the UAV system.

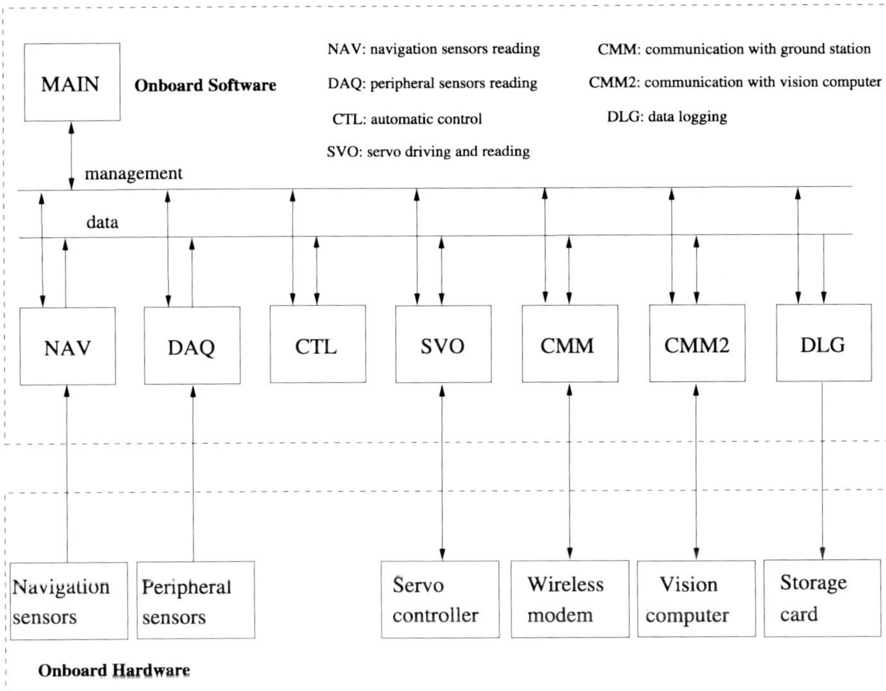

Fig. 4.1 Framework of the flight control software system

3. The CTL block is for realizing automatic flight control laws. In our current flight control software, this is the only block that does not interact with hardware devices. Its main functions include obtaining the UAV helicopter flight status from globally shared data generated by the NAV block, executing control algorithms based on the flight status, and generating control signals to drive servo actuators. We note that the control algorithms adopted include multiple levels such as navigation, task scheduling, and lower-level controllers that directly control the UAV dynamics.
4. The SVO block is for driving the UAV servo actuators and reading their corresponding control-surface deflections, which include $\delta_{\rm lat}$ for aileron, $\delta_{\rm lon}$ for elevator, $\delta_{\rm col}$ for collective pitch, and $\delta_{\rm ped}$ for rudder. The SVO block also records the data of the throttle and manual/automatic switching channels, for the purpose of constant-RPM governing and easy post-flight analysis.
5. The CMM block is for the communications between the avionic system and the ground control station through wireless modems, which perform necessary downloading of the inflight data and uploading of the user-defined command/trajectory.
6. The CMM2 block is for the data exchange between the flight-control and vision computers. Data transferred in this block include user command(s) for direct operation of the vision computer, inflight data (shared by the flight control com-

puter) needed for executing vision processing algorithms, and commands sent from the vision computer to the flight control computer.
7. The DLG block is for real-time data logging, which saves all necessary inflight data. As mentioned earlier in Chap. 3, a compact flash card is used as the storage medium.
8. Finally, the MAIN block is for managing all the tasks mentioned above. It is noted that data exchange within our software system is accomplished and centralized by a globally shared data scheme.

4.2.2 Task Management

We adopt a multiple-thread architecture in our flight control software system, in which every thread is designed for a specific task. To ensure their reliable and harmonious execution, the following task management scheme (program code) for task scheduling, individual task execution, and time allocation is proposed:

```
main program
  initialize task threads
  initialize timer
  loop {
    wait for timer signal
    read user command
    if command is for exit, exit loop
    send an activating pulse to task thread 1
    wait some time for task 1 accomplishment
    send an activating pulse to task thread 2
    wait some time for task 2 accomplishment
    ...
    send an activating pulse to task thread n,
    wait some time for task n accomplishment
  }
  send an exit pulse to task thread 1,
  send an exit pulse to task thread 2,
  ...
  send an exit pulse to task thread n.
  exit main program
task thread
  loop {
    wait for a pulse
    if pulse is for exit, exit loop
    do work
    ...
    set notification of accomplishment
    }
  exit task thread
```

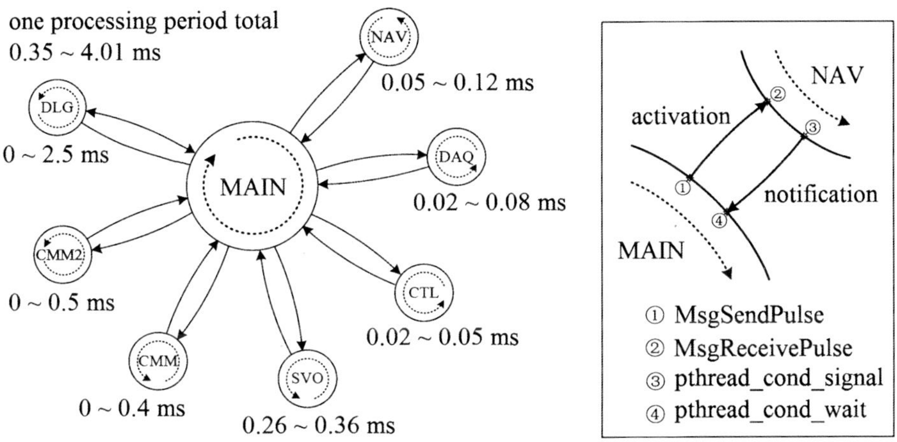

Fig. 4.2 Task scheduling of the flight control software system

4.2.2.1 Task Scheduling

Task scheduling is performed by the MAIN thread (see Fig. 4.2). In Fig. 4.2, the solid arrows represent either one of the following three actions: (i) task-thread activation, (ii) task-thread accomplishment notification, or (iii) process-exchange indication. The dotted round arrow in the MAIN thread denotes the execution sequence of the task threads depicted in the figure.

4.2.2.2 Individual Task Execution

The MAIN and the task threads follow different ways in terms of execution method. For the MAIN thread, its execution scheme is as in the program code presented earlier. Special attention should be paid to its two key components, i.e., the timer and the program loop. More specifically, to ensure the real-time property, a high-precision timer, which provides nanosecond-level accuracy and a variety of system functions for synchronization [102, 148], is used. It emits a pulse signal to activate the sequential loop of the MAIN thread in a predefined frequency. For SheLion, we choose a frequency of 50 Hz (a sampling rate of 20 ms), which coincides with that of the GPS-aided AHRS. The program loop keeps an idle status until a timer signal is captured. Once the timer signal is captured, the program loop reads user commands. All the task threads are terminated if an exit command is received. Otherwise, the program loop cooperates with the timer and iteratively sends pulse signals to activate the task threads for performing their functions and waits for the corresponding accomplishment notification signals, in accordance with the pre-scheduled sequence. The activation and notification methods are depicted in the diagram on the right side of Fig. 4.2, where the activation is implemented by a pair of messages (send/receive calls) and the notification is done by a pair of condition variables (signal/wait calls). After all the task threads have been executed, the program loop returns to idle status

Table 4.1 QNX run-time functions

Function	Description
pthread_create	create a new thread
pthread_mutex_lock	lock a mutex object
pthread_mutex_unlock	unlock a mutex object
pthread_cond_signal	signal a condition variable and unblock threads waiting for it
pthread_cond_wait	wait for a condition variable and block the calling thread on it
timer_create	create a timer
timer_settime	set property of a timer (starting time, repeating interval,...)
MsgSendPulse	send a pulse to a channel
MsgReceivePulse	receive a pulse from a channel
TimerTimeout	set expiration time for waiting function

and waits for the next pulse signal. We can effectively prevent the MAIN thread and other task threads from blocking each other with such an execution mechanism.

On the other hand, the execution of the task thread is carried out as follows. At the beginning of the task thread loop, it waits for a pulse signal from the MAIN thread. Once an execution signal is received, all steps in the thread are sequentially executed. The task thread loop is terminated when an exit pulse issued by the MAIN thread is received. After the task thread is successfully processed, an accomplishment notification signal is sent back to the main program loop. We provide the QNX-based library of threads, messages, and synchronization in Table 4.1 for ease of reference.

4.2.2.3 Time Allocation

In every execution cycle (20 ms for HeLion and SheLion), we need to allocate a time slot to each task thread for processing necessary functions. The time allocation is determined based on its task features.

1. For the NAV thread, the time consumption consists of two parts, one for retrieving the measurement data of the GPS-aided AHRS stored in the serial-port buffer, and the other for performing the measurement signal enhancement (see Chap. 5 for a more detailed description of the measurement signal enhancement techniques). According to our experiments and experience, the total time consumption for the NAV thread is within the range of 0.05 ms to 0.12 ms.
2. The DAQ thread is very similar to the NAV thread, but with less data processing involved. Its time consumption is roughly within 0.02 ms to 0.08 ms.
3. The CTL thread does not deal with any hardware component. Its time consumption totally depends on the complexity of the algorithms employed for the flight control system. For HeLion and SheLion, it is generally within the range of 0.02 ms to 0.05 ms.

Table 4.2 Time allocation for the task threads in the flight control software unit

NAV	DAQ	CTL	SVO	CMM	CMM2	DLG
1 ms	1 ms	1 ms	1 ms	1 ms	1 ms	5 ms

4. The SVO thread is rather different from the hardware-related NAV and DAQ threads. Its time consumption comes from those for servo actuator driving and deflection reading, in which the former sends control signal to a pre-assigned serial port buffer. For servo deflection reading, we note that the servo controller does not send its deflection reading signal until it receives a request from the serial port. The overall procedure from sending the request to receiving deflection reading in the serial port buffer requires an average time of 10 ms. As such, we decide to send the request in the current cycle but read the deflection signal in the next one. With such an arrangement, we have avoided unnecessary waiting time for the servo deflection signal in an individual cycle. The total time consumption for the SVO thread can thus be reduced to a range between 0.26 ms and 0.36 ms.
5. For the CMM thread, the time consumption similarly resides in both the serial-port reading and writing. The former is for checking and decoding user commands, whereas the latter covers data transferring to the serial port connected to the onboard wireless modem. Our experience has shown that a 1 Hz data-downloading rate is suitable for the monitoring purpose. As such, we choose to issue the downloading process in the CMM thread once every 50 cycles, i.e., once per second. For the CMM thread, its time consumption is less than 0.4 ms.
6. The CMM2 thread is similar compared to the CMM thread. Its executing frequency is set at 10 Hz, which coincides with the vision processing software module. Based on the experimental records, the time consumption of the CMM2 thread is generally less than 0.5 ms.
7. Finally, the time consumption of the DLG thread depends on the size of the data package and the speed of the selected storage device. To improve the efficiency of the flight control software module, the DLG thread is not issued until the data are piled up to certain predefined volume. In SheLion, inflight data are also stored at 1 Hz, that is, once every 50 cycles. The maximum time consumption is less than 2.5 ms, according to our experimental record.

Considering all necessary redundancies, we adopt a practical time-allocation solution as given in Table 4.2.

Figure 4.3 gives an illustration of the time allocations for both the flight control software and associated hardware processing. The stack on the left describes the software processing in every cycle, which totally consumes about 0.35 ms to 4.01 ms. The hardware processing runs in parallel with its software counterpart. As shown in the figure, the servo controller board begins to process data collection after a request command is received. The software part does not stop and wait for hardware processing to be finished. It continues processing other scheduled task threads. When the requested data are ready, they are to be read in the next cycle. After all the tasks are executed (within 4.01 ms in this case), the flight control software module

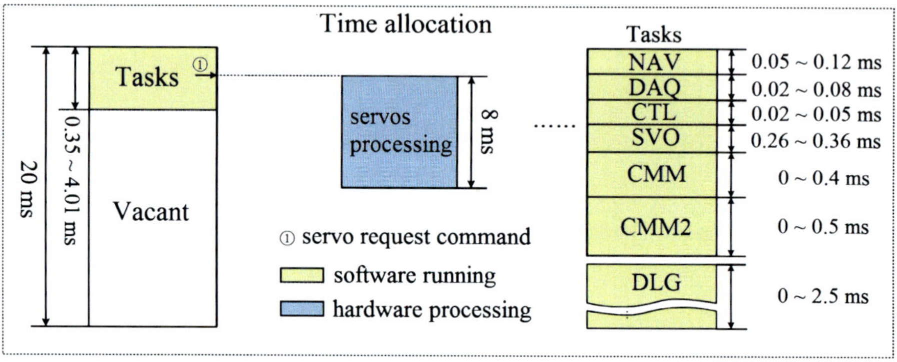

Fig. 4.3 Time allocation for task threads

enters a vacant status, which can be used for additional task thread(s), if any, developed in the future. It is desirable for the software system to keep a low CPU-working load. The description of the software processing is depicted by the stack on the right side of Fig. 4.3, along with the time-consumption list for the involved task threads.

4.2.3 Implementation of Automatic Control

The implementation of automatic control laws in the CTL thread is one of the most essential issues in the design of the flight control software module. It is well known that a comprehensive flight plan generally consists of multiple flight patterns or behaviors, which commonly have different control parameter settings and even require different control algorithms. These issues have been carefully considered in our software development procedure. We have employed a behavior-based flight scheduling block to organize various behaviors in a flight plan. In this architecture, any operation of the UAV helicopter is recognized as a specific behavior with specific parameters (such as control signal references and execution time limitations) and flag setting. A hierarchical control system block is created to receive the behavior and realize the corresponding automatic control algorithm. It consists of two layers with multiple control algorithms being integrated. We note that the proposed behavior-based architecture is rather different from the traditional behavior-based architecture introduced in [52, 115, 157] for robotic systems. In our work, the behavior is much like a concept standing for an action or a kind of operation of an agent. In what follows, we first document the detailed descriptions of the flight-scheduling and control system blocks and then use a specific example to demonstrate its implementation procedure.

The flight scheduling block hosts flight plans for the UAV helicopter. A specific flight plan can be either pre-stored in the avionic system or generated online by the ground control station. Figure 4.4 depicts a simple flight plan, in which the UAV helicopter is required to start with a hovering fly, then fly from one point to another,

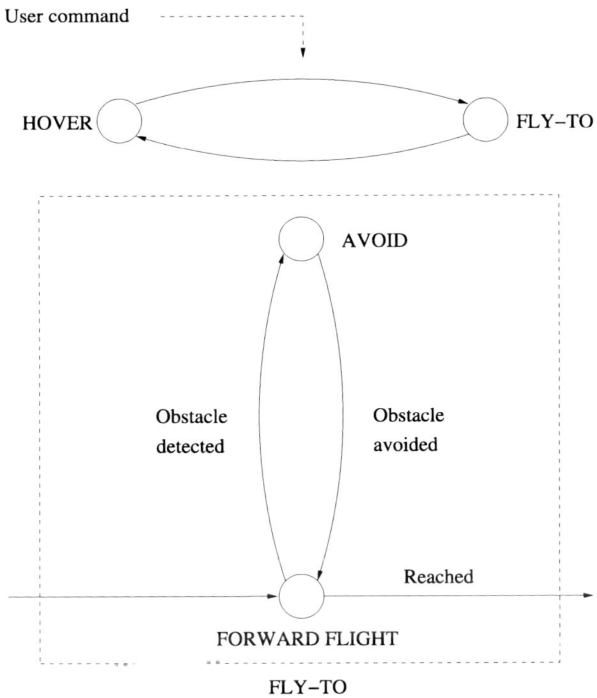

Fig. 4.4 A simple behavior-based flight plan

and eventually retrieve hover again. We note that the flight plan consists of multiple nodes and arrows. Each node represents a sub-plan or a specific behavior, whereas each arrow denotes an event. The flight scheduling block employs an event-driven mechanism. When an event is triggered, the flight scheduling block transfers from one behavior (or sub-plan) to another. An event can be triggered by either one of the following sources: (i) change on the surrounding environment (e.g., obstacle encountering and target detection), (ii) change in working situations (e.g., hardware component malfunction and data loss), (iii) change in inflight status (e.g., the achievement of a certain flight mission), or (iv) user command uploaded from the ground control station. Based on the triggered source, the flight scheduling block determines what behavior is to be performed and transmits it to the consequent control system block for further action. It is clear that the flight scheduling block acts like a decision maker or a commander.

The function of the control system block is to decode the behaviors received and to drive the servo actuators of the UAV. The control system block is developed based on a hierarchical structure (see Fig. 4.5). As introduced earlier in Chap. 1, the inner loop guarantees the asymptotic stability and good disturbance-rejection capacity of the UAV helicopter, and the outer loop follows the behavior request to guide the UAV helicopter for achieving a predefined flight trajectory with a desired orientation. The outer loop is responsible for generating control references for the inner loop. The two layers work together to realize the UAV automatic control. It is noted that the numerical values of the inner- and outer-loop control laws are pre-loaded

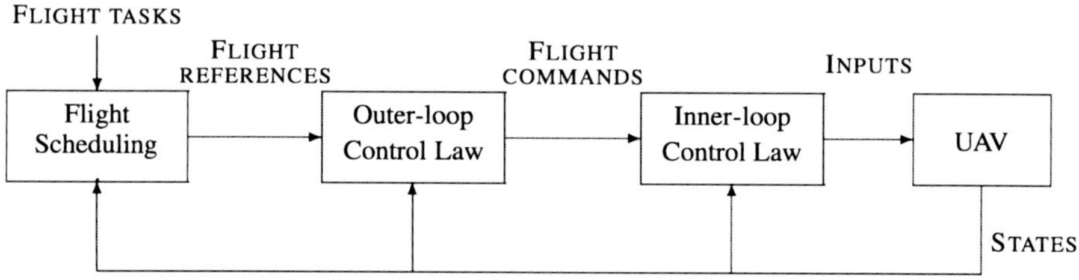

Fig. 4.5 Structure of the hierarchical flight control system

to the globally shared memory from an onboard text file when the MAIN thread is initialized. They can be directly imported for automatic control implementation in the CTL-thread execution.

The following is a specific example for illustrating the working principle of the CTL thread, in which SheLion is used as the experimental platform. We intend to evaluate the performance of the automatic control system, which is designed using the H_∞ control technique (for the inner loop) and the robust perfect tracking (RPT) technique (for the outer loop), and to collect camera images of the surrounding environment of the destination point for post-flight analysis (emulation of reconnaissance). The CTL thread is implemented as follows:

1. For the flight plan, SheLion starts with a stable hover. The consequent motion execution is triggered by a user command uploaded from the ground control station. After receiving the command, SheLion performs a forward flight with a maximum speed of 12 m/s. When it reaches the destination, SheLion is required to hover for 10 s and then make a head turning motion with a constant yaw rate. In this procedure, the onboard vision system is commanded to work simultaneously. When the event condition (i.e., two and a half rotations) is achieved, SheLion stays hovering at the destination, with its nose pointing to the starting point. The next behavior, triggered by the completion of hovering, is to fly back to the starting point. The overall flight plan ends again with a stable hover at the starting point. The whole process is depicted in Fig. 4.6. We note that FLY-TO and RECON are sub-plans, whereas the specific behavior set consists of HOVER, FORWARD FLIGHT, PHOTO, and HEAD TURN. Assuming that there is no obstacle between the starting and destination points, we can obtain the behavior sequence shown in the second part of Fig. 4.6, which is sequentially to be passed to the control system block.
2. For a specified behavior, information transferred from the flight scheduling block to the control system block consists of two parts, necessary control reference signals (see, e.g., position and velocity references) and flag signals for activating control algorithms. In this example, the flags are set to load the H_∞ inner-loop controller and the RPT outer-loop control law into the control system block, and the corresponding control gain matrices are then retrieved from the globally shared memory for generating the driving signals of the servo actuators. The detailed information on the control system design will be given later in Chaps. 7 and 8.

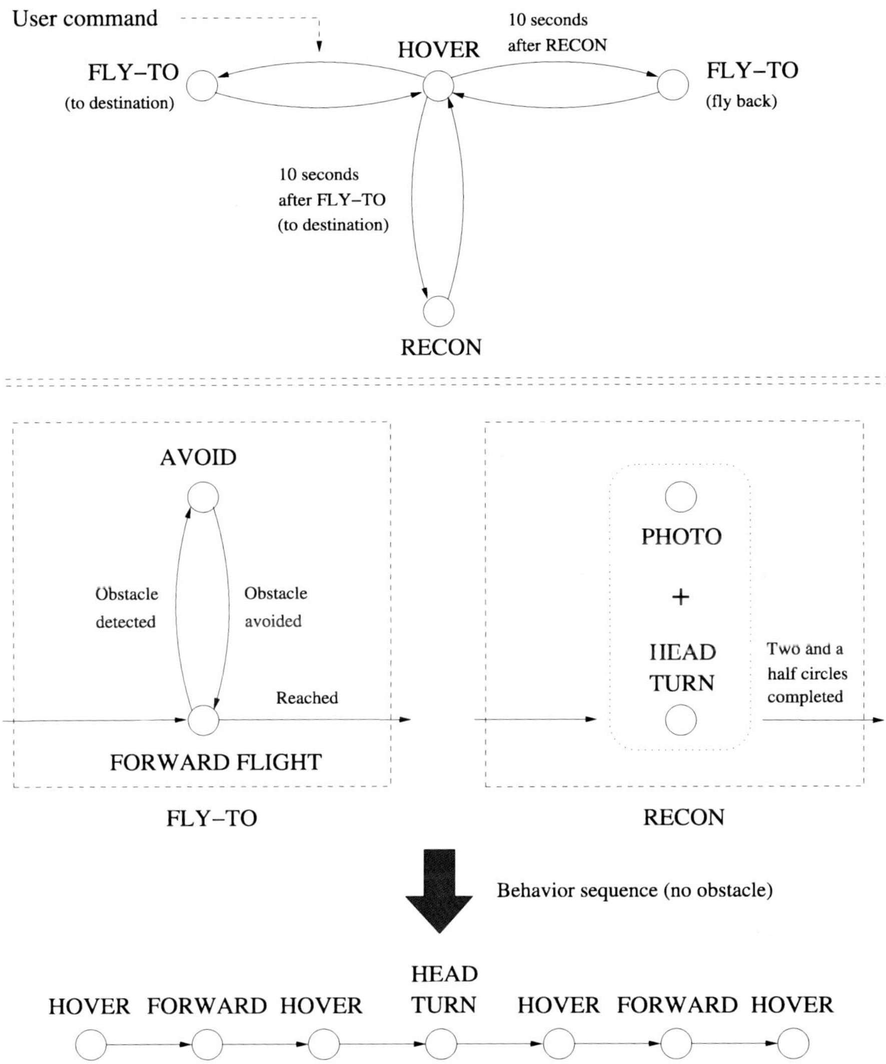

Fig. 4.6 Behavior-based flight plan and the resulting behavior sequence

4.2.4 Emergency Handling

This additional safety measure is a result that we learned from a couple of crash accidents of our UAV helicopters. There are many sources that would cause failures in the UAV control system, which include drastic changes in environment, hardware failure, GPS disorder, and problems in the control system design and software implementation. A mechanism for handling emergency situations is built in the onboard software system. At every cycle, before applying control action, the control task thread checks all data received from the NAV and other devices. Once any abnormality is observed, the emergency control function is called up immediately to sequentially (i) send an alert signal to the ground control station to inform the pilot to take over the control authority, (ii) drive and maintain all four input channels to their trimmed values in the hovering condition, and (iii) shut down the helicopter engine via the throttle input channel if the control authority is still at the automatic side after a predefined alert time (3 s for HeLion and SheLion).

It is also very important to keep logging data as much as possible in emergency situations, which can be used to identify problems causing the incidents. However, in an exceptional occasion when a crash occurs, all tasks run onboard including data logging are not terminated as expected. In a normal operation, the data logging file is opened at the beginning of a flight test, is amended with inflight data during the test, and finally is closed and saved at the end of the test. In an abnormal situation, the data logging file is likely to be corrupted, resulting in the loss of all data. If the process is interrupted, the whole file will be damaged. A mechanism similar to the black box in commercial aircraft is implemented to keep saving logged data even in a crash. Such a feature is illustrated in the program below:

initialization stage
    do nothing
logging stage
1st time (the first 50 cycles)
    open logging file in writing and appending mode
    write 1st pack of data to logging file
    close logging file
2nd time (the next 50 cycles)
    open logging file in writing and appending mode
    write 2nd pack of data to logging file
    close logging file
...
n-th time
    open logging file in writing and appending mode
    write n-th pack of data to logging file
    close logging file
...
final stage
    do nothing

As illustrated in the program above, the new feature enables the data logging process to complete data writing and saving every 50 cycles, i.e., 1 s. If an accident occurs, the data can still be saved until the last second before the breakdown of the flight control software system.

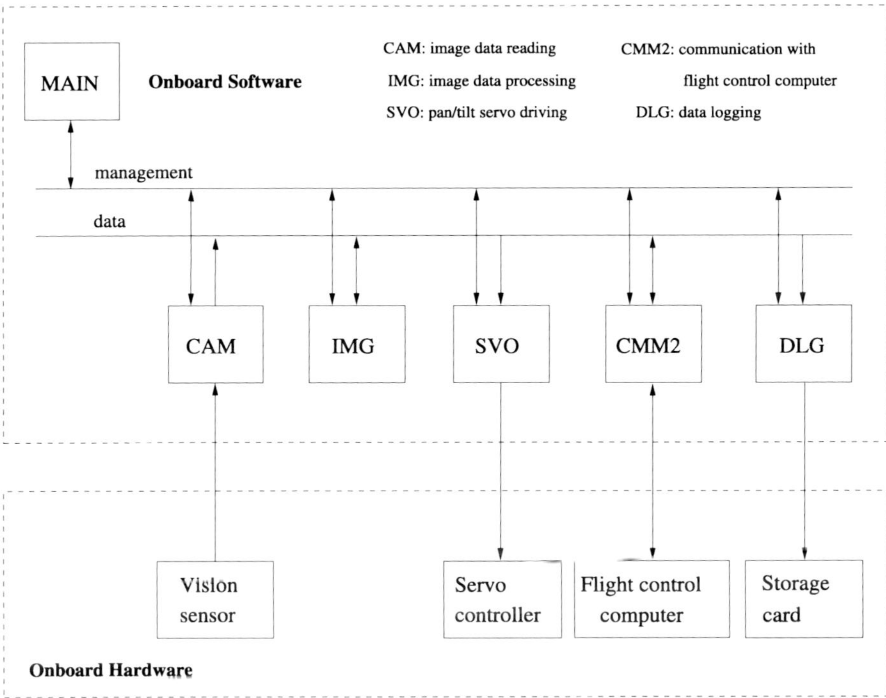

Fig. 4.7 Framework of vision software system

4.2.5 Vision Processing Software Module

The main functions of vision processing software include (i) capturing video signals, (ii) executing vision algorithms, (iii) controlling the pan/tilt servo mechanism, and (iv) communicating with flight control system and the ground control station. The vision processing software module adopts a similar framework and task management scheme to that for the flight control software module. The key difference between the flight control and vision processing parts resides in the looping time and the functions of the sub-task threads. Figure 4.7 shows the vision software framework, which runs on the vision processing unit with a period of 100 ms.

1. CAM BLOCK: The main purpose of the CAM block is to read the image data captured by the frame grabber and store them in the globally shared memory. In SheLion, the adopted frame grabber can support a sustained frame rate up to 30 FPS (frames per second) for the uncompressed RGB images with a resolution of 720×576 pixels. Such a capability is sufficient in real-time operation. In fact, the real capturing rate is limited by the complexity of the vision algorithms and other tasks involved, such as writing processed images to a storage device. In SheLion, the frame rate of the vision system is adjusted to 10 FPS.
2. IMG BLOCK: The vision processing algorithms are implemented in the IMG block, which is a critical part in the vision software module. The key functions

Table 4.3 Time allocation of the vision processing software unit

CAM	IMG	SVO	CMM2	DLG
8 ms	85 ms	1 ms	1 ms	15 ms

of the IMG block include processing image data and making decisions based on image data and information shared by the flight control unit (see, e.g., the GPS-aided AHRS measurement). The algorithms for image processing, object segmentation, feature extraction, pattern recognition, automatic tracking and camera control are to be executed in this block. The processed image is also saved to the globally shared memory. Vision information is to be used together with the outputs of other sensors to decide on actions to be taken for driving the pan/tilt servo mechanism and sending control commands to the flight control computer. The IMG block is time-consuming and its processing time is generally within the range of 50 ms to 80 ms. As such, the looping frequency of the vision processing software is greatly reduced to 10 Hz.

3. SVO BLOCK: The SVO block is utilized to control the rotation of the pan/tilt servo mechanism to keep the target in a certain location of the camera image, typically at its center. The control input of the pan/tilt servo mechanism is calculated based on the vision information of the target and the status of the UAV system.
4. CMM2 BLOCK: The CMM2 block interacts with its counterpart in the flight control software module to realize data exchange. More specifically, the flight control computer shares the inflight states of the unmanned helicopter with the vision computer. The vision unit, on the other hand, might send guidance information to the flight control unit. We have also programmed the flight control computer to act as a bridge between the vision computer and the ground control station.
5. DLG BLOCK: The DLG block saves the processed images to the compact flash card equipped in the vision computer. Due to the larger volume of the image data, the recording time is set to 15 ms. In each loop, the MAIN thread calculates the actual spare time and decides if the DLG block is to be executed.

Finally, we provide the time allocation for the task threads of the vision software system in Table 4.3.

4.3 Ground Control Station Software

The ground station plays a role as a terminal for end users to monitor and command the UAV helicopter through the wireless communication channel. In the flight tests, data of the UAV helicopter are transferred from the onboard system to the ground station and displayed. The task of the ground station is to provide a friendly and realistic interface for users to monitor the process of the flight tests. Different methods

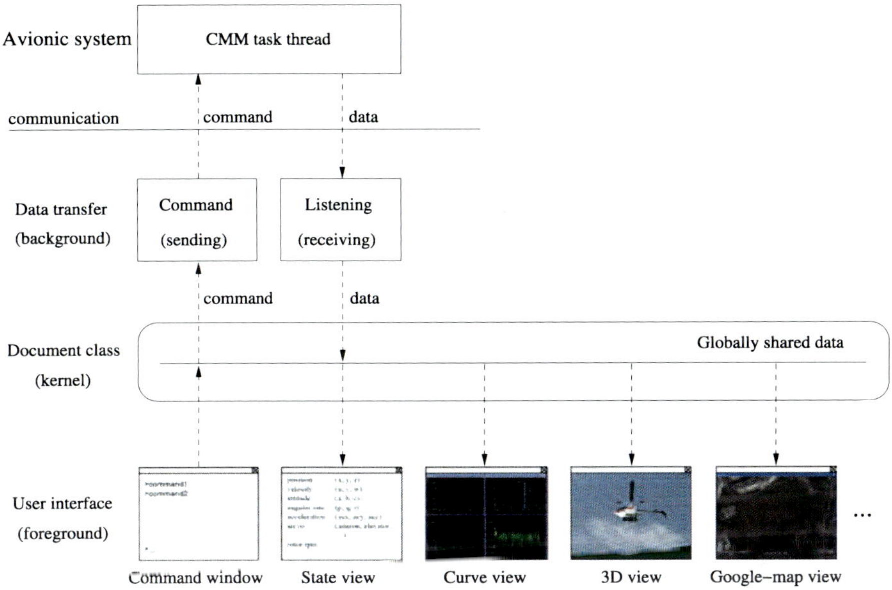

Fig. 4.8 Framework of the ground station software system

of data visualization are to be implemented. Our ground control station software is developed in the Windows XP Professional operating system with Visual C++ 6.0 as the primary developing environment. It features a document-view structure based on the MFC (Microsoft Foundation Class) library, which is recommenced by Microsoft for application oriented development [103]. In this section, we first introduce the framework for developing the ground station software system and then particularly highlight the development of a 3D view interface, which is capable of transforming the data received from the helicopter into a realistic 3D view in the ground station.

4.3.1 Framework of Ground Station Software Module

The framework of the ground station software is depicted in Fig. 4.8. It consists of three layers: the foreground layer, the background layer, and the kernel layer.

1. The foreground layer directly interacts with the end users. The main function of this layer is to provide a straightforward and user friendly GUI. Figure 4.9 illustrates the GUI that we have created for the ground station system. The main window contains a series of sub-windows for displaying inflight data. Five types of display (or view) modes have been adopted in the GUI design, as shown in Fig. 4.9. Among them, the state view displays inflight data in a two-column text list, with the left column showing the UAV state variables (such as position and

Fig. 4.9 Graphical user interface of the ground station software

velocity) and the right column displaying the numerical values. The curve view sub-windows display the inflight data of the selected variable(s) in the 2D form, whereas the 3D view function reconstructs the actual flight motions of the UAV helicopters in a virtual 3D environment. The Google-map view displays the data in a 2D satellite map, which is pre-loaded into the kernel layer. Lastly, the bottom part of the GUI is a command view, which consists of (i) an input bar for online issuing command(s) and (ii) a command-view panel for displaying the command(s) history and the received inflight-data packages.

2. The background layer is created for data transferring. This layer communicates with the software module of the avionic system (i.e., the CMM task thread) through the wireless communication channel. The data-receiving thread in the background layer continuously reads the serial port connected to the wireless modem attached to the ground station and updates the globally shared data storage created by the kernel layer. The data-transmitting thread of the background layer, on the other hand, keeps checking the globally shared data storage to capture the user command issued in the foreground layer and then writing it to the serial port.
3. The kernel layer is a document class that hosts the globally shared data. A variety of methods for data processing and accessing have been integrated. The document class links the communication threads and multiple view functions in the foreground and background layers. Two dynamical data chains are involved:

(i) when the data-receiving thread in the background layer receives a new inflight data package, the kernel layer updates its globally shared data storage and activates the foreground layers to call the display functions in the foreground layer; and (ii) when a new user command is issued in the foreground layer, the kernel layer similarly updates its data storage and activates the background layers to upload it to the onboard system.

The execution scheme for realizing the above software framework can be summarized in the following pseudo-code:

```
document class
init
    create communication threads
    (pass pointer to this document to threads as parameter)
    create views
    (pass pointer to this document to these views)
    create timer
ontimer
    loop {
        get pointer to next view
        if the pointer is null(end), exit loop
        call the updating function (onupdate) of that view
    }
view class
onupdate
    get the pointer pointing to the document
    get new data in the document
    draw view according to the new data
communication thread 1 (data-receiving)
    loop {
        read serial port for communications
        if new data received {
            store new data in the document
        }
    }
communication thread 2 (data-transmitting)
    loop {
        look up to the document if new command captured
        if there is new command {
            translate command and parameter in telegraph
            write telegraph to communication serial port
        }
    }
```

In this program, we should pay special attention to the following four key points:

1. The *init* function is called to initialize the document class and to create the communication threads, various data views, and updating timer. When these threads

and views are successfully created, a pointer of the document is assigned to them to establish the necessary links. The document class itself also stores an array of pointers.

2. The timer is created to send updating signals periodically. The timer message is handled by the member function *ontimer*. In every cycle, when the timer message is captured, the ontimer function searches all views linked to the document class and calls their updating functions (*onupdate*) sequentially.
3. A universal method is used to define all the view functions in the view class. The *onupdate* member function is called every time when the shared data storage is updated by new inflight data packages. In *onupdate*, it is programmed to first obtain the pointer, which points to the document class, and then access the associated stored content, before a drawing function is called to display new data.
4. The kernel layer is for decoding the user command to a format that can be recognized by the avionic software system. Once a new user command has been received, the kernel layer calls a function to obtain the string and translates it in an internal representation carrying command code and parameters, based on a predefined mapping table (or coding scheme). The translated code and parameters are then stored in the globally shared data storage.

4.3.2 3D View Development

The 3D virtual view of the helicopter is very useful when the UAV helicopter flies beyond the visible range of users. It provides users a straightforward impression of the helicopter status. The development of the 3D view in our ground station (see, e.g., [42]) involves (i) the creation of the 3D models of the helicopter and ground environment, (ii) the *OpenGL* (open graphical library) programs for kinematical transformations of these models, and lastly, (iii) the integration of the 3D view into the overall ground station software system.

4.3.2.1 3D Model Development

We create the 3D models of the UAV helicopter and the surrounding environment using a well-known software tool called *3ds max*, in which a 3D object is represented by a set of vertices and a series of polygons depicting its shape. Each vertex is described by three float numbers representing its position. Each polygon is described by a series of vertices representing its boundary points. In *OpenGL*, vertices and polygons are loaded from pre-saved files to construct models of the helicopter and the surrounding environment. Figure 4.10 shows the sample models in the development workspace. The left part of the figure is the 3D model of the helicopter, which is much like the physical helicopter. The right part is a hemisphere model to imitate the sky part of the surrounding environment.

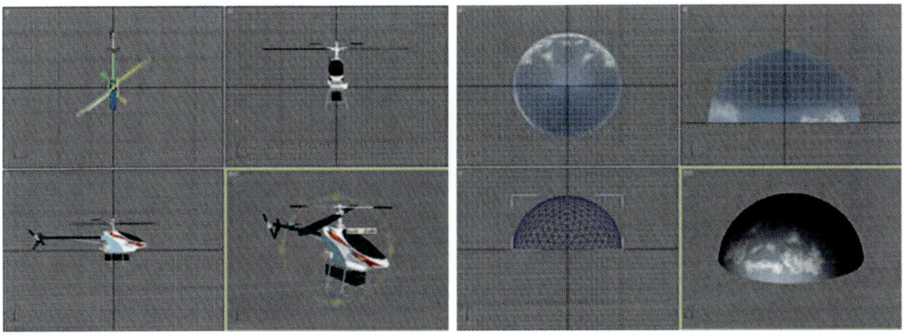

Fig. 4.10 Model development in *3ds max*

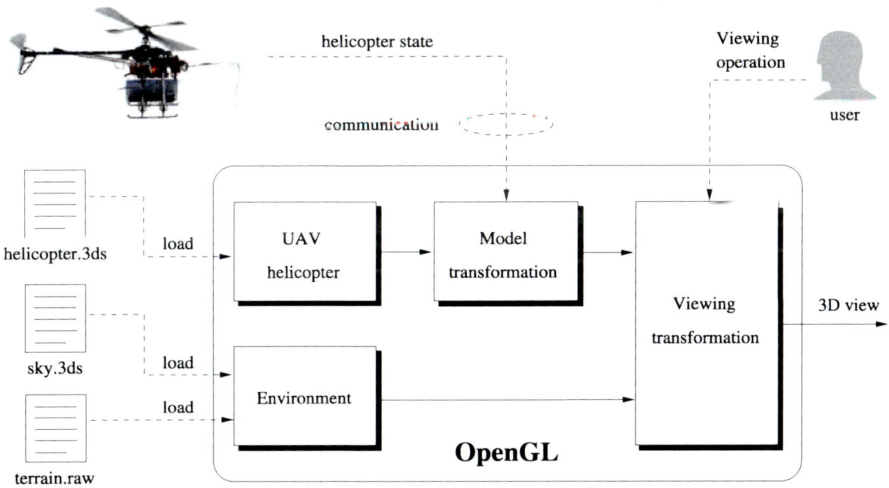

Fig. 4.11 OpenGL drawing

4.3.2.2 3D Drawing

OpenGL is adopted for drawing the UAV helicopter model in the virtual environment. It provides a rich collection of functions for drawing objects based on vertices and polygons. As mentioned earlier, objects are represented by a set of vertices and polygons. The drawing of an object is accomplished by going through all of its vertices and polygons. The framework of the 3D drawing is depicted in Fig. 4.11, which illustrates how the 3D models created offline, the data of the helicopter state, and view operations from users are integrated. Both the helicopter and the environment models are loaded from pre-created files. Real-time data required for the helicopter include positions and attitudes transferred from the onboard system. View operations from users include moving, rotating, and zooming of the virtual 3D view.

Fig. 4.12 The 3D view of the UAV helicopter

4.3.2.3 3D View

The 3D view is lastly encapsulated in a class in the C++ program. It is linked to the globally shared data (i.e., the kernel layer), in which the inflight data of the UAV helicopter are stored. To dynamically simulate the motion of the helicopter, the view is updated with a rate of 10 Hz. In every cycle, the newest values of the helicopter state are read from the global data pool and the content of the 3D view is repainted accordingly. The periodical update is scheduled in the program by a timer, which is created in the initialization stage of the 3D view. At the end of each cycle, a timer message will be sent to the view to request an update, which proceeds as in the following code.

```
CTDView::OnUpdate()
{
   double pos[3] = { _data[1], _data[2], _data[3] };    //Local NED position
   double att[3] = { _data[7], _data[8], _data[9] };    //Euler angles
   angle += rate*period/1000;    //incremental rotor angle
   SetPositionAttitudeRotor(pos, att, angle);    //calculate transformation
   GLDraw();    //redraw
}
```

The *OnUpdate* function is a member of the 3D view class *CTDView*, which is called whenever a timer message is captured. The front two lines get the local NED position and Euler angles of the unmanned system from the global data pool, which is updated from time to time as the flight test progresses. The angle of the rotor is added by an incremental value every period to emulate the rotating effect of the UAV main rotor. Then, the member function *SetPositionAttitudeRotor* is called to assign these states to the 3D view. This function calculates transformation matrices and applies them to the 3D objects. The *GLDraw* member function is finally called to redraw the content of the 3D view.

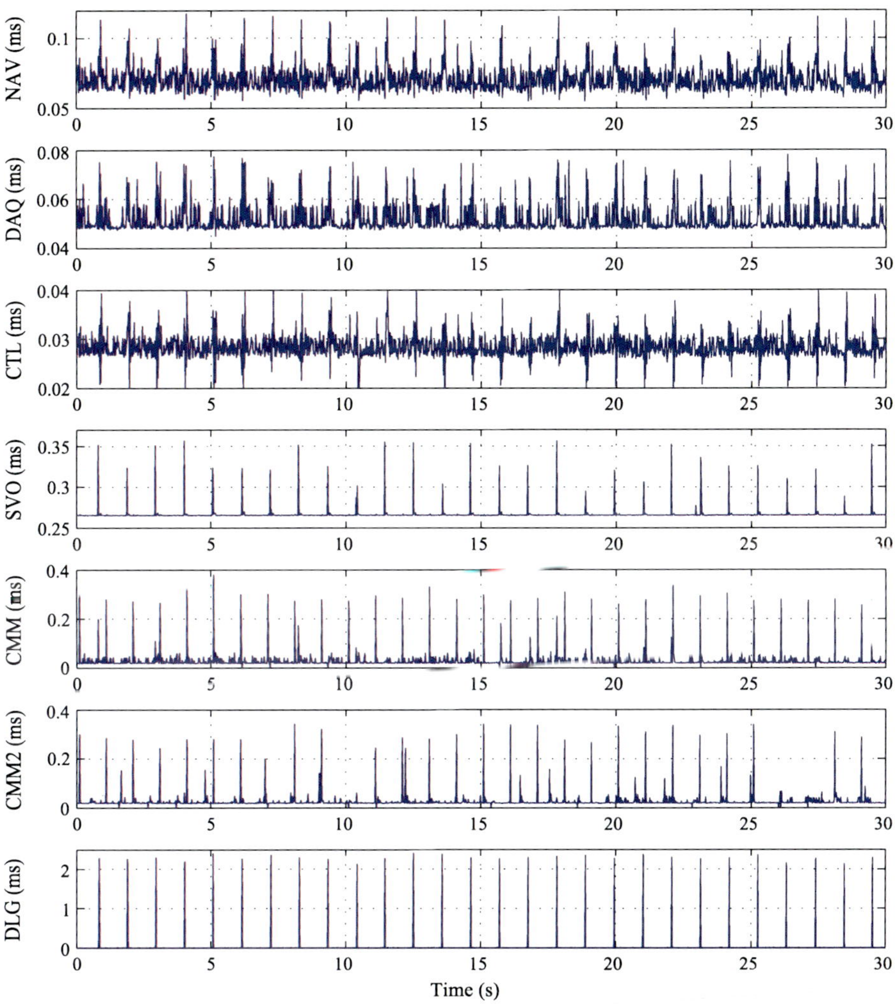

Fig. 4.13 Time consumption of the task threads in flight control software module

Finally, the appearance of the 3D view is illustrated in Fig. 4.12. Pictures are captured when SheLion is performing a wide envelope automatic flight test. Some virtual views obtained from different viewing points are listed in the left and right columns.

4.4 Software Evaluation

In this section, we present some initial evaluation results of the overall software system. Real-time properties and working load evaluation of the onboard system are

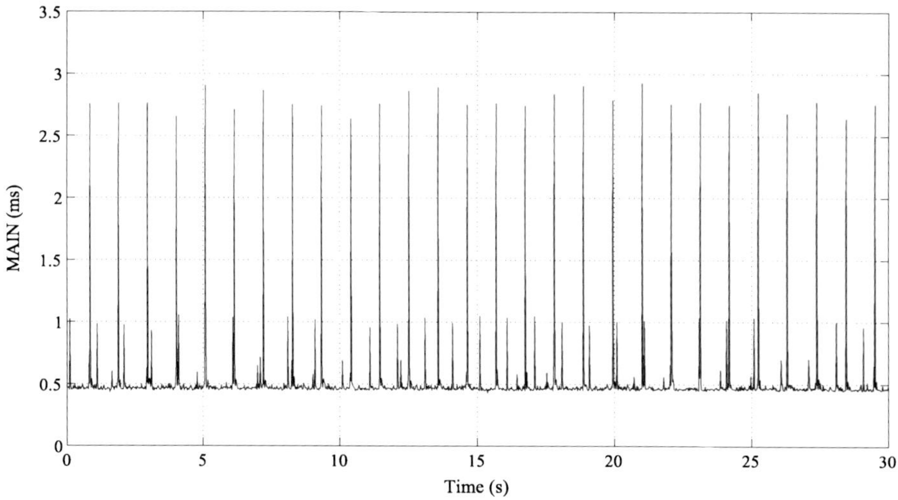

Fig. 4.14 Time consumption of the MAIN thread in the flight control software module

the primary focus in this procedure. To evaluate the working load of task threads, we record the time used by task threads in every cycle. Figure 4.13 shows a graph of the time consumptions by task threads of the flight control software module in an actual test flight. The times spent by all task threads are summed up in Fig. 4.14 as the CPU time consumption of the main thread. It is clear that the time consumption for each thread is within the pre-allocated range. The real-time operation for both task threads and the MAIN thread are well maintained during the whole experimental procedure. We note that the communication task is scheduled to transfer data once every second due to the bandwidth limitation of the serial wireless communications. The data logging thread is also operated at a slower pace to enhance the overall software efficiency.

The CPU usage rate of the flight control computer can be calculated as the ratio of the total time consumption and the length of the period, which is 20 ms in the actual test. From the results shown in Fig. 4.14, we can determine that the lowest usage rate of the flight control computer is

$$\text{usage}_{\text{min}} = \tau_{\text{min}} / T_{\text{cyc}} = 0.45/20 = 2.25\%$$

and the highest peak is

$$\text{usage}_{\text{max}} = \tau_{\text{max}} / T_{\text{cyc}} = 2.93/20 = 14.65\%,$$

where τ_{min} and τ_{max} stand respectively for the minimum and maximum total time consumptions by all threads in one cycle, and T_{cyc} is the 20 ms period of the cycle. The peak of the CPU usage rate comes with the large working load of data logging and communications, which occurs once every 50 cycles, i.e., 1 second. The average time consumption in each cycle is about 0.49 ms. Thus, the average CPU usage rate is

$$\text{usage}_{\text{ave}} = \tau_{\text{ave}} / T = 0.49/20 = 2.45\%.$$

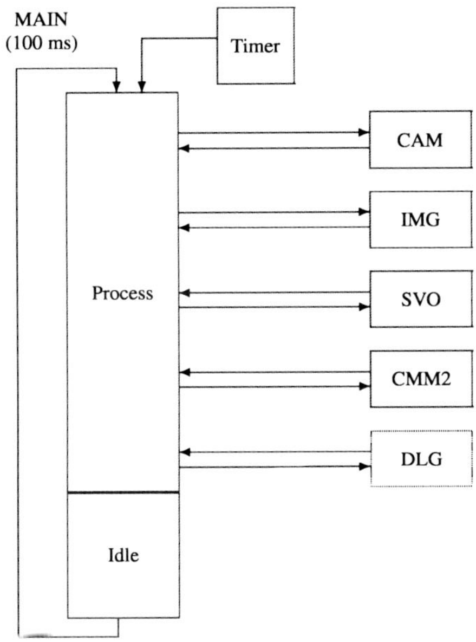

	Time limit (ms)	Execution time (ms)
CAM	8	5–7
IMG	85	50–80
SVO	1	0.2–0.3
CMM2	1	0.1–0.2
DLG	15	15–25

Fig. 4.15 Multiple-task execution of the vision software system

The low usage rate of the flight control computer is ideal for guaranteeing the reliability of the overall software system. To balance the working load, the communication and data logging threads have to be scheduled in a more distributive and efficient way.

Similarly, for the vision processing software unit, the time consumption of each thread is measured in the same experiment and shown in Fig. 4.15. It is observed that the vision algorithms consume more than 80% of the computational resource, and the complexity of the vision algorithms severely affects the real-time application of the vision system.

Chapter 5
Measurement Signal Enhancement

5.1 Introduction

A sophisticated attitude and heading reference system (AHRS) is a key element in modeling and control of unmanned aerial vehicles. It consists of necessary sensors to provide measurement signals for the UAVs. Typical sensors used to form an AHRS are accelerometers to measure the proper acceleration along the three axes of the UAV body coordinate, gyroscopes to provide the three-axis angular rates, and magnetometers to capture the magnet values of the three axes, and some sensors provide reliable position and velocity information of the UAV for navigation, trajectory tracking, autonomous flight control, and mission completion.

To obtain accurate position and navigation signals, it is common to employ a GPS sensor unit together with an inertial navigation system (INS), which are capable of providing position and orientation information of the UAV. Unfortunately, performance obtained from low-cost inertial sensors available on the market is pretty poor due to error sources such as random noise, biases, and scaling factor errors. Even though expensive and bulky INS is able to provide accurate navigation data, its performance degrades gradually with time (see, e.g., [12] and [180]). Similarly, navigation data generated by GPS sensors carry bounded errors. The GPS signals for positions available for public use have an accuracy of only about 3 m (CEP) with a sampling frequency of 4 Hz. The velocities, determined from the Doppler effect, have an accuracy of about 0.3 m/s. These measurement errors are rather bad for mini-scale UAVs flying at low speed. As such, it is necessary to introduce some filtering schemes to smooth and improve the AHRS and INS measurements for control purposes and for the attenuation of high frequency noises.

There have been a number of approaches recently introduced for improving the performance of GPS and low-cost INS integration. For example, Jang and Liccardo [87] have utilized the well-known EKF technique to construct an effective GPS-aided AHRS for a hand-launched fixed-wing UAV. Nassar et al. [129] have investigated the improvement of the accuracy of an inertial measurement unit (IMU) using an autoregressive modeling approach. Different integration filters have also been investigated in the literature. For instance, Shin and El-Sheimy [164] and

G. Cai et al., *Unmanned Rotorcraft Systems*, Advances in Industrial Control,
DOI 10.1007/978-0-85729-635-1_5,

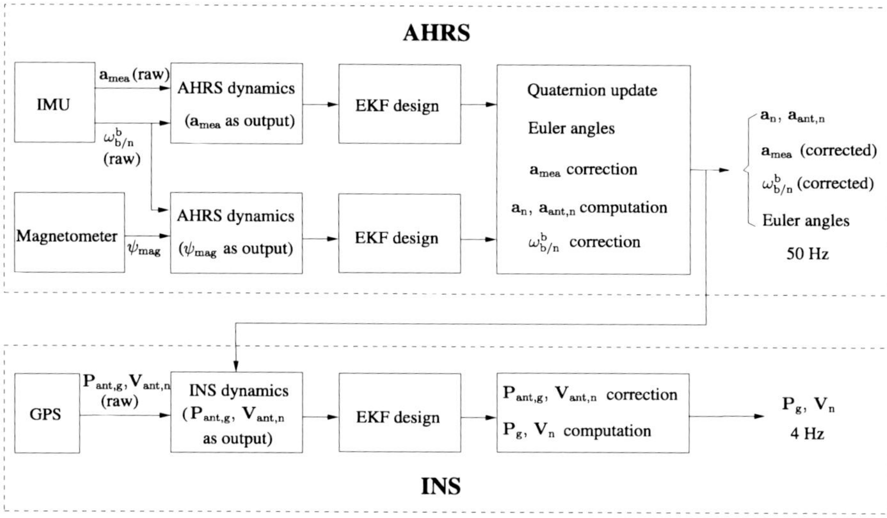

Fig. 5.1 Signal enhancement for the GPS-aided attitude and heading reference system

van der Merwe and Wan [187] have studied the use of unscented Kalman filters, Noureldin et al. [133] have considered solutions using neural networks, and more recently, Yun et al. [220] have introduced a measurement enhancement scheme using a robust H_∞ filtering technique.

The goal of this chapter is to introduce an integration of a low-cost inertial attitude and position reference system for a mini UAV helicopter by utilizing the well-known EKF technique. More specifically, we propose a systematic signal enhancement procedure, which is for MNAV sensors adopted to yield more accurate measurement of the Euler angles, angular rates, positions, and velocities of the unmanned aerial vehicles. The overall procedure is summarized and depicted in Fig. 5.1, which consists of two independent parts, one for the AHRS and one for the INS.

5.2 Extended Kalman Filtering

The extended Kalman filter is developed to deal with systems with nonlinear dynamics or is regarded as the nonlinear version of the Kalman filter. It is widely used in signal processing and estimation. We recall in the following a general procedure for designing an extended Kalman filter for discrete-time nonlinear systems. More specifically, we consider a given system characterized by

$$\mathbf{x}_k = \mathbf{f}(\mathbf{x}_{k-1}, \mathbf{u}_{k-1}) + \mathbf{w}_{k-1} \tag{5.1}$$

and

$$\mathbf{y}_k = \mathbf{h}(\mathbf{x}_k) + \mathbf{v}_k, \tag{5.2}$$

where $\mathbf{x}_k$ is the true state vector at time k, $\mathbf{f}(\mathbf{x}_{k-1}, \mathbf{u}_{k-1})$ and $\mathbf{h}(\mathbf{x}_k)$ are some non-linear (or linear) functions, and $\mathbf{w}$ and $\mathbf{v}$ are the process and measurement noises. The noises, $\mathbf{w}$ and $\mathbf{v}$, are assumed to be independent, zero mean, and with normal probability distributions

$$\mathbf{w}_k \sim \mathrm{N}(0, \mathbf{Q}_{\mathrm{cov}}) \tag{5.3}$$

and

$$\mathbf{v}_k \sim \mathrm{N}(0, \mathbf{R}_{\mathrm{cov}}), \tag{5.4}$$

where $\mathbf{Q}_{\mathrm{cov}}$ and $\mathbf{R}_{\mathrm{cov}}$ are the process and measurement noise covariance matrices, respectively. The initial state and the noise vectors at each step are also assumed to be mutually independent [203]. It should be noted that although $\mathbf{Q}_{\mathrm{cov}}$ and $\mathbf{R}_{\mathrm{cov}}$ might be time variant, they are assumed to be time invariant in our work.

The extended Kalman filter design for the given discrete-time nonlinear system consists of the following two stages:

1. Prediction stage: The prediction stage uses the previous time step to produce an estimation of the state at the current time step [203]. More specifically, in this stage we have the predicted state estimate and the predicted estimated covariance as follows:

$$\hat{\mathbf{x}}_{k|k-1} = \mathbf{f}(\hat{\mathbf{x}}_{k-1|k-1}, \mathbf{u}_{k-1}) \tag{5.5}$$

and

$$\mathbf{P}_{k|k-1} = \mathbf{F}_{k-1}\mathbf{P}_{k-1|k-1}\mathbf{F}_{k-1}^{\mathrm{T}} + \mathbf{Q}_{\mathrm{cov}}, \tag{5.6}$$

where $\mathbf{F}_{k-1}$, the Jacobian matrix of the partial derivative of the function $\mathbf{f}$ with respect to $\mathbf{x}$ at time $k-1$, is defined by

$$\mathbf{F}_{k-1} = \left.\frac{\partial \mathbf{f}}{\partial \mathbf{x}}\right|_{\hat{\mathbf{x}}_{k-1|k-1}, \mathbf{u}_{k-1}}. \tag{5.7}$$

2. Correction stage: In the correction stage, the current predicted estimate is combined with current observation information to refine the state estimate [203]. More specifically, at time point k, we need to compute the Kalman gain $\mathbf{K}_k$ and then update the state estimate and error covariance as follows:

$$\mathbf{K}_k = \mathbf{P}_{k|k-1}\mathbf{H}_k^{\mathrm{T}}(\mathbf{H}_k\mathbf{P}_{k|k-1}\mathbf{H}_k^{\mathrm{T}} + \mathbf{R}_{\mathrm{cov}})^{-1}, \tag{5.8}$$

$$\hat{\mathbf{x}}_{k|k} = \hat{\mathbf{x}}_{k|k-1} + \mathbf{K}_k(\mathbf{y}_k - \mathbf{H}_k\hat{\mathbf{x}}_{k|k-1}), \tag{5.9}$$

and

$$\mathbf{P}_{k|k} = (\mathbf{I} - \mathbf{K}_k\mathbf{H}_k)\mathbf{P}_{k|k-1}, \tag{5.10}$$

where $\mathbf{H}_k$, the Jacobian matrix of the partial derivative of the function $\mathbf{h}$ with respect to $\mathbf{x}$ at time k, is defined by

$$\mathbf{H}_k = \left.\frac{\partial \mathbf{h}}{\partial \mathbf{x}}\right|_{\hat{\mathbf{x}}_{k|k-1}}. \tag{5.11}$$

As pointed out in [203], the extended Kalman filter in general is not an optimal estimator. If the initial estimate of the state is wrong, or if the given dynamical model is inaccurate, the filter may quickly diverge, owing to its linearization. Nonetheless, the extended Kalman filter can generally give reasonable performance and is intensively used in navigation systems and GPS.

5.3 Dynamics Models of the GPS-Aided AHRS

In this section, we establish necessary dynamical models for the AHRS and for the INS, which can be utilized to enhance the measurement signals of each system through properly chosen filtering algorithms, i.e., the extended Kalman filtering adopted in our work.

5.3.1 AHRS Dynamics Model

The primary function of the AHRS is to provide reliable estimation for Euler angles. Furthermore, drift correction for the angular velocity and acceleration measurements are also required. A natural selection for constructing the AHRS dynamics model is to use the inverse relationship of (2.36) and (2.38). However, as mentioned in Chap. 2, the transformations are singular when $\theta = \pm 90^\circ$. Such a problem can be resolved by using the so-called quaternion formulation.

The concept of quaternion was introduced by Hamilton in 1843. The quaternion form, which is a 4D vector space over the real numbers, is given by

$$x_0\mathbf{1} + x_\mathrm{i}\mathbf{i} + x_\mathrm{j}\mathbf{j} + x_\mathrm{k}\mathbf{k} \tag{5.12}$$

with

$$\mathbf{i}^2 = \mathbf{j}^2 = \mathbf{k}^2 = \mathbf{ijk} = -1, \qquad \mathbf{ij} = \mathbf{k} = -\mathbf{ji} \tag{5.13}$$

and

$$\mathbf{jk} = \mathbf{i} = -\mathbf{kj}, \qquad \mathbf{ki} = \mathbf{j} = -\mathbf{ik}, \tag{5.14}$$

where x_0, x_i, x_j, and x_k are real numbers, and $\mathbf{1}$, $\mathbf{i}$, $\mathbf{j}$, and $\mathbf{k}$ are the elements of the basis. Interested readers are referred to [171, 210] for more detailed descriptions on the properties of the quaternion concept. In our work, the quaternion-based AHRS dynamics model is expressed by

$$\dot{\mathbf{x}}_\mathrm{A} = \mathbf{f}_\mathrm{A}(\mathbf{x}_\mathrm{A}, \boldsymbol{\omega}^\mathrm{b}_{\mathrm{b/n}}) + \mathbf{w}_\mathrm{A} \tag{5.15}$$

and

$$\mathbf{y}_\mathrm{A} = \mathbf{h}_\mathrm{A}(\mathbf{x}_\mathrm{A}) + \mathbf{v}_\mathrm{A} \tag{5.16}$$

where $\mathbf{x}_\mathrm{A}$ is the state vector, $\mathbf{w}_\mathrm{A}$ is the process noise vector, $\mathbf{y}_\mathrm{A}$ is the measurement output vector, $\mathbf{v}_\mathrm{A}$ is the measurement noise vector, and $\mathbf{f}_\mathrm{A}(\mathbf{x}_\mathrm{A}, \boldsymbol{\omega}^\mathrm{b}_{\mathrm{b/n}})$ and $\mathbf{h}_\mathrm{A}(\mathbf{x}_\mathrm{A})$ are nonlinear functions of the AHRS dynamics model.

We further note that the state vector $\mathbf{x}_{\mathrm{A}}$ is given by

$$\mathbf{x}_{\mathrm{A}} = [\,\mathbf{q} \quad \mathbf{b}_{\omega}\,]^{\mathrm{T}} = [\,q_0 \quad q_1 \quad q_2 \quad q_3 \quad b_{\mathrm{p}} \quad b_{\mathrm{q}} \quad b_{\mathrm{r}}\,]^{\mathrm{T}}, \tag{5.17}$$

where $\mathbf{q} = [\,q_0 \quad q_1 \quad q_2 \quad q_3\,]^{\mathrm{T}}$ is the quaternion with a unity form, i.e.,

$$\|\mathbf{q}\| = q_0^2 + q_1^2 + q_2^2 + q_3^2 = 1. \tag{5.18}$$

It is particularly determined for the 3–2–1 Euler rotation sequence with the involved items being given by

$$q_0 = \cos\frac{\phi}{2}\cos\frac{\theta}{2}\cos\frac{\psi}{2} + \sin\frac{\phi}{2}\sin\frac{\theta}{2}\sin\frac{\psi}{2}, \tag{5.19}$$

$$q_1 = \sin\frac{\phi}{2}\cos\frac{\theta}{2}\cos\frac{\psi}{2} - \cos\frac{\phi}{2}\sin\frac{\theta}{2}\sin\frac{\psi}{2}, \tag{5.20}$$

$$q_2 = \cos\frac{\phi}{2}\sin\frac{\theta}{2}\cos\frac{\psi}{2} + \sin\frac{\phi}{2}\cos\frac{\theta}{2}\sin\frac{\psi}{2}, \tag{5.21}$$

$$q_3 = \cos\frac{\phi}{2}\cos\frac{\theta}{2}\sin\frac{\psi}{2} - \sin\frac{\phi}{2}\sin\frac{\theta}{2}\cos\frac{\psi}{2} \tag{5.22}$$

and $\mathbf{b}_{\omega} = [\,b_{\mathrm{p}} \quad b_{\mathrm{q}} \quad b_{\mathrm{r}}\,]^{\mathrm{T}}$ is the gyro biases. The nonlinear function $\mathbf{f}_{\mathrm{A}}(\mathbf{x}_{\mathrm{A}}, \boldsymbol{\omega}_{\mathrm{b/n}}^{\mathrm{b}})$ is expressed by

$$\mathbf{f}_{\mathrm{A}}(\mathbf{x}_{\mathrm{A}}, \boldsymbol{\omega}_{\mathrm{b/n}}^{\mathrm{b}}) = \begin{pmatrix} \dfrac{1}{2}\begin{bmatrix} 0 & b_{\mathrm{p}} - p & b_{\mathrm{q}} - q & b_{\mathrm{r}} - r \\ p - b_{\mathrm{p}} & 0 & r - b_{\mathrm{r}} & b_{\mathrm{q}} - q \\ q - b_{\mathrm{q}} & b_{\mathrm{r}} - r & 0 & p - b_{\mathrm{p}} \\ r - b_{\mathrm{r}} & q - b_{\mathrm{q}} & b_{\mathrm{p}} - p & 0 \end{bmatrix}\begin{pmatrix} q_0 \\ q_1 \\ q_2 \\ q_3 \end{pmatrix} \\ 0 \\ 0 \\ 0 \end{pmatrix}, \tag{5.23}$$

where the first four rows are also known as quaternion kinematical equations that are widely used to express the relative orientations between two coordinate systems with relative angular motion. The last three zero rows mean that the gyro biases $\mathbf{b}_{\omega}$ in our work are simply modeled as slowly varying constants (without considering the process noises). We note that such an approximation is only reasonable for the systems such as the miniature UAV helicopters with very limited working time.

Two signal sources that are independent of the gyroscope can be selected to form the output measurement vector $\mathbf{y}_{\mathrm{A}}$: the accelerometer (50 Hz) and the magnetometer (10 Hz). Due to the different sampling rates, we need to construct the AHRS dynamics model and the associated filtering design separately (see also Fig. 5.1).

For the accelerometer, the output measurement vector that corresponds to the proper acceleration measured, i.e., $\mathbf{a}_{\mathrm{mea}}$, is given by

$$\mathbf{y}_{\mathrm{A},\mathbf{a}_{\mathrm{mea}}} = \mathbf{h}_{\mathrm{A},\mathbf{a}_{\mathrm{mea}}}(\mathbf{x}_{\mathrm{A}}) + \mathbf{v}_{\mathrm{A},\mathbf{a}_{\mathrm{mea}}}, \tag{5.24}$$

where

$$\mathbf{h}_{\mathrm{A},\mathbf{a}_{\mathrm{mea}}} = \begin{pmatrix} -2g(q_1 q_3 - q_0 q_2) \\ -2g(q_0 q_1 + q_2 q_3) \\ -g(q_0^2 - q_1^2 - q_2^2 + q_3^2) \end{pmatrix} \tag{5.25}$$

and where g is the acceleration of gravity. It should be noted that the above equation is valid only for the small-scale aircraft performing normal maneuvers with very small acceleration change (compared to the gravitational acceleration projections).

For the magnetometer we have the output measurement vector

$$\mathbf{y}_{\mathrm{A},\psi_{\mathrm{mag}}} = \mathbf{h}_{\mathrm{A},\psi_{\mathrm{mag}}}(\mathbf{x}_{\mathrm{A}}) + \mathbf{v}_{\mathrm{A},\psi_{\mathrm{mag}}} \tag{5.26}$$

with

$$\mathbf{h}_{\mathrm{A},\psi_{\mathrm{mag}}} = \mathrm{atan2}\left[\frac{2(q_1 q_2 + q_0 q_3)}{q_0^2 + q_1^2 - q_2^2 - q_3^2}\right], \tag{5.27}$$

where atan2 is the four-quadrant inverse tangent function. More detailed derivation of the above expressions can be found in [171].

5.3.2 INS Dynamics Model

The INS is responsible for the navigation task in the local NED coordinate system. The INS model to be established in this section is based on the physical location and received signals of the GPS antenna. More specifically, it depends on the geodetic position,

$$\mathbf{P}_{\mathrm{ant,g}} = \begin{pmatrix} \boldsymbol{\lambda}_{\mathrm{ant}} \\ \varphi_{\mathrm{ant}} \\ h_{\mathrm{ant}} \end{pmatrix}, \tag{5.28}$$

the local NED velocity,

$$\mathbf{V}_{\mathrm{ant,n}} = \begin{pmatrix} u_{\mathrm{ant,n}} \\ v_{\mathrm{ant,n}} \\ w_{\mathrm{ant,n}} \end{pmatrix}, \tag{5.29}$$

and the local NED proper acceleration,

$$\mathbf{a}_{\mathrm{ant,n}} = \begin{pmatrix} a_{\mathrm{ax,n}} \\ a_{\mathrm{ay,n}} \\ a_{\mathrm{az,n}} \end{pmatrix}, \tag{5.30}$$

measured at the location of the GPS antenna.

The INS dynamics model can be expressed as

$$\dot{\mathbf{x}}_{\mathrm{I}} = \mathbf{f}_{\mathrm{I}}(\mathbf{x}_{\mathrm{I}}, \mathbf{a}_{\mathrm{ant,n}}) + \mathbf{w}_{\mathrm{I}} \tag{5.31}$$

and

$$\mathbf{y}_{\mathrm{I}} = \mathbf{h}_{\mathrm{I}}(\mathbf{x}_{\mathrm{I}}) + \mathbf{v}_{\mathrm{I}}, \tag{5.32}$$

where $\mathbf{x}_{\mathrm{I}}$ is the state vector, $\mathbf{w}_{\mathrm{I}}$ is the process noise vector, $\mathbf{y}_{\mathrm{I}}$ is the measurement output vector, $\mathbf{v}_{\mathrm{I}}$ is the measurement noise vector, and $\mathbf{f}_{\mathrm{I}}(\mathbf{x}_{\mathrm{I}}, \mathbf{a}_{\mathrm{ant,n}})$ and $\mathbf{h}_{\mathrm{I}}(\mathbf{x}_{\mathrm{I}})$ are appropriate linear or nonlinear functions associated with the INS dynamics. More specifically, the state vector has nine elements, which are given by

$$
\begin{aligned}
\mathbf{x}_{\mathrm{I}} &= [\,\mathbf{P}_{\mathrm{ant,g}} \quad \mathbf{V}_{\mathrm{ant,n}} \quad \mathbf{b}_{\mathbf{a}_{\mathrm{ant,n}}}\,]^{\mathrm{T}} \\
&= [\,\lambda_{\mathrm{ant}} \quad \varphi_{\mathrm{ant}} \quad h_{\mathrm{ant}} \quad u_{\mathrm{ant,n}} \quad v_{\mathrm{ant,n}} \quad w_{\mathrm{ant,n}} \quad b_{a_{\mathrm{ax,n}}} \quad b_{a_{\mathrm{ay,n}}} \quad b_{a_{\mathrm{az,n}}}\,]^{\mathrm{T}},
\end{aligned}
\tag{5.33}
$$

where $\mathbf{b}_{\mathbf{a}_{\mathrm{ant,n}}}$ is the acceleration bias. The nonlinear function $\mathbf{f}_{\mathrm{I}}(\mathbf{x}_{\mathrm{I}}, \mathbf{a}_{\mathrm{ant,n}})$ is given by

$$
\mathbf{f}_{\mathrm{I}}(\mathbf{x}_{\mathrm{I}}, \mathbf{a}_{\mathrm{ant,n}}) = \begin{pmatrix}
\dfrac{v_{\mathrm{ant,n}}}{(N_{\mathrm{E}} + h_{\mathrm{ant}})\cos\varphi_{\mathrm{ant}}} \\
\dfrac{u_{\mathrm{ant,n}}}{M_{\mathrm{E}} + h_{\mathrm{ant}}} \\
-w_{\mathrm{ant,n}} \\
-\dfrac{v_{\mathrm{ant,n}}^2 \sin\varphi_{\mathrm{ant}}}{(N_{\mathrm{E}} + h_{\mathrm{ant}})\cos\varphi_{\mathrm{ant}}} + \dfrac{u_{\mathrm{ant,n}} w_{\mathrm{ant,n}}}{M_{\mathrm{E}} + h_{\mathrm{ant}}} + a_{\mathrm{ax,n}} - b_{a_{\mathrm{ax,n}}} \\
\dfrac{u_{\mathrm{ant,n}} v_{\mathrm{ant,n}} \sin\varphi_{\mathrm{ant}}}{(N_{\mathrm{E}} + h)\cos\varphi_{\mathrm{ant}}} + \dfrac{v_{\mathrm{ant,n}} w_{\mathrm{ant,n}}}{N_{\mathrm{E}} + h_{\mathrm{ant}}} + a_{\mathrm{ay,n}} - b_{a_{\mathrm{ay,n}}} \\
\dfrac{v_{\mathrm{ant,n}}^2}{N_{\mathrm{E}} + h_{\mathrm{ant}}} - \dfrac{u_{\mathrm{ant,n}}^2}{M_{\mathrm{E}} + h_{\mathrm{ant}}} + g + a_{\mathrm{az,n}} - b_{a_{\mathrm{az,n}}} \\
0 \\
0 \\
0
\end{pmatrix}, \tag{5.34}
$$

where the first six rows are based on (2.25) and (2.28), with additional acceleration biases. As in the gyro biases, we model the acceleration biases $\mathbf{b}_{\mathbf{a}_{\mathrm{ant,n}}}$ to be slowly varying constants.

For the majority of the commercial GPS receivers (including those adopted in HeLion and SheLion), $\mathbf{P}_{\mathrm{ant,g}}$ and $\mathbf{V}_{\mathrm{ant,n}}$ are automatically provided by the system. As such, we define $\mathbf{h}_{\mathrm{I}}(\mathbf{x}_{\mathrm{I}})$ simply as

$$
\mathbf{h}_{\mathrm{I}}(\mathbf{x}_{\mathrm{I}}) = [\,\mathbf{I}_6 \quad \mathbf{0}_{6\times 3}\,]\,\mathbf{x}_{\mathrm{I}}, \tag{5.35}
$$

where $\mathbf{I}_6$ is a 6×6 identity matrix and $\mathbf{0}_{6\times 3}$ is a 6×3 zero matrix.

5.4 Design of Extended Kalman Filters

Although recently there are various algorithms available for the filtering design (see, e.g., the autoregressive modeling approach [129], the unscented Kalman filtering or UKF technique [164, 187], neural networks [133], and the H_∞ filtering technique [220], to name a few), the EKF design is still playing a dominant role both in practical applications and in the research community. In this section, we follow the mainstream of current practices in aerospace to employ the well-known EKF technique as described in Sect. 5.2 to realize our aim for the measurement signal enhancement.

5.4.1 EKF for AHRS with Accelerometer Measurement

Based on the AHRS dynamics model with the accelerometer measurement as given by (5.23) and (5.24), we present the detailed EKF parameters for the accelerometer-based AHRS signal enhancement below. We should note that we have used the result of [87] as the initial baseline of our design.

1. Initial setting: Initialization is required for the state vector and error covariance matrix. The numerical values adopted for the EKF of the accelerometer-based AHRS are respectively given as
$$\mathbf{x}_{\text{A},0} = [\,1 \quad 0 \quad 0 \quad 0 \quad 0 \quad 0 \quad 0\,]^{\text{T}} \tag{5.36}$$
and
$$\mathbf{P}_{\text{A},0} = \text{diag}\{0.1,\ 0.1,\ 0.1,\ 0.1,\ 0.1,\ 0.1,\ 0.1\}. \tag{5.37}$$
2. Noise covariance matrices setting: $\mathbf{Q}_{\text{cov}}$ and $\mathbf{R}_{\text{cov}}$ are respectively set as
$$\mathbf{Q}_{\text{cov}} = \text{diag}\{10^{-8},\ 10^{-8},\ 10^{-8},\ 10^{-8},\ 0.8\times 10^{-12},\ 0.8\times 10^{-12},\ 0.8\times 10^{-12}\} \tag{5.38}$$
and
$$\mathbf{R}_{\text{cov}} = \text{diag}\{0.9781,\ 0.9781,\ 0.9781\}. \tag{5.39}$$
3. Prediction and correction: The prediction and correction stages follow the iterative procedure given in (5.5) to (5.10). The execution frequencies for both stages are 50 Hz. The measurement output vector $\mathbf{y}_{\text{A},k}$ at time k is the raw acceleration data delivered by the integrated accelerometer.
4. Signal update: In each loop, we need to update the following parameters: (i) the Euler angles ϕ, θ, and ψ, (ii) the corrected angular velocities $\boldsymbol{\omega}^{\text{b}}_{\text{b/n}}$, (iii) the body frame proper acceleration at the CG, $\mathbf{a}_{\text{mea,b}}$, (iv) the local NED acceleration at the CG, $\mathbf{a}_{\text{n}}$, and (v) the local NED proper acceleration at the GPS antenna location, $\mathbf{a}_{\text{ant,n}}$. We note that items (iii) to (v) are subjected to the lever arm effect (see Chap. 3). Also, item (v) will be used in the EKF for the INS. More specifically, we compute
$$\begin{pmatrix} \phi \\ \theta \\ \psi \end{pmatrix} = \begin{pmatrix} \text{atan2}\left[\frac{2(q_2q_3+q_0q_1)}{1-2(q_1^2+q_2^2)}\right] \\ \sin^{-1}[-2(q_1q_3-q_0q_2)] \\ \text{atan2}\left[\frac{2(q_1q_2+q_0q_3)}{1-2(q_2^2+q_3^2)}\right] \end{pmatrix} \tag{5.40}$$
and
$$\boldsymbol{\omega}^{\text{b}}_{\text{b/n}} = \boldsymbol{\omega}^{\text{b}}_{\text{b/n,mea}} - \mathbf{b}_{\omega}, \tag{5.41}$$
where $\boldsymbol{\omega}^{\text{b}}_{\text{b/n,mea}}$ is the raw signal measurement of the gyroscope. We also compute
$$\mathbf{a}_{\text{n}} = \mathbf{R}_{\text{n/b}}\mathbf{a}_{\text{mea,b}} + \mathbf{g}, \tag{5.42}$$

where

$$\mathbf{g} = \begin{pmatrix} 0 \\ 0 \\ g \end{pmatrix} \tag{5.43}$$

and

$$\mathbf{a}_{\mathrm{mea,b}} = \mathbf{a}_{\mathrm{mea}} - \dot{\boldsymbol{\omega}}^{\mathrm{b}}_{\mathrm{b/n}} \times \mathbf{d}_{\mathrm{CG}} - \boldsymbol{\omega}^{\mathrm{b}}_{\mathrm{b/n}} \times (\boldsymbol{\omega}^{\mathrm{b}}_{\mathrm{b/n}} \times \mathbf{d}_{\mathrm{CG}}), \tag{5.44}$$

and where $\mathbf{a}_{\mathrm{mea}}$ is the raw measurement of the accelerometer, $\mathbf{d}_{\mathrm{CG}} = [\,0 \quad 0 \quad 0.21\,]^{\mathrm{T}}$ is the position offset of the MNAV with respect to the CG in SheLion and HeLion in the body frame, and $\dot{\boldsymbol{\omega}}^{\mathrm{b}}_{\mathrm{b/n}}$ is the derivative of the corrected angular velocity that can be easily obtained based on the measurements at time k and $k-1$. Lastly, we calculate

$$\mathbf{a}_{\mathrm{ant,n}} = \mathbf{R}_{\mathrm{n/b}}\mathbf{a}_{\mathrm{ant,b}} = \mathbf{R}_{\mathrm{n/b}}\big[\mathbf{a}_{\mathrm{mea,b}} + \dot{\boldsymbol{\omega}}^{\mathrm{b}}_{\mathrm{b/n}} \times \mathbf{d}_{\mathrm{off}} + \boldsymbol{\omega}^{\mathrm{b}}_{\mathrm{b/n}} \times (\boldsymbol{\omega}^{\mathrm{b}}_{\mathrm{b/n}} \times \mathbf{d}_{\mathrm{off}})\big], \tag{5.45}$$

where $\mathbf{a}_{\mathrm{ant,b}}$ is the body frame proper acceleration at the GPS antenna location, and $\mathbf{d}_{\mathrm{off}} = [\,-1.035 \quad 0 \quad -0.172\,]^{\mathrm{T}}$ is the position offset of the GPS antenna with respect to the CG of HeLion and SheLion in the body frame. We note that in both (5.44) and (5.45), "×" denotes a cross product.

5.4.2 *EKF for AHRS with Magnetometer Measurement*

The AHRS dynamics model with the magnetometer measurement is characterized by (5.23) and (5.26). The EKF for the magnetometer-based AHRS is partially overlapped with that of the accelerometer-based AHRS. In software realization, we may encounter a situation in which both filters are ready to execute in the same loop. In such a case, we set a lower priority for the EKF for the magnetometer-based AHRS.

1. Initial setting: The initial values for the state vector and error covariance matrix are identical to those given in (5.36) and (5.37), respectively.
2. Noise covariance matrices setting: $\mathbf{Q}_{\mathrm{cov}}$ is chosen to be identical to that in (5.38) and $\mathbf{R}_{\mathrm{cov}}$ is set to

$$\mathbf{R}_{\mathrm{cov}} = 0.0145. \tag{5.46}$$

3. Prediction and correction: The prediction and correction stages, given in (5.5) to (5.10), for the EKF associated with the magnetometer-based AHRS are executed at different frequencies. More specifically, the prediction stage, shared with the EKF for the accelerometer-based AHRS, runs at a frequency of 50 Hz, whereas the correction stage is executed at a frequency of 10 Hz, which is the sampling frequency of the magnetometer. The measurement output vector $\mathbf{y}_{\mathrm{A},k}$ at time k is the measurement of the heading angle provided by the magnetometer and is computed based on the measured magnetic field strength characterized by

$$\psi_{\mathrm{mag}} = \mathrm{atan2}\left(-\frac{H_{\mathrm{y,n}} - H_{\mathrm{y,off}}}{H_{\mathrm{x,n}} - H_{\mathrm{x,off}}}\right), \tag{5.47}$$

where $H_{x,n}$ and $H_{y,n}$ are the horizontal magnetic field measurement in the local NED frame, and $H_{x,off}$ and $H_{y,off}$ are the magnetic field measurement offsets projected onto the local NED frame, which are mainly caused by some minor EMI of the unmanned system and can be determined by hard- and soft-iron calibration. They are obtained via the following coordinate transformation:

$$H_{x,n} = H_x \cos\theta + H_y \sin\phi \sin\theta + H_z \cos\phi \sin\theta \tag{5.48}$$

and

$$H_{y,n} = H_y \cos\phi - H_z \sin\phi, \tag{5.49}$$

where H_x, H_y, and H_z are the raw magnetometer readings in the body frame. The Euler angles used in the above two equations are those updated in the previous time step, that is, time $k-1$.

4. Signal update: The signal update step is identical to its counterpart in the EKF associated with the accelerometer-based AHRS.

5.4.3 EKF for INS

The dynamics model for the INS is given earlier in (5.34) and (5.35). We summarize in what follows the parameters selected for the EKF associated with the INS.

1. Initial setting: Initial values for the state vector and the error covariance matrix are respectively given by

$$\mathbf{x}_{I,0} = [\,\boldsymbol{\lambda}_{ref} \quad \boldsymbol{\varphi}_{ref} \quad h_{ref} \quad 0 \quad 0 \quad 0 \quad 0 \quad 0 \quad 0\,]^{T} \tag{5.50}$$

and

$$\mathbf{P}_{I,0} = \mathbf{I}_9. \tag{5.51}$$

2. Noise covariance matrices setting: For the INS, $\mathbf{Q}_{cov}$ and $\mathbf{R}_{cov}$ are respectively selected as

$$\mathbf{Q}_{cov} = \text{diag}\{0, 0, 0, 0.04, 0.04, 0.04, 0.8 \times 10^{-8}, 0.8 \times 10^{-8}, 0.8 \times 10^{-8}\} \tag{5.52}$$

and

$$\mathbf{R}_{cov} = \text{diag}\left\{\left(\frac{\pi}{180}\right)^2 \times 10^{-10}, \left(\frac{\pi}{180}\right)^2 \times 10^{-10}, 4, 0.01, 0.01, 0.025\right\}. \tag{5.53}$$

3. Prediction and correction: The recursive procedure as in (5.5) to (5.10) is executed with a frequency of 4 Hz. The measurement output vector $\mathbf{y}_{I,k}$ at time k is directly obtained from the GPS receiver.
4. Signal update: The EKF for the INS updates the position and velocity information in the local NED frame. We note that the position of the GPS antenna in the local NED frame can be determined by the coordinate transformations given

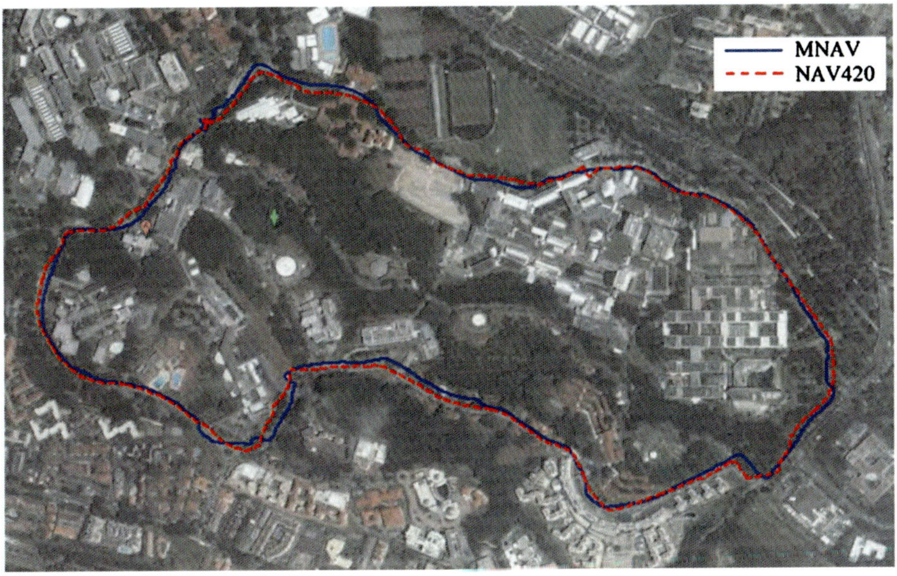

Fig. 5.2 Local NED positions recorded in the ground experiment

in (2.22) and (2.23). The local NED position and velocity at the CG of the unmanned system can be determined by eliminating the lever arm effect from those obtained at the location of the GPS antenna. Both the position and velocity signals are updated at a frequency of 4 Hz.

We would like to highlight that the development of MEMS-based navigation sensors is still a challenging problem, especially for long endurance utilization. Extra works, such as modeling the error dynamics, adjusting the covariance matrices of the process and observation noises, and adding empirical or experimental corrections, are needed if one wants to enhance the working envelope and/or performance of the overall measurement system.

5.5 Performance Evaluation

We have carried out numerous ground experiments to evaluate the efficiency and reliability of the signal enhancement scheme presented in the previous sections. We present in this section the results of a representative experiment that we have conducted. More specifically, in this ground experiment, the trajectory is chosen to be the main road surrounding the campus of the National University of Singapore (see Fig. 5.2), with the starting and ending points being overlapped with each other. Two UAV helicopters, HeLion and SheLion, are fastened at the rear side of a van, which has its cover removed to strengthen signals received by the GPS receivers integrated on the UAVs.

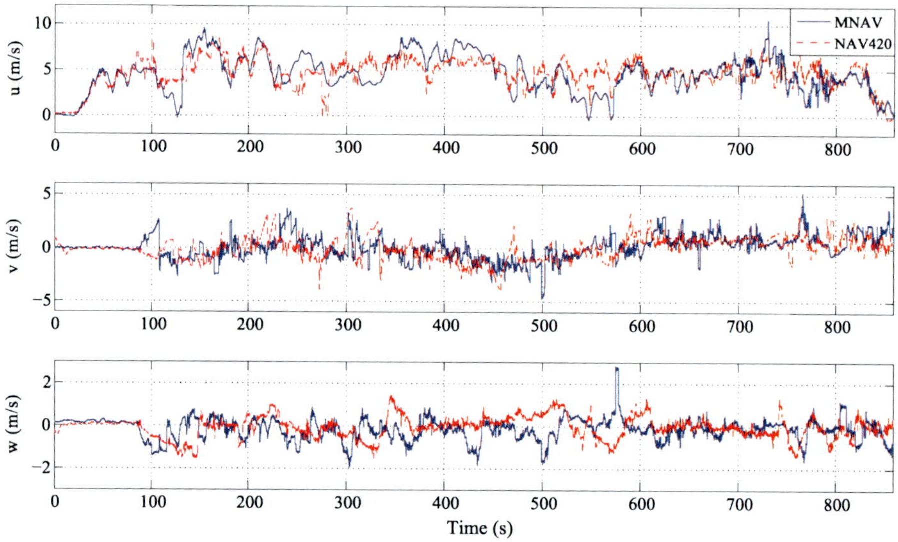

Fig. 5.3 Body frame velocities recorded in the ground experiment

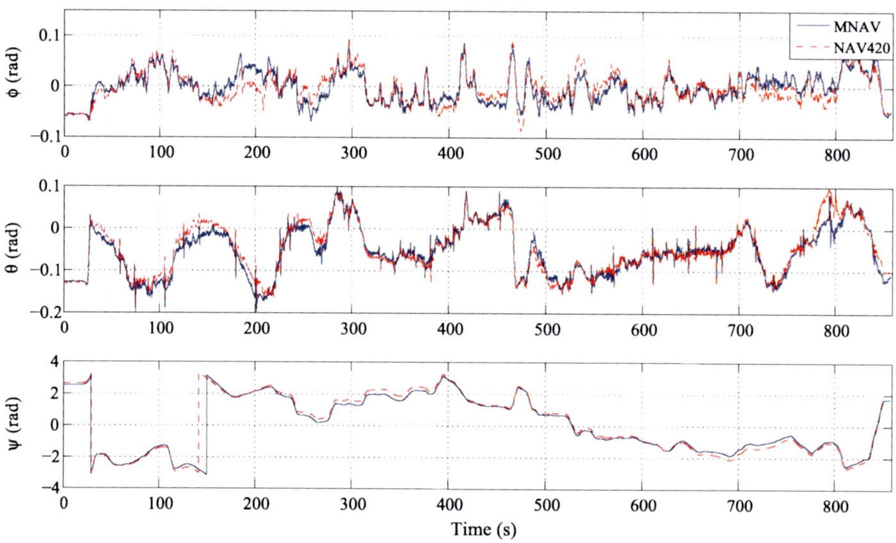

Fig. 5.4 Euler angles recorded in the ground experiment

We should note that for this experiment setup, SheLion carries an MNAV and HeLion is equipped with a high-performance commercial GPS-aided AHRS unit, NAV420 [130], which is used as a reference and for comparison. During the experiment, the van is driven with a moderate speed ranging from 4 to 8 m/s, with some random stops at traffic lights. Such a procedure is a good simulation of a cruising

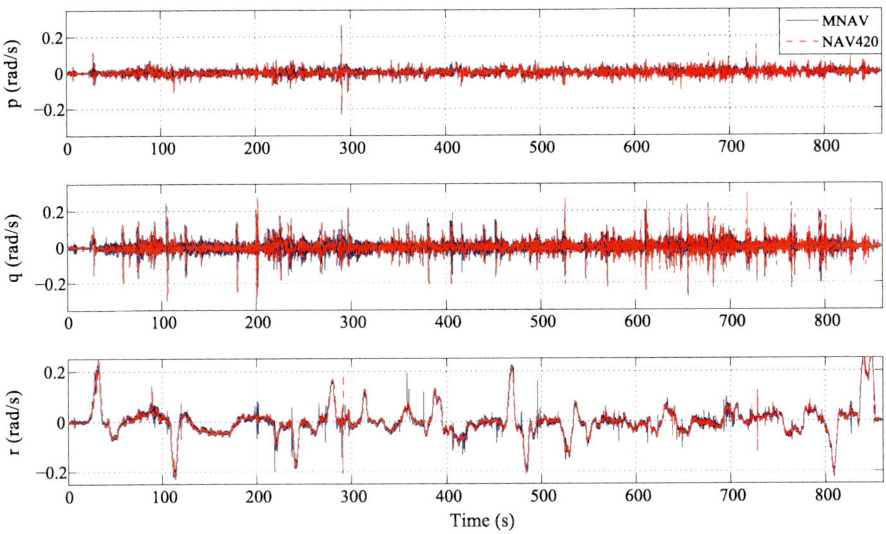

Fig. 5.5 Angular velocities recorded in the ground experiment

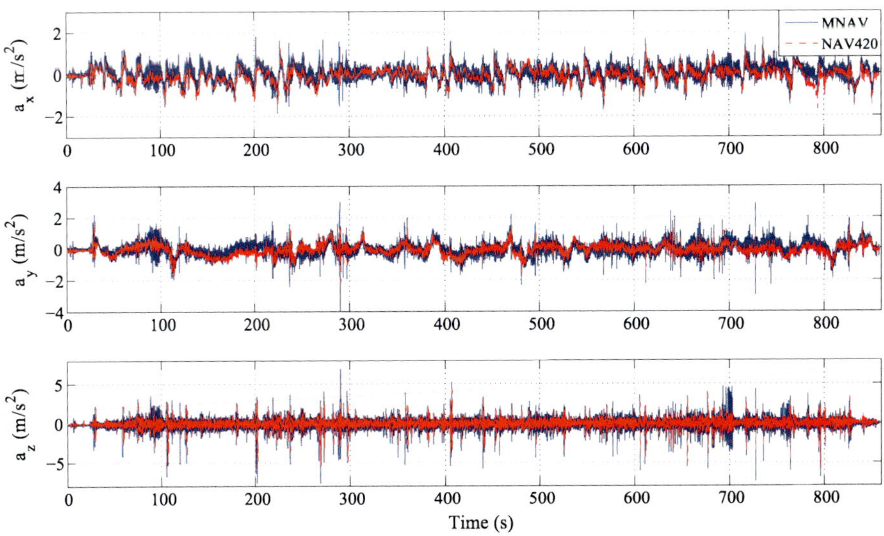

Fig. 5.6 Body-frame accelerations recorded in the ground experiment

mission performed by a miniature UAV helicopter. It takes about 8 to 9 minutes to complete the experiment.

The experimental results, which include the local NED position, velocity, Euler angles, angular velocities, and body-frame accelerations for both NAV420 (HeLion) and MNAV (SheLion), are shown in Figs. 5.2 to 5.6. Generally, the result produced by the commercial unit, NAV420, and that of our own integrated MNAV are very closely matched. Using the result of the NAV420 as a reference, the average deviations for the results shown in Figs. 5.2 to 5.6 are 2.5 m, 0.5 m/s, 0.025 rad (1.5 deg), 0.005 rad/s (0.3 deg/s), and 0.05 m/s^2, respectively. It is clear that the MNAV with the signal enhancement process has a very good performance within the predefined working time.

We note that at several locations, the matches between the recorded NED-based positions and the background satellite map are not as good as we expected. Such mismatches, mainly caused by trees along the road, happen to both the NAV420 and MNAV. On the other hand, we do not observe any divergence in the NED-based velocities, which are well matched all the way. During the whole experiment, Euler angles are accurately estimated by the MNAV and are almost perfectly matched with those obtained from the NAV420 unit.

During the test, the engines of both UAVs are running at 60% of the normal speed (1850 RPM). It can be observed that the vibration, after signal enhancement processing, has been greatly reduced. The van bumps quite a few times, which can be regarded as environmental disturbances or equivalently as wind gust disturbances in actual flight tests. These disturbances have been accurately recorded in the Euler angles and angular velocities in both the NAV420 and MNAV units.

Chapter 6
Flight Dynamics Modeling

6.1 Introduction

A mathematical model that can accurately reflect the flight dynamics of an aircraft is necessary if one wishes to design a flight control system using advanced control techniques. Many works related to flight dynamics modeling of miniature rotorcraft have been conducted since the early 1990s and some successful results have been achieved based on either the first-principles modeling approach (see, e.g., [33, 67, 139]) or the system identification method (see, e.g., [30, 98, 122, 162]). However, research on such a topic is still in its initial stage. The main challenges and key issues on the dynamics modeling of the miniature rotorcraft are as follows:

1. MODELING METHODOLOGY: Although both the first-principles modeling and the system identification approaches have shown their success, we note that using either of the two methods alone is difficult to generate a model with good fidelity in the wide flight envelope. Theoretically, the first-principles-based nonlinear model should cover any operation point in a flight envelope. However, in practical implementation, it needs to be tuned iteratively in terms of structure and aerodynamic parameters, based on practical ground/flight databases or empirical experience. On the other hand, the system identification method is very suitable for deriving a linear model for certain operating points or small flight regions. To cover a wide envelope, it needs to be repeatedly implemented in multiple flight conditions. The efficiency generally decreases in situations associated with higher speed or more aggressiveness, since performing the data collection is extremely difficult or even impractical (see, e.g., [121]).
2. STRUCTURE DETERMINATION: The selected structure directly determines the complexity and validity of the flight dynamics model. For example, main rotor flapping, due to the existence of the stabilizer bar, is required to be included in the modeling process. If the purpose of building up the dynamics model is for flight control law design, the flapping motion can be modeled by two coupled first-order equations instead of the high-order structure proposed in [78].
3. PARAMETER IDENTIFICATION: This issue, which is also referred to as model identifiability, is tightly coupled with the determined model structure. Obviously,

G. Cai et al., *Unmanned Rotorcraft Systems*, Advances in Industrial Control,
DOI 10.1007/978-0-85729-635-1_6,

a model with too much complexity makes its associated parameters difficult or even impossible to be identified accurately. It should be noted that an oversimplified structure also causes bias to the modeling procedure. Using the above rotor flapping example, if the rotor flapping dynamics model is omitted, the associated effect is lumped into the helicopter fuselage dynamics. As a result, the obtained parameters have poor physical meaning and the overall model is highly inaccurate.

In this chapter, we aim to present a comprehensive modeling procedure for the miniature rotorcraft. More specifically, a minimum-complexity model structure, which covers all of the important dynamic features necessary for flight control law design, is proposed and analyzed in Sect. 6.2. Based on this structured model, we carefully develop a five-step procedure in Sect. 6.3, which is a systematic combination of the first-principles and system identification approaches, to determine all of the associated model parameters. We then carry out a thorough time-domain validation in Sect. 6.4 to guarantee the fidelity of the flight dynamics model in the wide flight envelope. Finally, in Sect. 6.5, we proceed to determine the flight envelope of the obtained flight dynamics model, which is essential before proceeding to conduct the flight control law design and flight experiment.

6.2 Model Structure

The main purpose of our modeling work is to obtain a model that can capture the wide-envelope flight dynamics for control law design. With this in mind, we decide to adopt the nonlinear model depicted in Fig. 6.1. We note that there are four key components included in the model: (i) kinematics, (ii) 6-DOF rigid-body dynamics, (iii) main rotor flapping dynamics, and (iv) factory-installed yaw rate feedback controller dynamics. This flight dynamics model features minimum complexity and contains fifteen states and four inputs, which are summarized in Table 6.1.

6.2.1 Kinematics

The kinematical part, which focuses on the translational and rotational motions between the local NED and the body coordinate systems, is derived from the analysis given in Chap. 2. In the kinematical modeling process, the local NED frame is assumed to be inertial and the NED directions of the miniature UAV rotorcraft are constantly aligned with those of the local NED frame.

For the translational motion, we consider (2.9), (2.32) and (2.39), and have

$$\dot{\mathbf{P}}_{\mathrm{n}} = \mathbf{V}_{\mathrm{n}} = \mathbf{R}_{\mathrm{n/b}}\mathbf{V}_{\mathrm{b}}, \tag{6.1}$$

where the rotation matrix $\mathbf{R}_{\mathrm{n/b}}$ is given by

$$\mathbf{R}_{\mathrm{n/b}} = \begin{bmatrix} \mathbf{c}_\theta \mathbf{c}_\psi & \mathbf{s}_\phi \mathbf{s}_\theta \mathbf{c}_\psi - \mathbf{c}_\phi \mathbf{s}_\psi & \mathbf{c}_\phi \mathbf{s}_\theta \mathbf{c}_\psi + \mathbf{s}_\phi \mathbf{s}_\psi \\ \mathbf{c}_\theta \mathbf{s}_\psi & \mathbf{s}_\phi \mathbf{s}_\theta \mathbf{s}_\psi + \mathbf{c}_\phi \mathbf{c}_\psi & \mathbf{c}_\phi \mathbf{s}_\theta \mathbf{s}_\psi - \mathbf{s}_\phi \mathbf{c}_\psi \\ -\mathbf{s}_\theta & \mathbf{s}_\phi \mathbf{c}_\theta & \mathbf{c}_\phi \mathbf{c}_\theta \end{bmatrix} \tag{6.2}$$

with $\mathbf{s}_\star = \sin(\star)$ and $\mathbf{c}_\star = \cos(\star)$.

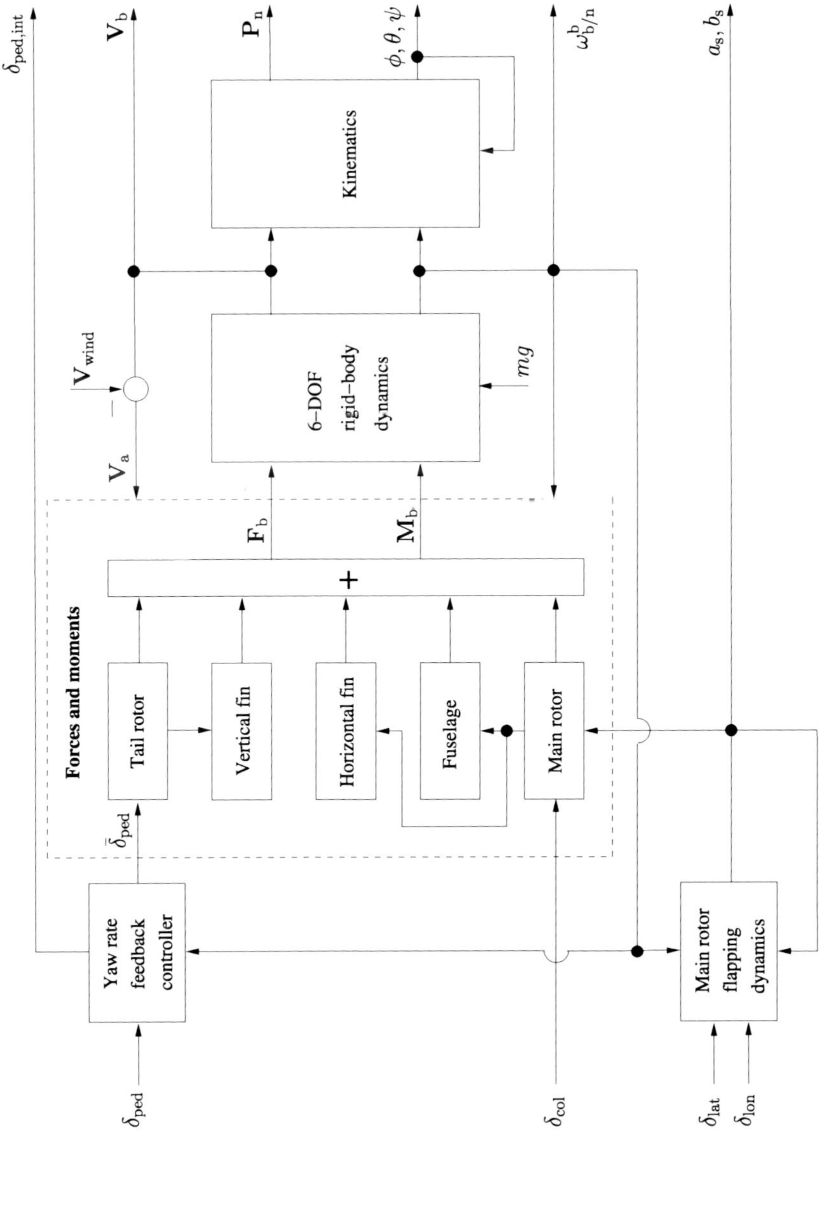

Fig. 6.1 Structure of the flight dynamics model

Table 6.1 State and input variables of the flight dynamics model

Variable	Physical description	Unit
$\mathbf{P}_{\mathrm{n}} = (x_{\mathrm{n}} \quad y_{\mathrm{n}} \quad z_{\mathrm{n}})'$	Position vector in the local NED frame	m
$\mathbf{V}_{\mathrm{b}} = (u \quad v \quad w)'$	Local NED velocity projected onto the body frame	m/s
$\boldsymbol{\omega}^{\mathrm{b}}_{\mathrm{b/n}} = (p \quad q \quad r)'$	Angular velocity vector	rad/s
ϕ, θ, ψ	Euler angles	rad
$a_{\mathrm{s}}, b_{\mathrm{s}}$	Tip-path-plane (TPP) flapping angles of the main rotor	rad
$\delta_{\mathrm{ped,int}}$	Intermediate state of yaw rate feedback controller	NA
δ_{lat}	Normalized aileron servo input $(-1, 1)$	NA
δ_{lon}	Normalized elevator servo input $(-1, 1)$	NA
δ_{col}	Normalized collective pitch servo input $(-1, 1)$	NA
δ_{ped}	Normalized rudder servo input $(-1, 1)$	NA

The rotational motion, based on (2.36) to (2.38), is given by

$$\begin{pmatrix} \dot{\phi} \\ \dot{\theta} \\ \dot{\psi} \end{pmatrix} = \mathbf{S}^{-1} \boldsymbol{\omega}^{\mathrm{b}}_{\mathrm{b/n}}, \tag{6.3}$$

where the lumped transformation matrix $\mathbf{S}^{-1}$ is given in (2.38), i.e.,

$$\mathbf{S}^{-1} = \begin{bmatrix} 1 & \sin\phi \tan\theta & \cos\phi \tan\theta \\ 0 & \cos\phi & -\sin\phi \\ 0 & \sin\phi / \cos\theta & \cos\phi / \cos\theta \end{bmatrix}. \tag{6.4}$$

Note that here we do not use the quaternion expression since $\theta = \pm 90^\circ$ does not appear in our focused flight envelope, which consists of normal helicopter maneuvers.

The wind coordinate system, which is used in many aerodynamic analyses, is not included in our work. However, the wind velocity cannot be ignored since it is essential for aerodynamic forces computation. Our solution is to project the wind velocity onto the body frame. Thus, we have the velocity vector (denoted by a subscript "wind") as

$$\mathbf{V}_{\mathrm{wind}} = \begin{pmatrix} u_{\mathrm{wind}} \\ v_{\mathrm{wind}} \\ w_{\mathrm{wind}} \end{pmatrix}. \tag{6.5}$$

Following the above definition, we get the rotorcraft velocity relative to the air expressed in the body frame, $\mathbf{V}_{\mathrm{a}}$, as

$$\mathbf{V}_{\mathrm{a}} = \begin{pmatrix} u_{\mathrm{a}} \\ v_{\mathrm{a}} \\ w_{\mathrm{a}} \end{pmatrix} = \begin{pmatrix} u - u_{\mathrm{wind}} \\ v - v_{\mathrm{wind}} \\ w - w_{\mathrm{wind}} \end{pmatrix}. \tag{6.6}$$

6.2.2 Rigid-Body Dynamics

The 6-DOF rigid-body dynamics of the helicopter fuselage is derived based on the assumption that the local NED frame is inertial. It can be expressed by the following Newton-Euler equations:

$$\dot{\mathbf{V}}_{\mathrm{b}} = -\boldsymbol{\omega}^{\mathrm{b}}_{\mathrm{b/n}} \times \mathbf{V}_{\mathrm{b}} + \frac{\mathbf{F}_{\mathrm{b}}}{m} + \frac{\mathbf{F}_{\mathrm{b,g}}}{m} \tag{6.7}$$

and

$$\dot{\boldsymbol{\omega}}^{\mathrm{b}}_{\mathrm{b/n}} = \mathbf{J}^{-1}\left[\mathbf{M}_{\mathrm{b}} - \boldsymbol{\omega}^{\mathrm{b}}_{\mathrm{b/n}} \times \left(\mathbf{J}\boldsymbol{\omega}^{\mathrm{b}}_{\mathrm{b/n}}\right)\right], \tag{6.8}$$

where "$\times$" denotes the cross-product of two vectors, m is the mass of the helicopter,

$$\mathbf{F}_{\mathrm{b,g}} = \begin{pmatrix} -mg\sin\theta \\ mg\sin\phi\cos\theta \\ mg\cos\phi\cos\theta \end{pmatrix} \tag{6.9}$$

is the gravity force vector projected onto the body frame, $\mathbf{J} = \mathrm{diag}\{J_{\mathrm{xx}}, J_{\mathrm{yy}}, J_{\mathrm{zz}}\}$ is the moment of inertia matrix (note that the off-axis moment of inertia for our small-scale UAV helicopters is very small and thus ignored for simplicity), $\mathbf{F}_{\mathrm{b}}$ is the aerodynamic force vector, and $\mathbf{M}_{\mathrm{b}}$ is the aerodynamic moment vector. Furthermore, the last two terms are given by

$$\mathbf{F}_{\mathrm{b}} = \begin{pmatrix} X_{\mathrm{mr}} + X_{\mathrm{fus}} \\ Y_{\mathrm{mr}} + Y_{\mathrm{fus}} + Y_{\mathrm{tr}} + Y_{\mathrm{vf}} \\ Z_{\mathrm{mr}} + Z_{\mathrm{fus}} + Z_{\mathrm{hf}} \end{pmatrix} \tag{6.10}$$

and

$$\mathbf{M}_{\mathrm{b}} = \begin{pmatrix} L_{\mathrm{mr}} + L_{\mathrm{vf}} + L_{\mathrm{tr}} \\ M_{\mathrm{mr}} + M_{\mathrm{hf}} \\ N_{\mathrm{mr}} + N_{\mathrm{vf}} + N_{\mathrm{tr}} \end{pmatrix}, \tag{6.11}$$

where $(\cdot)_{\mathrm{mr}}$, $(\cdot)_{\mathrm{tr}}$, $(\cdot)_{\mathrm{fus}}$, $(\cdot)_{\mathrm{vf}}$, and $(\cdot)_{\mathrm{hf}}$ stand, respectively, for the main rotor, tail rotor, fuselage, vertical fin, and horizontal fin of the helicopter, which are the five sources for generating aerodynamic forces and moments. Correspondingly, we can categorize the force and moment components into five groups:

1. Main rotor forces and moments: X_{mr}, Y_{mr}, Z_{mr}, L_{mr}, M_{mr}, N_{mr}
2. Tail rotor force and moments: Y_{tr}, L_{tr}, N_{tr}
3. Fuselage forces: X_{fus}, Y_{fus}, Z_{fus}
4. Vertical fin force and moments: Y_{vf}, L_{vf}, N_{vf}
5. Horizontal fin force and moment: Z_{hf}, M_{hf}

In what follows, we detail the expressions of these force and moment items. It is noted that these expressions are mainly based on the results presented in [79] with some minor modifications.

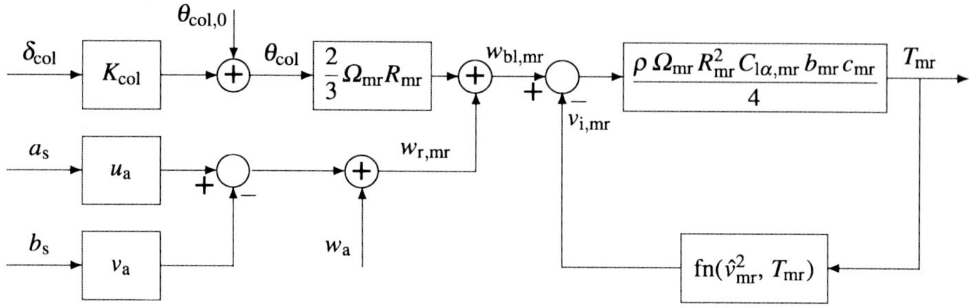

Fig. 6.2 Block diagram of main rotor thrust computation

6.2.2.1 Main Rotor Force and Moment

The main rotor is the source of lift. The forces and moments generated by the main rotor are computed based on the classical momentum theory, under the assumption of uniform inflow distribution. As introduced in [91], momentum theory is an efficient global analysis regarding the overall flow velocities and the total thrust and power. This is in line with the focus of the main rotor dynamics that we intend to examine.

Shown in Fig. 6.2 is the block diagram of the main rotor thrust computation, which was initially proposed by Heffley in [79] and was partially modified to suit the flight dynamics of the small-scale UAV helicopters. This process features a recursion scheme, which is able to effectively interpret the main rotor behavior and achieve a quick convergence of two key terms: the main rotor thrust T_{mr} and the induced velocity $v_{i,mr}$, which are given by

$$T_{mr} = \frac{\rho\Omega_{mr}R_{mr}^2 C_{l\alpha,mr}b_{mr}c_{mr}}{4}(w_{bl,mr} - v_{i,mr}) \tag{6.12}$$

and

$$v_{i,mr}^2 = \sqrt{\left(\frac{\hat{v}_{mr}^2}{2}\right)^2 + \left(\frac{T_{mr}}{2\rho\pi R_{mr}^2}\right)^2} - \frac{\hat{v}_{mr}^2}{2}, \tag{6.13}$$

where

$$\hat{v}_{mr}^2 = u_a^2 + v_a^2 + w_{r,mr}(w_{r,mr} - 2v_{i,mr}), \tag{6.14}$$

$$w_{r,mr} = w_a + a_s u_a - b_s v_a, \tag{6.15}$$

$$w_{bl,mr} = w_{r,mr} + \frac{2}{3}\Omega_{mr}R_{mr}\theta_{col}, \tag{6.16}$$

$$\theta_{col} = K_{col}\delta_{col} + \theta_{col,0}, \tag{6.17}$$

and ρ is the local air density, Ω_{mr} is the rotation speed of the main rotor, R_{mr} is the radius of the main rotor disc, $C_{l\alpha,mr}$ is the lift curve slope of the main rotor blade, b_{mr} is the blade number, c_{mr} is the chord length of the main rotor blade, $w_{bl,mr}$ is the net vertical velocity relative to the main rotor blade, $\hat{v}_{mr}^2$ is an intermediate variable in main rotor thrust calculation, $w_{r,mr}$ is the net vertical velocity through the main rotor disc (note that this item includes the main rotor flapping angles a_s and b_s that are detailed later in this chapter), and θ_{col} is the collective pitch angle of the main rotor blade. It should be noted that the change of θ_{col} results from the collective pitch servo input δ_{col}. We note that their relationship is linear and can be expressed in terms of the scaling factor K_{col} and the offset value of the collective pitch angle $\theta_{col,0}$ (when δ_{col} is zero).

In the above computation procedure, we note that (i) the blade twist angle (θ_{tw}) and shaft incidence (i_s) are both zeros for our UAV helicopters, HeLion and SheLion, and thus not involved in the recursive expression, (ii) for any flight condition, the iteration scheme starts with the associated trim values of T_{mr} and $v_{i,mr}$, and (iii) we run the iteration scheme for ten loops for each computational process to ensure the result convergence for T_{mr} and $v_{i,mr}$, following the experimental result introduced in [79].

The force components generated by the main rotor are computed as follows:

$$X_{mr} = -T_{mr} \sin a_s, \qquad Y_{mr} = T_{mr} \sin b_s, \qquad Z_{mr} = -T_{mr} \cos a_s \cos b_s. \tag{6.18}$$

The moments generated by the main rotor are given by

$$L_{mr} = (K_\beta + T_{mr} H_{mr}) \sin(b_s), \tag{6.19}$$

$$M_{mr} = (K_\beta + T_{mr} H_{mr}) \sin(a_s), \tag{6.20}$$

and

$$N_{mr} = -P_{mr}/\Omega_{mr}, \tag{6.21}$$

where K_β is the effective main rotor spring constant, H_{mr} is the main rotor hub location above the CG of the helicopter, and P_{mr} is the total power consumption which is the sum of four components, including the main rotor profile power P_{pr}, main rotor induced power P_i, parasite power P_{pa}, and climbing power P_c, i.e.,

$$P_{mr} = P_{pr} + P_i + P_{pa} + P_c \tag{6.22}$$

with

$$P_{pr} = \frac{\rho \Omega_{mr} R_{mr}^2 C_{D0} b_{mr} c_{mr}}{8} \left[(\Omega_{mr} R_{mr})^2 + 4.6 \left(u_a^2 + v_a^2 \right) \right], \tag{6.23}$$

$$P_i = T_{mr} v_{i,mr}, \tag{6.24}$$

$$P_{pa} = |X_{fus} u_a| + |Y_{fus} v_a| + |Z_{fus} (w_a - v_{i,mr})|, \tag{6.25}$$

and

$$P_c = \begin{cases} -mg w_a, & \text{if } w_a < 0, \\ 0, & \text{if } w_a \geq 0. \end{cases} \tag{6.26}$$

In (6.23), C_{D0} is the main rotor blade drag coefficient, and in (6.25), X_{fus}, Y_{fus}, and Z_{fus} are the fuselage drag forces to be addressed later.

6.2.2.2 Tail Rotor Force and Moment

The tail rotor generates a thrust to counter the fuselage torque arising from the rotation of the main rotor. Similar to the main rotor, the tail rotor thrust $T_{\rm tr}$ and the induced velocity $v_{\rm i,tr}$ can be calculated using the above ten-loop recursive scheme. Since the size of the tail rotor blade is very small, its flapping effect is negligible and the recursive procedure is modified as follows:

$$T_{\rm tr} = \frac{\rho \Omega_{\rm tr} {R_{\rm tr}}^2 C_{{\rm l}\alpha,{\rm tr}} b_{\rm tr} c_{\rm tr}}{4} (w_{\rm bl,tr} - v_{\rm i,tr}) \tag{6.27}$$

and

$$v_{\rm i,tr}^2 = \sqrt{\left(\frac{\hat{v}_{\rm tr}^2}{2}\right)^2 + \left(\frac{T_{\rm tr}}{2\rho\pi R_{\rm tr}^2}\right)^2} - \frac{\hat{v}_{\rm tr}^2}{2}, \tag{6.28}$$

where

$$\hat{v}_{\rm tr}^2 = (w_{\rm a} + q D_{\rm tr})^2 + u_{\rm a}^2 + w_{\rm r,tr}(w_{\rm r,tr} - 2v_{\rm i,tr}), \tag{6.29}$$

$$w_{\rm r,tr} = v_{\rm a} - r D_{\rm tr} + p H_{\rm tr}, \tag{6.30}$$

$$w_{\rm bl,tr} = w_{\rm r,tr} + \frac{2}{3}\Omega_{\rm tr} R_{\rm tr} \theta_{\rm ped}, \tag{6.31}$$

$$\theta_{\rm ped} = K_{\rm ped}\bar{\delta}_{\rm ped} + \theta_{\rm ped,0}, \tag{6.32}$$

and $\Omega_{\rm tr}$ is the rotation speed of the tail rotor, $R_{\rm tr}$ is the radius of the tail rotor disc, $C_{{\rm l}\alpha,{\rm tr}}$ is the lift curve slope of the tail rotor blade, $b_{\rm tr}$ is the tail rotor blade number, $c_{\rm tr}$ is the chord length of the tail rotor blade, $w_{\rm bl,tr}$ is the net vertical velocity relative to the tail rotor disc, $\hat{v}_{\rm tr}^2$ is an intermediate variable in the recursive calculation, $D_{\rm tr}$ is the tail rotor hub location behind the CG of the helicopter, $w_{\rm r,tr}$ is the net vertical velocity through the tail rotor disc, $H_{\rm tr}$ is the tail rotor hub location above the CG of the helicopter, and $\theta_{\rm ped}$ is the collective pitch angle of the tail rotor blade. Similar to the main rotor configuration, the relationship between $\theta_{\rm ped}$ and the rudder servo actuator deflection $\bar{\delta}_{\rm ped}$ is linear and denoted by the scaling factor $K_{\rm ped}$ and the offset value of $\theta_{\rm ped,0}$ (when $\bar{\delta}_{\rm ped}$ is zero). It should be noted that for this linear relationship, the input is $\bar{\delta}_{\rm ped}$ instead of $\delta_{\rm ped}$, due to the existence of the yaw rate feedback controller as shown in Fig. 6.1.

The force component $Y_{\rm tr}$ is then calculated by

$$Y_{\rm tr} = -T_{\rm tr}. \tag{6.33}$$

We note that Raptor 90 SE helicopters have clockwise rotation (from top view), and thus $Y_{\rm tr}$ is negative-valued as a result of the definition of the axes of the body frame.

The above tail rotor force generates two moments, i.e., $L_{\rm tr}$ and $N_{\rm tr}$. The first item is caused by the vertical distance between the helicopter CG and the tail rotor hub, and the latter is responsible for countering the torque $N_{\rm mr}$ generated by the main rotor. They are given by

$$L_{\rm tr} = Y_{\rm tr} H_{\rm tr} \quad \text{and} \quad N_{\rm tr} = -Y_{\rm tr} D_{\rm tr}, \tag{6.34}$$

respectively.

6.2.2.3 Fuselage Forces

The fuselage causes drags along three body-frame directions during the flight. In the modeling process, we consider the main rotor downwash effect and express the drag forces in horizontal and vertical directions as the following.

For the horizontal directions, the main rotor downwash is deflected by u_a or v_a. In the situation when u_a (or v_a) is less than $v_{i,mr}$, such a deflection effect is required to be taken into account. While u_a (or v_a) exceeds $v_{i,mr}$, the downwash effect is relatively weak and can be ignored. The fuselage is then considered as a 3D virtual flag plate and the drag forces are represented by a quadratic form. As such, the horizontal fuselage forces are defined by

$$X_{fus} = \begin{cases} -\frac{\rho}{2} S_{fx} u_a v_{i,mr}, & \text{if } |u_a| \le v_{i,mr}, \\ -\frac{\rho}{2} S_{fx} u_a |u_a|, & \text{if } |u_a| > v_{i,mr} \end{cases} \tag{6.35}$$

and

$$Y_{fus} = \begin{cases} -\frac{\rho}{2} S_{fy} v_a v_{i,mr}, & \text{if } |v_a| \le v_{i,mr}, \\ -\frac{\rho}{2} S_{fy} v_a |v_a|, & \text{if } |v_a| > v_{i,mr} \end{cases} \tag{6.36}$$

where S_{fx} and S_{fy} are the effective drag area along the body-frame X- and Y-axes, respectively.

For the vertical direction, the fuselage is constantly exposed to the main rotor downwash. We thus use a uniform quadratic equation to express the vertical fuselage drag force Z_{fus}:

$$Z_{fus} = -\frac{\rho}{2} S_{fz}(w_a - v_{i,mr})|w_a - v_{i,mr}|, \tag{6.37}$$

where S_{fz} is the effective drag area along the body-frame Z-axis.

We note that the fuselage moment components are generally very small due to the symmetric mechanical structure of our HeLion and SheLion. As such, they are not included in the flight dynamics model.

6.2.2.4 Vertical Fin Force and Moment

The main function of the vertical fin is to increase the yaw motion stability via the generated sideforce. For the sideforce computation, the following three points should be noted:

1. The vertical fin equipped in the small-scale helicopters is generally within 2 to 3 mm thickness only. Consequently, the sideforce arising from the camber is sufficiently small and not considered.
2. The mounting position of the vertical fin affects the local lateral airspeed v_{vf} at the vertical fin. We define a parameter λ_{vf} to indicate whether the vertical fin is exposed to the tail rotor wake ($\lambda_{vf} = 1$ if the vertical fin is exposed to the tail rotor, otherwise $\lambda_{vf} = 0$).

3. Stall, in which the sideforce reduces with respect to the increase of the angle of attack (AOA), is required to be considered. The critical AOA, defined by α_{st}, should be determined via practical experiment data. The tangential value of α_{st} is the ratio of local lateral speed magnitude to the longitudinal airspeed magnitude. In stall, we assume that the sideforce is caused only by the dynamic pressure perpendicular to the vertical fin.

Next, we proceed to model the vertical fin force. We define v_{vf}, the local lateral airspeed at the vertical fin, as

$$v_{\mathrm{vf}} = v_{\mathrm{a}} - r D_{\mathrm{vf}} - \lambda_{\mathrm{vf}} v_{\mathrm{i,tr}}, \tag{6.38}$$

where D_{vf} is the vertical fin location behind the CG of the helicopter. The vertical fin force is then given by

$$Y_{\mathrm{vf}} = \begin{cases} -\frac{\rho}{2} C_{\mathrm{l}\alpha,\mathrm{vf}} S_{\mathrm{vf}} v_{\mathrm{vf}} |u_{\mathrm{a}}|, & \text{if } |\frac{v_{\mathrm{vf}}}{u_{\mathrm{a}}}| \leq \tan(\alpha_{\mathrm{st}}), \\ -\frac{\rho}{2} S_{\mathrm{vf}} v_{\mathrm{vf}} |v_{\mathrm{vf}}|, & \text{if } |\frac{v_{\mathrm{vf}}}{u_{\mathrm{a}}}| > \tan(\alpha_{\mathrm{st}}) \text{ (surface stalled)} \end{cases} \tag{6.39}$$

where $C_{\mathrm{l}\alpha,\mathrm{vf}}$ is the lift curve slope of the vertical fin, and S_{vf} is the vertical fin area.

Similar to the tail rotor, the vertical fin generates two moment components, L_{vf} and N_{vf}, along the body-frame X-axis and Z-axis, respectively. They are given by

$$L_{\mathrm{vf}} = Y_{\mathrm{vf}} H_{\mathrm{vf}} \quad \text{and} \quad N_{\mathrm{vf}} = -Y_{\mathrm{vf}} D_{\mathrm{vf}}, \tag{6.40}$$

where H_{vf} is the vertical fin location above the CG of the helicopter.

6.2.2.5 Horizontal Fin Force and Moment

The horizontal fin is equipped to provide extra stability for the pitching motion. Its force computation is quite similar to the counterpart of the vertical fin. We note that (i) the minor force component arising from the horizontal fin camber is omitted, (ii) for our HeLion and SheLion, the horizontal fin is fully immersed in the main rotor downwash, which should be considered in the local vertical airspeed w_{hf} computation, and (iii) the α_{st} is identical to the counterpart for the vertical fin since both fins share similar size and shape. We define w_{hf} as

$$w_{\mathrm{hf}} = w_{\mathrm{a}} + q D_{\mathrm{hf}} - v_{\mathrm{i,mr}}, \tag{6.41}$$

where D_{hf} is the horizontal fin location behind the helicopter CG. The horizontal fin force is computed as follows:

$$Z_{\mathrm{hf}} = \begin{cases} -\frac{\rho}{2} C_{\mathrm{l}\alpha,\mathrm{hf}} S_{\mathrm{hf}} w_{\mathrm{hf}} |u_{\mathrm{a}}|, & \text{if } |\frac{w_{\mathrm{hf}}}{u_{\mathrm{a}}}| \leq \tan(\alpha_{\mathrm{st}}), \\ -\frac{\rho}{2} S_{\mathrm{hf}} w_{\mathrm{hf}} |w_{\mathrm{hf}}|, & \text{if } |\frac{w_{\mathrm{hf}}}{u_{\mathrm{a}}}| > \tan(\alpha_{\mathrm{st}}) \text{ (surface stalled)} \end{cases} \tag{6.42}$$

where $C_{\mathrm{l}\alpha,\mathrm{hf}}$ is the lift curve slope of the horizontal fin, and S_{hf} is the horizontal fin area. Lastly, the moment generated by Z_{hf} provides the above mentioned stability for the pitching motion and is given by

$$M_{\mathrm{hf}} = Z_{\mathrm{hf}} D_{\mathrm{hf}}. \tag{6.43}$$

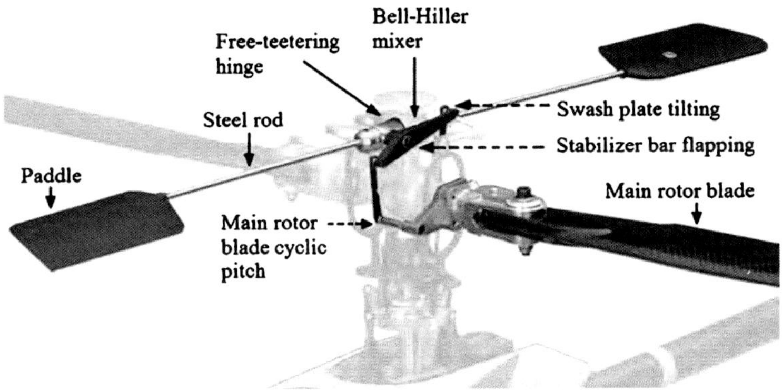

Fig. 6.3 Main rotor augmented by the stabilizer bar

6.2.3 Main Rotor Flapping Dynamics

The majority of RC helicopters (such as the Raptor 90 SE) were originally designed for 3D acrobatic flight. The on-axis angular rate dynamics, that is, from δ_{lat} (or δ_{lon}) to angular rate velocity p (or q), is a tight coupling between the fuselage inertia response and the main rotor flapping response. We refer interested readers to [40, 78, 177] for more detailed information on the rotor-fuselage coupling characteristics and derivations.

The main rotor disc deflection is naturally modeled as flapping motions in longitudinal and lateral directions, i.e., the tip-path-plane (TPP) flapping dynamics. We need to highlight that the stabilizer bar, a unique feature of almost all RC helicopters, plays an important role in the TPP flapping response. In our work, we have included this important feature but lumped its effect into the bare main rotor disc flapping motion. The feasibility of such a simplification has been proven by many successful research studies such as [33, 121, 122].

6.2.3.1 Stabilizer Bar Dynamics

The stabilizer bar, as shown in Fig. 6.3, consists of a steel rod and two plastic paddles with a small aerodynamic surface. It is attached to the main rotor shaft via a free-teetering hinge and can be regarded as a second rotor. For the small-scale helicopters, its primary function is to provide enhanced stability to make the helicopter more stable (or manually controllable) and meanwhile reduce the effect of wind gust and turbulence.

The TPP flapping angles of the stabilizer bar disc are defined as c_{s} and d_{s} for longitudinal and lateral directions, respectively. As introduced in [121], the flapping dynamics from the stabilizer bar cyclic pitch to the flapping angle can be represented by two coupled first-order differential equations:

$$\dot{c}_{\text{s}} = -q - \frac{1}{\tau_{\text{sb}}} c_{\text{s}} + \frac{C_{\text{lon}}}{\tau_{\text{sb}}} \delta_{\text{lon}} \tag{6.44}$$

and

$$\dot{d}_{\mathrm{s}} = -p - \frac{1}{\tau_{\mathrm{sb}}} d_{\mathrm{s}} + \frac{D_{\mathrm{lat}}}{\tau_{\mathrm{sb}}} \delta_{\mathrm{lat}}, \tag{6.45}$$

where C_{lon} (and D_{lat}) is the ratio of stabilizer bar longitudinal (and lateral) cyclic pitch to servo input δ_{lon} (and δ_{lat}), τ_{sb} is the rotor flapping time constant of the stabilizer bar and defined by

$$\tau_{\mathrm{sb}} = \frac{16}{\gamma_{\mathrm{sb}} \Omega_{\mathrm{mr}}}, \tag{6.46}$$

and γ_{sb} is the stabilizer bar Lock number and given as

$$\gamma_{\mathrm{sb}} = \frac{\rho c_{\mathrm{sb}} C_{\mathrm{l}\alpha,\mathrm{sb}} (R_{\mathrm{sb,out}}^4 - R_{\mathrm{sb,in}}^4)}{I_{\beta,\mathrm{sb}}}. \tag{6.47}$$

Here, c_{sb} is the chord length of the stabilizer bar paddle, $C_{\mathrm{l}\alpha,\mathrm{sb}}$ is the lift curve slope of the stabilizer bar paddle, $R_{\mathrm{sb,out}}$ and $R_{\mathrm{sb,in}}$ are the outer and inner radiuses of the stabilizer bar rotor disc, and $I_{\beta,\mathrm{sb}}$ is the inertia moment of the stabilizer bar paddle, with the rotation axis at the main shaft. It should be noted that due to the free-teetering feature, there is no coupling effect between the longitudinal and lateral flapping motions.

6.2.3.2 Bare Main Rotor

The bare main rotor has more complicated flapping dynamics compared with the stabilizer bar. The TPP flapping angles of the main rotor disc are defined as a_{s} and b_{s} for longitudinal and lateral directions, respectively. Similarly, the flapping dynamics from the main rotor blade cyclic pitch to the rotor flapping angle can be represented by two coupled first-order differential equations,

$$\dot{a}_{\mathrm{s}} = -q - \frac{1}{\tau_{\mathrm{mr}}} a_{\mathrm{s}} + A_{b_{\mathrm{s}}} b_{\mathrm{s}} + \frac{1}{\tau_{\mathrm{mr}}} \theta_{\mathrm{cyc},a_{\mathrm{s}}} \tag{6.48}$$

and

$$\dot{b}_{\mathrm{s}} = -p + B_{a_{\mathrm{s}}} a_{\mathrm{s}} - \frac{1}{\tau_{\mathrm{mr}}} b_{\mathrm{s}} + \frac{1}{\tau_{\mathrm{mr}}} \theta_{\mathrm{cyc},b_{\mathrm{s}}}, \tag{6.49}$$

with all of its parameters given as in the following:

1. τ_{mr} is the time constant of the main rotor flapping motion. We note that for a rotor with a flapping hinge offset, the time constant computation given in (6.46) should be modified accordingly. As described in [78], it can be defined by

$$\tau_{\mathrm{mr}} = \frac{16}{\gamma_{\mathrm{mr}} \Omega_{\mathrm{mr}}} \left(1 - \frac{8 e_{\mathrm{mr}}}{3 R_{\mathrm{mr}}}\right)^{-1}, \tag{6.50}$$

   where e_{mr} is the effective hinge offset of the main rotor, γ_{mr} is the main rotor blade Lock number, which is given as

$$\gamma_{\mathrm{mr}} = \frac{\rho c_{\mathrm{mr}} C_{\mathrm{l}\alpha,\mathrm{mr}} R_{\mathrm{mr}}^4}{I_{\beta,\mathrm{mr}}}, \tag{6.51}$$

and $I_{\beta,\mathrm{mr}}$ is the main rotor blade inertia moment with the rotation axis at the main shaft.

2. $\theta_{\mathrm{cyc,a_s}}$ and $\theta_{\mathrm{cyc,b_s}}$, the longitudinal and lateral cyclic pitch of the main rotor blade, are calculated as follows:

$$\theta_{\mathrm{cyc,a_s}} = A_{\mathrm{lon}}\delta_{\mathrm{lon}} + K_{\mathrm{sb}}c_{\mathrm{s}} \tag{6.52}$$

and

$$\theta_{\mathrm{cyc,b_s}} = B_{\mathrm{lat}}\delta_{\mathrm{lat}} + K_{\mathrm{sb}}d_{\mathrm{s}}, \tag{6.53}$$

which indicate how the stabilizer bar affects the main rotor flapping motion. We note that the cyclic pitch is driven by both the servo input (coming from the swash plate) and stabilizer bar flapping angle. It is mechanically realized by a Bell-Hiller mixer (shown in Fig. 6.3). With such a configuration, we are able to keep the instant response coming directly from the swash plate tilting, meanwhile gaining the desired stability from the flapping movement of the stabilizer bar. Adjusting the lever length of the Bell-Hiller mixer results in a different speed of the main rotor flapping response. We note that for a fixed Bell-Hiller mixer, the following three parameters, including A_{lon} (ratio of $\theta_{\mathrm{cyc,a_s}}$ to δ_{lon}), B_{lat} (ratio of $\theta_{\mathrm{cyc,b_s}}$ to δ_{lat}), and K_{sb} (ratio of main rotor blade cyclic pitch to stabilizer bar flapping), can be completely determined.

3. $A_{\mathrm{b_s}}$ and $B_{\mathrm{a_s}}$ represent the coupling effect between longitudinal and lateral flapping motions. As introduced in [121], their theoretical expression is given by

$$A_{\mathrm{b_s}} = -B_{\mathrm{a_s}} = \frac{8k_\beta}{\gamma_{\mathrm{mr}}\Omega^2 I_\beta}. \tag{6.54}$$

However, they are generally not aligned with the obtained results via the practical flight test. Actually identifying these off-axis coupling effects is one of the most challenging issues in the aerodynamic analysis for both full- and small-scale rotorcraft. In our later work, we will further tune their values based on the real flight test data.

6.2.3.3 Complete Main Rotor Flapping Dynamics

To derive the complete main rotor flapping dynamics, we need to integrate the above two TPP flapping dynamics together with the following four steps:

1. Apply the Laplace transform to (6.48) and (6.49).
2. Apply the Laplace transform to (6.44) and (6.45).
3. Insert $c_{\mathrm{s}}(s)$ and $d_{\mathrm{s}}(s)$ into the expression of $a_{\mathrm{s}}(s)$ and $b_{\mathrm{s}}(s)$.
4. Conduct Laplace inverse transform and ignore the items related to the derivatives $\dot{p}$, $\dot{q}$, $\dot{\delta}_{\mathrm{lat}}$, and $\dot{\delta}_{\mathrm{lon}}$.

The complete main rotor flapping dynamics is given by

$$\dot{a}_{\rm s} = -\frac{\tau_{\rm mr} + K_{\rm sb}\tau_{\rm sb}}{\tau_{\rm mr} + \tau_{\rm sb}} q - \frac{1}{\tau_{\rm mr} + \tau_{\rm sb}} a_{\rm s} + \frac{\tau_{\rm mr} A_{b_{\rm s}}}{\tau_{\rm mr} + \tau_{\rm sb}} b_{\rm s} + \frac{A_{\rm lon} + K_{\rm sb} C_{\rm lon}}{\tau_{\rm mr} + \tau_{\rm sb}} \delta_{\rm lon} \tag{6.55}$$

and

$$\dot{b}_{\rm s} = -\frac{\tau_{\rm mr} + K_{\rm sb}\tau_{\rm sb}}{\tau_{\rm mr} + \tau_{\rm sb}} p + \frac{\tau_{\rm mr} B_{a_{\rm s}}}{\tau_{\rm mr} + \tau_{\rm sb}} a_{\rm s} - \frac{1}{\tau_{\rm mr} + \tau_{\rm sb}} b_{\rm s} + \frac{B_{\rm lat} + K_{\rm sb} D_{\rm lat}}{\tau_{\rm mr} + \tau_{\rm sb}} \delta_{\rm lat}, \tag{6.56}$$

which are to be used to form the complete flight dynamics model.

6.2.4 Yaw Rate Feedback Controller

Yaw motion control was once a very tricky and challenging issue in RC helicopters as the yaw moment is extremely sensitive and is hard to be controlled by human pilots. To overcome this problem, almost all RC helicopter products available nowadays are equipped with a yaw rate gyro, which consists of a gyro sensor and a feedback controller to facilitate the human pilot in controlling the yaw rate and/or the heading angle. Ideally, these components should be removed in the unmanned rotorcraft systems. They are, however, commonly reserved for ease of the manual control backup. As such, the dynamics of the yaw rate feedback controller should be included in our model.

The framework of the yaw rate feedback controller is depicted in Fig. 6.4. The joystick input signal, $\delta_{\rm ped}$, is amplified (by a factor of $K_{\rm a}$) by a proportional amplifier circuit and then compared with the feedback yaw angular velocity r measured by the yaw rate gyro. The resulting difference signal is then sent to the embedded controller to generate the tail rotor servo deflection $\bar{\delta}_{\rm ped}$. The yaw rate feedback controllers equipped in HeLion and SheLion are uniform and are known to be a PI compensator, whose dynamics can be expressed by

$$\bar{\delta}_{\rm ped} = \left(K_{\rm P} + \frac{K_{\rm I}}{s} \right) (K_{\rm a}\delta_{\rm ped} - r), \tag{6.57}$$

where $K_{\rm P}$ and $K_{\rm I}$ are the proportional and integral gains of the embedded controller. We define an intermediate state $\delta_{\rm ped,int}$, which is the integration of the error between the amplified yaw channel input signal and the yaw rate feedback. Expression in (6.57) can then be rewritten as

$$\dot{\delta}_{\rm ped,int} = K_{\rm a}\delta_{\rm ped} - r \tag{6.58}$$

and

$$\bar{\delta}_{\rm ped} = K_{\rm P}(K_{\rm a}\delta_{\rm ped} - r) + K_{\rm I}\delta_{\rm ped,int}. \tag{6.59}$$

The above four subsections cover all the key dynamic components. We finally complete this section with a 15th order nonlinear dynamical model by combining

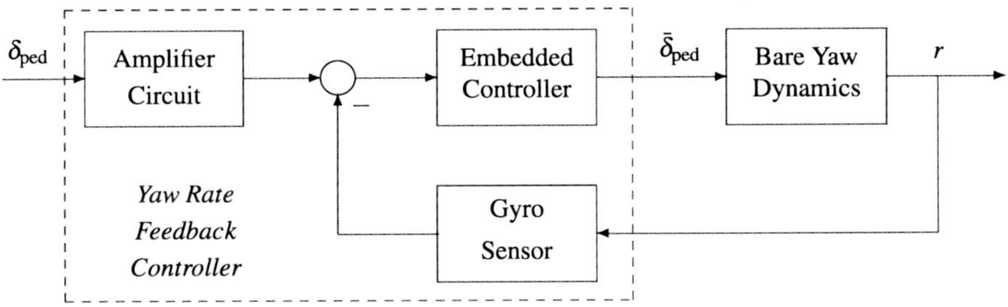

Fig. 6.4 Configuration of the yaw channel

(6.1) to (6.59). Among them, (6.1), (6.3), (6.7), (6.8), (6.55), (6.56), and (6.58) contain the fifteen states listed in Table 6.1. With such a structural dynamical model for the miniature helicopters, we proceed in the next section to identify all the associated model parameters for our unmanned rotorcraft systems.

6.3 Parameter Determination

In this section, we present a complete parameter determination method to identify all the parameters associated with the structural model obtained in the previous section. The parameter identification procedure includes the following five parts:

1. Direct measurement
2. Ground test
3. Estimation based on wind-tunnel data
4. Flight test
5. Fine tuning

We use SheLion as an example for illustration and demonstrate how the above procedure is utilized to determine the model parameters for a small-scale UAV helicopter.

6.3.1 Direct Measurement

We first deal with the parameters that can be directly measured. They are generally related to the environment, platform geometry, and loading. The measurement is completed via direct observation or using simple devices (such as rulers, weighing machines, and tachometers). Parameters belonging to this group are listed in Table 6.2, along with the determined results for SheLion. Note that in the table, I_{mr} and I_{sb} are computed based on the mass values of the main rotor blade and stabilizer bar (m_{mr} and m_{sb}). The equation used for inertia calculation can be easily found in many physics texts.

Table 6.2 Parameters determined via direct measurement

Parameter	Physical meaning
$I_{\mathrm{mr}} = 0.055\ \mathrm{kg{\cdot}m^2}$	Main rotor blade inertia w.r.t. rotor hub
$I_{\mathrm{sb}} = 0.004\ \mathrm{kg{\cdot}m^2}$	Stabilizer bar inertia w.r.t. rotor hub
$R_{\mathrm{mr}} = 0.705\ \mathrm{m}$	Main rotor radius
$R_{\mathrm{sb,in}} = 0.231\ \mathrm{m}$	Stabilizer bar inner radius
$R_{\mathrm{sb,out}} = 0.312\ \mathrm{m}$	Stabilizer bar outer radius
$R_{\mathrm{tr}} = 0.128\ \mathrm{m}$	Tail rotor radius
$S_{\mathrm{fx}} = 0.103\ \mathrm{m^2}$	Effective longitudinal fuselage drag area
$S_{\mathrm{fy}} = 0.900\ \mathrm{m^2}$	Effective lateral fuselage drag area
$S_{\mathrm{fz}} = 0.084\ \mathrm{m^2}$	Effective vertical fuselage drag area
$S_{\mathrm{hf}} = 0.011\ \mathrm{m^2}$	Horizontal fin area
$S_{\mathrm{vf}} = 0.007\ \mathrm{m^2}$	Vertical fin area
$b_{\mathrm{mr}} = 2$	Main rotor blade number
$b_{\mathrm{tr}} = 2$	Tail rotor blade number
$c_{\mathrm{mr}} = 0.062\ \mathrm{m}$	Main rotor blade chord length
$c_{\mathrm{sb}} = 0.059\ \mathrm{m}$	Stabilizer bar chord length
$c_{\mathrm{tr}} = 0.029\ \mathrm{m}$	Tail rotor chord length
$e_{\mathrm{mr}} = 0.07\ \mathrm{m}$	Effective hinge offset of the main rotor
$g = 9.781\ \mathrm{N{\cdot}kg^{-1}}$	Acceleration of gravity
$m = 9.750\ \mathrm{kg}$	Helicopter mass
$n_{\mathrm{tr}} = 4.650$	Gear ratio of the tail rotor to the main rotor
$\Omega_{\mathrm{mr}} = 193.73\ \mathrm{rad}$	Main rotor rotating speed
$\Omega_{\mathrm{tr}} = 900.85\ \mathrm{rad}$	Main rotor rotating speed
$\rho = 1.290\ \mathrm{kg/m^3}$	Air density

6.3.2 *Ground Tests*

We have performed a series of ground tests to determine some unknown parameters, which include (i) CG location determination, (ii) measurement of inertia moment, (iii) airfoil deflection test, and (iv) collective pitch curve examination.

6.3.2.1 CG Location Determination

The CG location experiment is required to be conducted at least three times. For each time, SheLion is suspended at an arbitrarily selected point. For our case, we choose the suspending points located at the main rotor hub, fuselage nose, and fuselage left side, respectively. The general idea is that the intersection of these three (or more) suspending lines is the CG of SheLion. High-resolution pictures are taken to ensure that the intersection position can be precisely determined. For SheLion, the overall CG is located 0.21 m above the CG of the MNAV along the body-frame

Table 6.3 Parameters determined in CG location experiment

Parameter	Physical meaning
$D_{\mathrm{hf}} = 0.751$ m	Horizontal fin location behind the CG
$D_{\mathrm{tr}} = 1.035$ m	Tail rotor hub location behind the CG
$D_{\mathrm{vf}} = 0.984$ m	Vertical fin location behind the CG
$H_{\mathrm{mr}} = 0.337$ m	Main rotor hub location above the CG
$H_{\mathrm{tr}} = 0.172$ m	Tail rotor hub location above the CG
$H_{\mathrm{vf}} = 0.184$ m	Vertical fin location above the CG

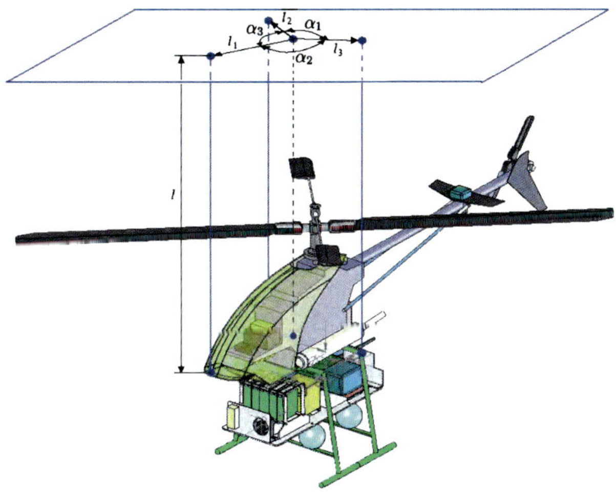

Fig. 6.5 Illustration of trifilar pendulum method

Z-axis. This measurement result is actually a good verification of our avionic system layout design addressed in Chap. 3. Based on the obtained CG location, we can determine another six parameters listed in Table 6.3.

6.3.2.2 Measurement of Inertia Moment

A simple and efficient method, which is named the trifilar pendulum method and proposed in [76], is implemented to obtain the numerical values of the inertia moments of SheLion. The experiment procedure is illustrated in Fig. 6.5. SheLion is suspended by three flexible lines with equal length l. The distances between the attached points and CG are l_1, l_2, and l_3, respectively. We then swing SheLion around the body-frame axis, which is parallel to these three suspended lines (Z-axis in this example), and record the torsional oscillation period t_1. The moment of inertia along this axis is given by

$$J_{\mathrm{xx/yy/zz}} = \frac{mgl_1l_2l_3t_1^2}{4\pi^2 l} \cdot \frac{l_1\sin\alpha_1 + l_2\sin\alpha_2 + l_3\sin\alpha_3}{l_2l_3\sin\alpha_1 + l_1l_3\sin\alpha_2 + l_1l_2\sin\alpha_3}. \tag{6.60}$$

Table 6.4 Moments of inertia determined via trifilar pendulum experiment

Parameter	Physical meaning
$J_{xx} = 0.251$ kg·m^2	Rolling moment of inertia
$J_{yy} = 0.548$ kg·m^2	Pitching moment of inertia
$J_{zz} = 0.787$ kg·m^2	Yawing moment of inertia

The above trifilar pendulum experiment has been repeated three times to obtain the moments of inertia, which are listed in Table 6.4.

6.3.2.3 Airfoil Deflection Test

Airfoil deflection tests aim for determining the parameters related to the Bell-Hiller mixer. For both longitudinal and lateral directions, three sub-experiments should be conducted. Here, we use the longitudinal direction as an example, whose experiment procedure is depicted in Fig. 6.6. Parameters that need to be determined include $A_{\rm lon}$, $C_{\rm lon}$, and $K_{\rm sb}$.

1. For $A_{\rm lon}$ determination (see Fig. 6.6.a), we need to
   a. adjust and maintain the stabilizer bar to be level;
   b. issue $\delta_{\rm lon}$ to tilt the swash plate longitudinally;
   c. record the cyclic pitch deflection $\theta_{\rm cyc,a_s}$ of the main rotor blade; and
   d. note that $A_{\rm lon}$ is the ratio of $\theta_{\rm cyc,a_s}$ to $\delta_{\rm lon}$.
2. For $C_{\rm lon}$ (see Fig. 6.6.b), we have to
   a. adjust the stabilizer bar to be level;
   b. keep the cyclic pitch of the main rotor blade unchanged in this experiment;
   c. inject $\delta_{\rm lon}$ to tilt the swash plate longitudinally;
   d. record the corresponding stabilizer bar deflection; and
   e. note that $C_{\rm lon}$ is the ratio of the stabilizer bar deflection to $\delta_{\rm lon}$.
3. For $K_{\rm sb}$ (see Fig. 6.6.c), we proceed to
   a. adjust the stabilizer bar to be level;
   b. keep the swash plate balanced in this experiment;
   c. manually change the stabilizer bar flapping angle $c_{\rm s}$ and record the corresponding $\theta_{\rm cyc,a_s}$; and
   d. note that $K_{\rm sb}$ is the ratio of $\theta_{\rm cyc,a_s}$ to $c_{\rm s}$.

The same experiments are applied to the lateral direction to identify $B_{\rm lat}$ and $D_{\rm lat}$. We note that we can obtain another numerical value for $K_{\rm sb}$ through a lateral deflection experiment. The two results are almost identical, which coincides with our expectation that $K_{\rm sb}$ is a Bell-Hiller mixer setting that is strictly symmetrical to both directions. The obtained results for SheLion (and HeLion) in this step are listed in Table 6.5.

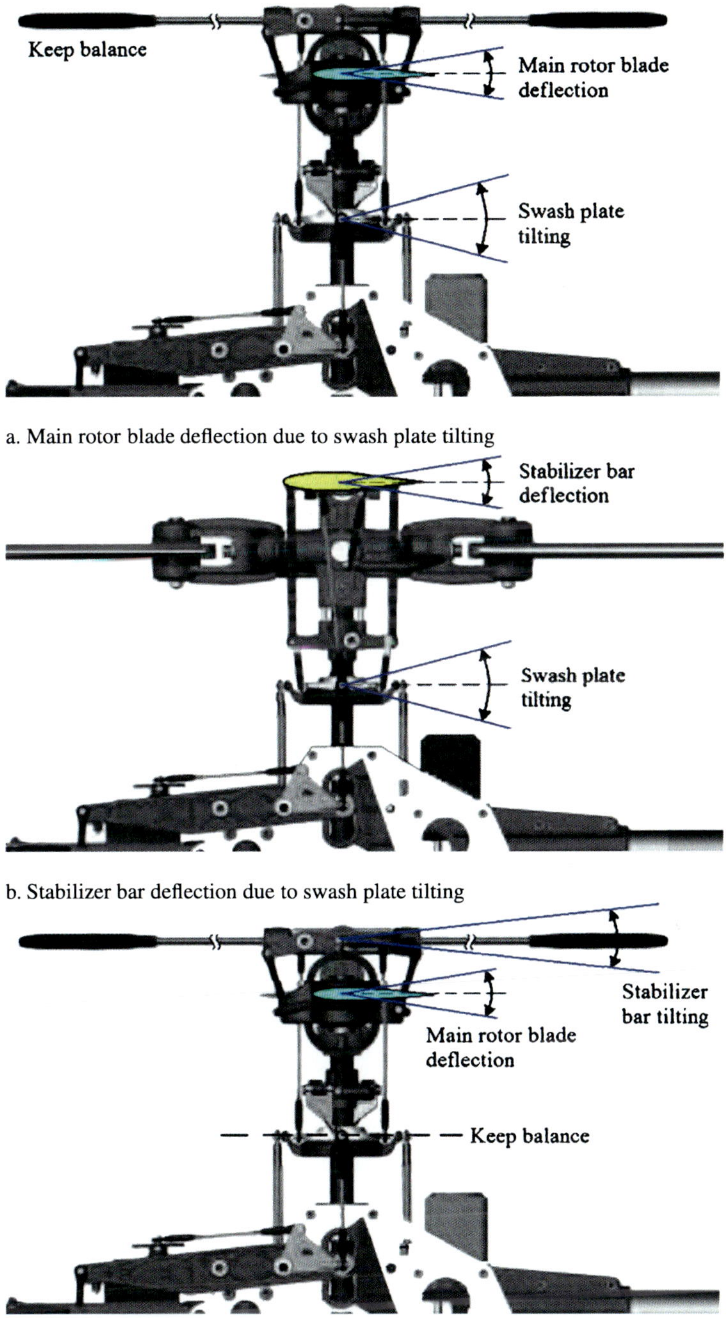

a. Main rotor blade deflection due to swash plate tilting

b. Stabilizer bar deflection due to swash plate tilting

c. Main rotor blade deflection due to stabilizer bar TPP tilting

Fig. 6.6 Airfoil deflection tests applied to SheLion

Table 6.5 Parameters measured in airfoil deflection test

Parameter	Physical meaning
$A_{\text{lon}} = 0.210$ rad	Linkage gain ratio of $\theta_{\text{cyc,a}_s}$ to δ_{lon}
$B_{\text{lat}} = 0.200$ rad	Linkage gain ratio of $\theta_{\text{cyc,b}_s}$ to δ_{lat}
$C_{\text{lon}} = 0.560$ rad	Linkage gain ratio of stabilizer bar cyclic change to δ_{lon}
$D_{\text{lat}} = 0.570$ rad	Linkage gain ratio of stabilizer bar cyclic change to δ_{lat}
$K_{\text{sb}} = 1$	Ratio of $\theta_{\text{cyc,a}_s}$ (or $\theta_{\text{cyc,b}_s}$) to c_s (or d_s)

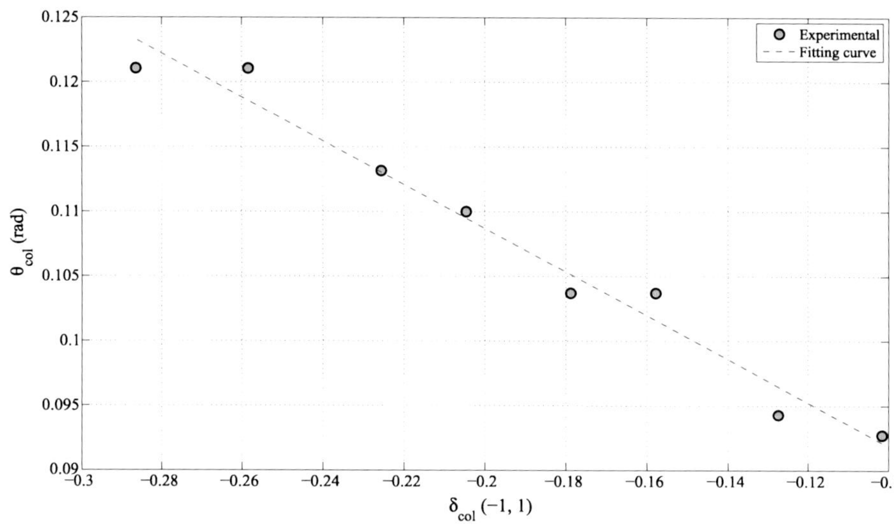

Fig. 6.7 Collective pitch curve of the main rotor

6.3.2.4 Collective Pitch Curve Examination

The collective pitch curve experiments are applied to both the main rotor and tail rotor. The main purpose is to determine the relationship between the servo actuator input δ_{col} (δ_{ped} for tail rotor) and the blade collective pitch angle θ_{col} (θ_{ped} for tail rotor).

1. For the main rotor, we choose seven input values within the effective working range of δ_{col} and record the corresponding collective pitch angles. The linear relationship is reflected by Fig. 6.7. We then adopt the least square curve fitting method to determine K_{col} and $\theta_{\text{col},0}$.
2. For the tail rotor, due to the existence of the yaw rate feedback controller, we need to divide the dynamics from δ_{ped} to θ_{ped} into two parts (i.e., $\bar{\delta}_{\text{ped}}$ to θ_{ped} and δ_{ped} to $\bar{\delta}_{\text{ped}}$) and analyze them individually. The former part has a similar linear relationship as depicted in Fig. 6.8. We can easily obtain the numerical results for the associated parameters K_{ped} and $\theta_{\text{ped},0}$. For the latter part, the focus is to estimate the parameters for the amplification circuit and the proportional and

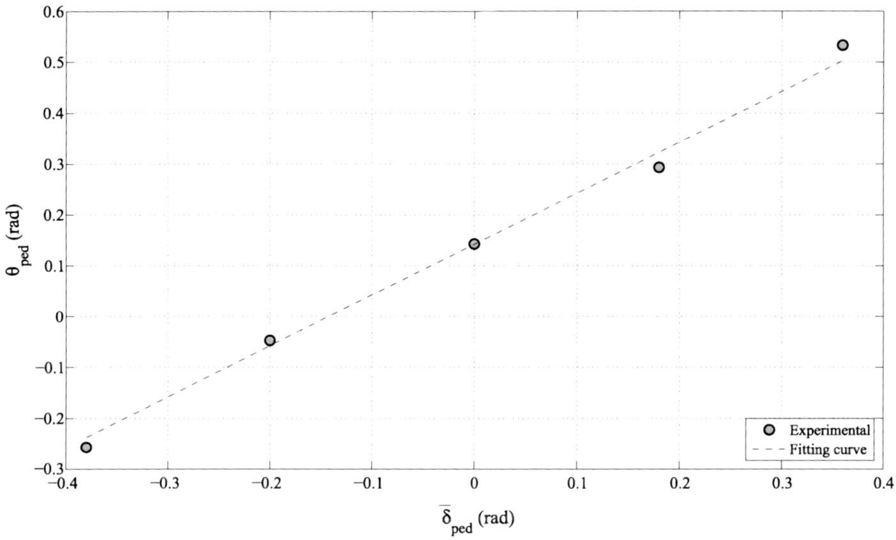

Fig. 6.8 Collective pitch curve of the tail rotor

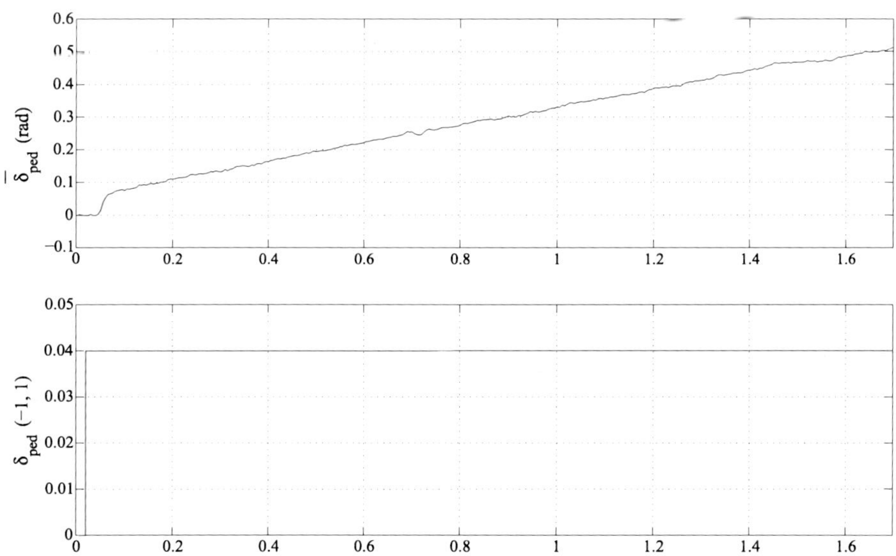

Fig. 6.9 Response of tail rotor servo actuator using step input signal

integration gains that are set in the feedback controller. A step input signal is injected in $\delta_{\rm ped}$ and the associated response of $\bar{\delta}_{\rm ped}$ is recorded and shown in Fig. 6.9. Using least square curve fitting, we can obtain the results for the lumped $K_{\rm P}K_{\rm a}$ and $K_{\rm I}K_{\rm a}$. Note that the results for the individual $K_{\rm P}$, $K_{\rm I}$, and $K_{\rm a}$ are to

Table 6.6 Parameters determined by collective pitch curve examination

Parameter	Physical meaning
$K_\mathrm{I} K_\mathrm{a} = 8.499$ rad	Lumped result of integral gain K_I and scaling factor K_a
$K_\mathrm{P} K_\mathrm{a} = 1.608$ rad	Lumped result of proportional gain K_P and scaling factor K_a
$K_\mathrm{col} = -0.165$ rad	Ratio of θ_col to δ_col
$K_\mathrm{ped} = 1$	Ratio of θ_ped to $\bar{\delta}_\mathrm{ped}$
$\theta_\mathrm{col,0} = 0.075$ rad	Offset of θ_col when δ_col is zero
$\theta_\mathrm{ped,0} = 0.143$ rad	Offset of θ_ped when $\bar{\delta}_\mathrm{ped}$ is zero

be determined after K_a is identified via another flight experiment reported in Sect. 6.3.4.

Table 6.6 lists six parameters obtained in the collective pitch curve examination.

6.3.3 Estimation Based on Wind-Tunnel Data

For our flight dynamics model, wind-tunnel data are essential to determine the lift curve slopes (and drag coefficient) for various airfoils, including (i) main rotor blade, (ii) tail rotor blade, (iii) stabilizer bar paddle, (iv) horizontal fin, and (v) vertical fin. Although it is impractical for us to conduct the wind-tunnel experiment, numerous wind-tunnel databases are available in the open literature and can be adopted as the baseline for the parameter estimation of our unmanned systems. Generally, we follow the following three important rules when selecting the wind-tunnel data:

1. Airfoil shape: Obviously, this is the most fundamental baseline for evaluating the data suitability.
2. Reynolds number (R_e): Miniature helicopters commonly work in a regime with a low Reynolds number.
3. Aspect ratio: With the same airfoil and Reynolds number, low aspect ratio can greatly reduce the lift coefficient.

Table 6.7 provides the three features for each of the involved airfoils. The Reynolds number computation can be found in many fluid dynamics texts. We need to highlight that (i) for the main rotor blade, tail rotor blade, and stabilizer bar, the computation points are located at the middle of these airfoils, with the chord lengths of 0.353 m, 0.064 m, and 0.272 m, and (ii) the horizontal and vertical fins are assumed to be isosceles triangles and the traveled length of the calculated Reynolds number is taken at half the height of the associated isosceles triangle. Based on this table, we choose the wind-tunnel data reported in [86, 141, 154] (as depicted in Figs. 6.10, 6.11, 6.12) for our estimation. The determined lift curve slopes and the drag coefficients are summarized in Table 6.8. We note that the lift curve slopes of the main rotor blade, tail rotor blade, and stabilizer bar can be further tuned in the last part of the proposed parameter determination procedure.

Table 6.7 Airfoil features of SheLion

Airfoil	Airfoil shape	Reynolds number	Aspect ratio
Main rotor blade	NACA 0012	3.07×10^5	11.37
Tail rotor blade	NACA 0012	1.21×10^6	4.41
Stabilizer bar	NACA 0012	2.25×10^5	1.37
Horizontal fin	Flat plate	1.02×10^5	3.11
Vertical fin	Flat plate	0.92×10^5	2.52

6.3.4 Flight Test

System identification methodology is integrated in this step. Through practical flight test data, we aim to determine more unknown parameters and validate some parameters obtained in the previous steps. We note that only the data collected in the hover and the near hover conditions are utilized because in flight conditions with moderate speed and aggressiveness, it is generally difficult for a human pilot to perturb a miniature rotorcraft consistently while maintaining the desired trimmed status. We have carried out the following flight experiments in this step.

6.3.4.1 Pirouette Flight

Pirouette motion requires the pilot to fly SheLion following a 10 m-radius circle, with the nose pointing to the circle center, whereas its yaw angular rate has to be kept to a constant. It follows from (6.57) that K_{a} is the ratio between the yaw angular rate and the injected input δ_{ped}. The recorded input and output responses for a successfully conducted flight experiment are shown in Fig. 6.13. Based on this experimental result, we can obtain K_{a} and further isolate K_{I} and K_{P}, which are given in Table 6.9.

6.3.4.2 Frequency Sweeping

Frequency-sweep technique with a typical input of chirp signals as shown in Fig. 6.14 is widely used in both full- and small-scale rotorcraft modeling (see, e.g., [30, 121, 177]). We note that for small-scale helicopters, this technique has the highest fidelity in modeling the linear angular-rate dynamics at hover condition. Inspired by this, we combine (6.8), (6.55), and (6.56) to form the following state-space dynamic structure:

$$\dot{\mathbf{x}} = \begin{bmatrix} 0 & 0 & 0 & L_{\mathrm{b_s}} \\ 0 & 0 & M_{\mathrm{a_s}} & 0 \\ 0 & -1 & -\frac{1}{\tau_{\mathrm{mr}}+\tau_{\mathrm{sb}}} & \frac{\tau_{\mathrm{mr}} A_{\mathrm{b_s}}}{\tau_{\mathrm{mr}}+\tau_{\mathrm{sb}}} \\ -1 & 0 & \frac{\tau_{\mathrm{mr}} B_{\mathrm{a_s}}}{\tau_{\mathrm{mr}}+\tau_{\mathrm{sb}}} & -\frac{1}{\tau_{\mathrm{mr}}+\tau_{\mathrm{sb}}} \end{bmatrix} \mathbf{x} + \begin{bmatrix} 0 & 0 \\ 0 & 0 \\ 0 & A_{\mathrm{lon,eff}} \\ B_{\mathrm{lat,eff}} & 0 \end{bmatrix} \mathbf{u} \qquad (6.61)$$

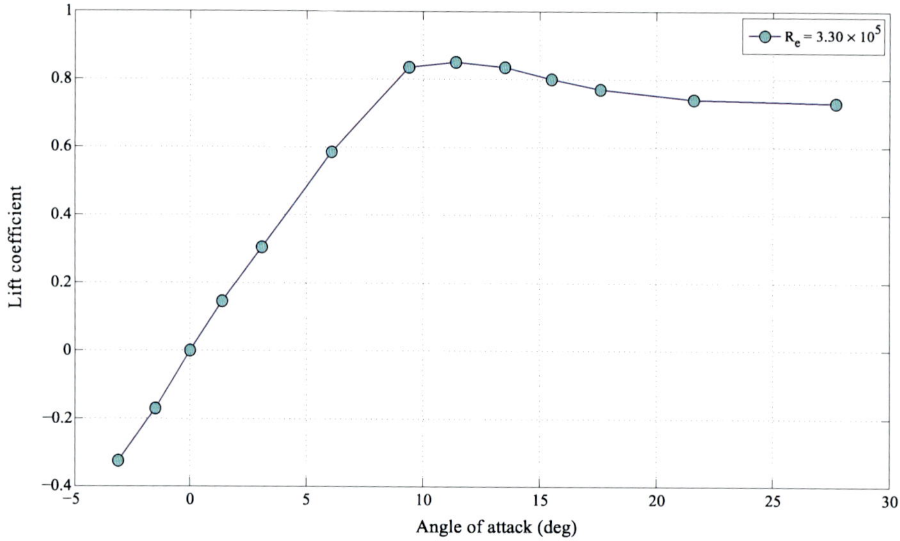

a. Wind-tunnel result for the lift curve slope of the main rotor blade (aspect ratio = 6)

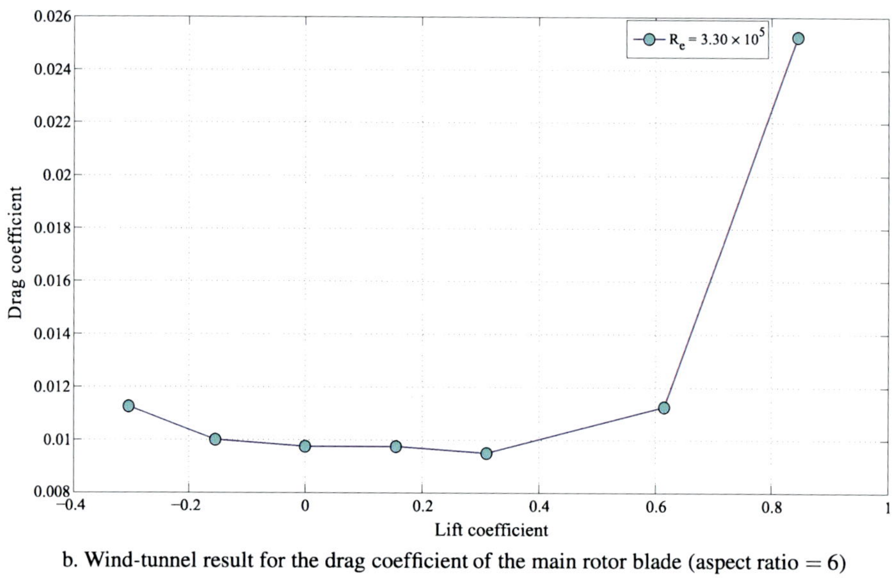

b. Wind-tunnel result for the drag coefficient of the main rotor blade (aspect ratio = 6)

Fig. 6.10 Wind-tunnel data related to the main rotor blade

and

$$\mathbf{y} = \begin{bmatrix} 1 & 0 & 0 & 0 \\ 0 & 1 & 0 & 0 \end{bmatrix} \mathbf{x}, \tag{6.62}$$

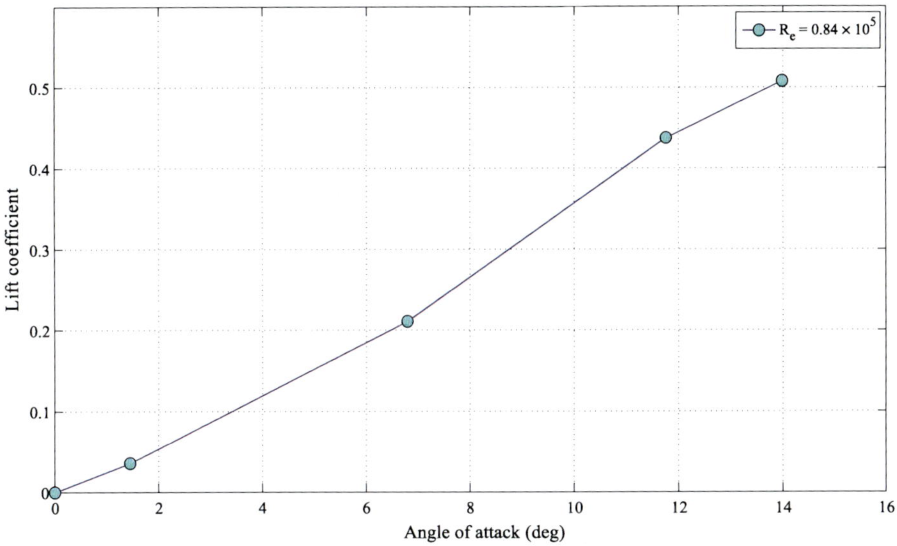

Fig. 6.11 Wind tunnel data related to the stabilizer bar and tail rotor blade (aspect ratio = 1)

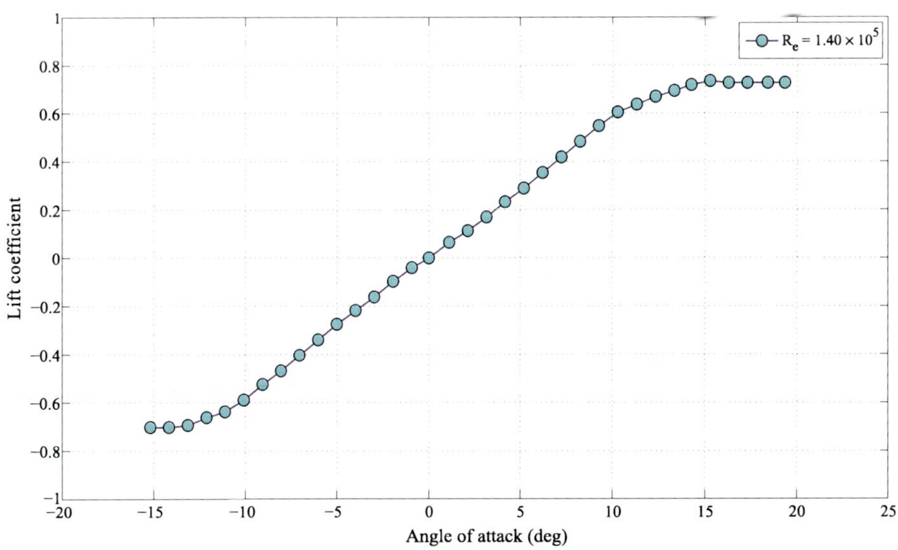

Fig. 6.12 Wind-tunnel data related to the flat-plate horizontal and vertical fins (aspect ratio = 2)

where

$$\mathbf{x} = \begin{pmatrix} p \\ q \\ a_s \\ b_s \end{pmatrix}, \qquad \mathbf{u} = \begin{pmatrix} \delta_{\text{lat}} \\ \delta_{\text{lon}} \end{pmatrix}, \qquad \mathbf{y} = \begin{pmatrix} p \\ q \end{pmatrix} \tag{6.63}$$

Table 6.8 Parameters identified by wind-tunnel data

Parameter	Physical meaning
$C_{\mathrm{D0}} = 0.01$	Drag coefficient of main rotor blade
$C_{\mathrm{l}\alpha,\mathrm{hf}} = 2.85\ \mathrm{rad}^{-1}$	Horizontal fin lift curve slope
$C_{\mathrm{l}\alpha,\mathrm{mr0}} = 5.73\ \mathrm{rad}^{-1}$	Initial setting of the main rotor blade lift curve slope
$C_{\mathrm{l}\alpha,\mathrm{sb0}} = 2.08\ \mathrm{rad}^{-1}$	Initial setting of the stabilizer bar lift curve slope
$C_{\mathrm{l}\alpha,\mathrm{tr0}} = 2.08\ \mathrm{rad}^{-1}$	Initial setting of the tail rotor blade lift curve slope
$C_{\mathrm{l}\alpha,\mathrm{vf}} = 2.85\ \mathrm{rad}^{-1}$	Vertical fin lift curve slope

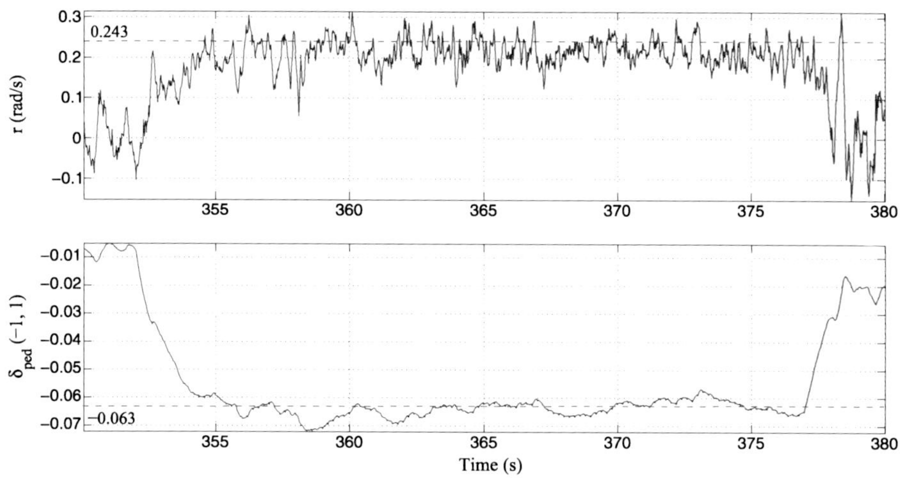

Fig. 6.13 Pirouette flight test for K_{a} determination

and the lumped derivatives $L_{\mathrm{b_s}}$, $M_{\mathrm{a_s}}$, $A_{\mathrm{lon,eff}}$, and $B_{\mathrm{lat,eff}}$ are respectively given by

$$L_{\mathrm{b_s}} = \frac{mgH_{\mathrm{mr}} + K_{\beta}}{J_{\mathrm{xx}}}, \qquad M_{\mathrm{a_s}} = \frac{mgH_{\mathrm{mr}} + K_{\beta}}{J_{\mathrm{yy}}} \tag{6.64}$$

and

$$A_{\mathrm{lon,eff}} = \frac{A_{\mathrm{lon}} + K_{\mathrm{sb}}C_{\mathrm{lon}}}{\tau_{\mathrm{mr}} + \tau_{\mathrm{sb}}}, \qquad B_{\mathrm{lat,eff}} = \frac{B_{\mathrm{lat}} + K_{\mathrm{sb}}D_{\mathrm{lat}}}{\tau_{\mathrm{mr}} + \tau_{\mathrm{sb}}}. \tag{6.65}$$

We note that the parameters A_{lon}, B_{lat}, C_{lon}, D_{lat}, K_{sb}, J_{xx}, and J_{yy} have already been determined earlier in Sect. 6.3.2. Among them, the first five are related to the mechanical design of the Bell-Hiller mixer, and the obtained parameters have sufficient reliability. For the last two parameters, we are to examine their validity using the system identification process below.

The identification process is assisted by an identification toolkit called CIFER, developed by the NASA Ames Research Center for military-based rotorcraft systems. It first converts the collected input-output data to frequency-domain responses. The parameters in the structured model of (6.61) are then identified by minimizing

Table 6.9 Parameters identified by pirouette experiment

Parameter	Physical meaning
$K_\mathrm{I} = 2.2076$	Integral gain of the embedded PI controller
$K_\mathrm{P} = 0.4177$	Proportional gain of the embedded PI controller
$K_\mathrm{a} = -3.85$ rad	Scaling factor of the amplifier circuit

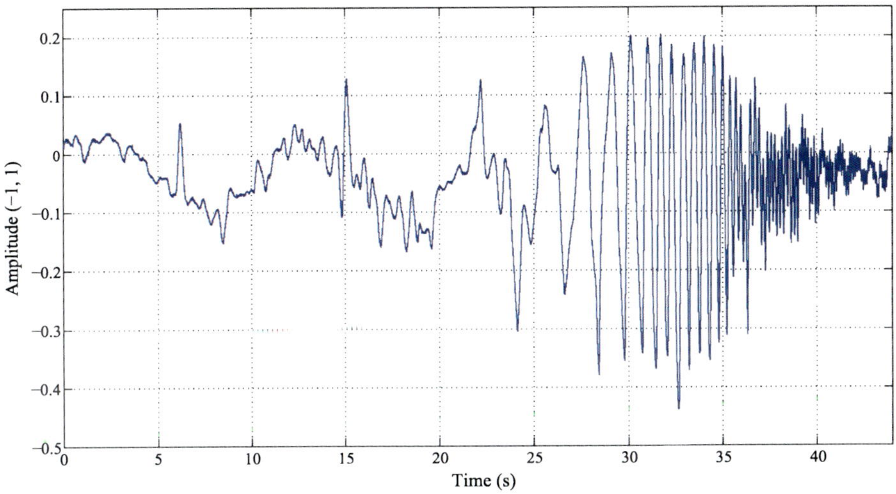

Fig. 6.14 Frequency-sweep input signal

the difference between the actual and simulation frequency responses, which are shown in Figs. 6.15, 6.16, 6.17, 6.18. We note that the coherence value of the frequency domain matching is an important index to indicate whether the system can be well characterized as a linear process in the interested frequency range [177]. Following the successful experience in [177], we take the threshold value as 0.6, for a very close matching can be achieved. The associated numerical values are listed in Table 6.10, together with two statistics provided by CIFER for evaluating the identification fidelity: (i) Cramer-Rao bound (%), which indicates the level of the parameter identifiability, and (ii) insensitivity (%), which indicates whether the parameter is important to the selected model structure [177]. According to [177], parameters identified are considered to be accurate and acceptable if their associated Cramer-Rao bound is less than 20% and insensitivity is less than 10%. It is clear from Table 6.10 that all the parameters identified using CIFER are highly accurate.

Based on the obtained intermediate results, we proceed to further perform the following:

1. Validation of J_xx and J_yy: From (6.64), we have

$$\frac{L_{\mathrm{b_s}}}{M_{\mathrm{a_s}}} = \frac{J_\mathrm{yy}}{J_\mathrm{xx}}. \tag{6.66}$$

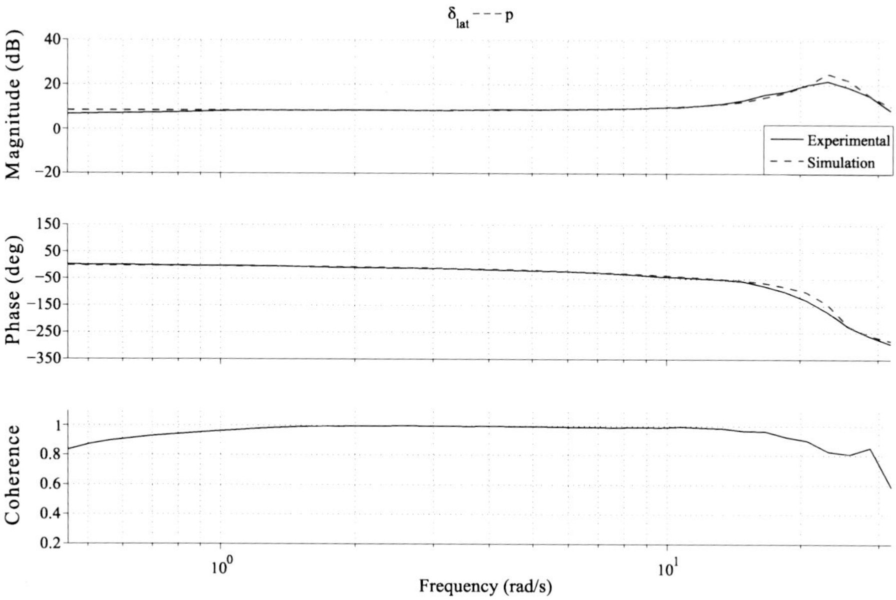

Fig. 6.15 Response comparison using frequency-sweep input (δ_{lat}—p)

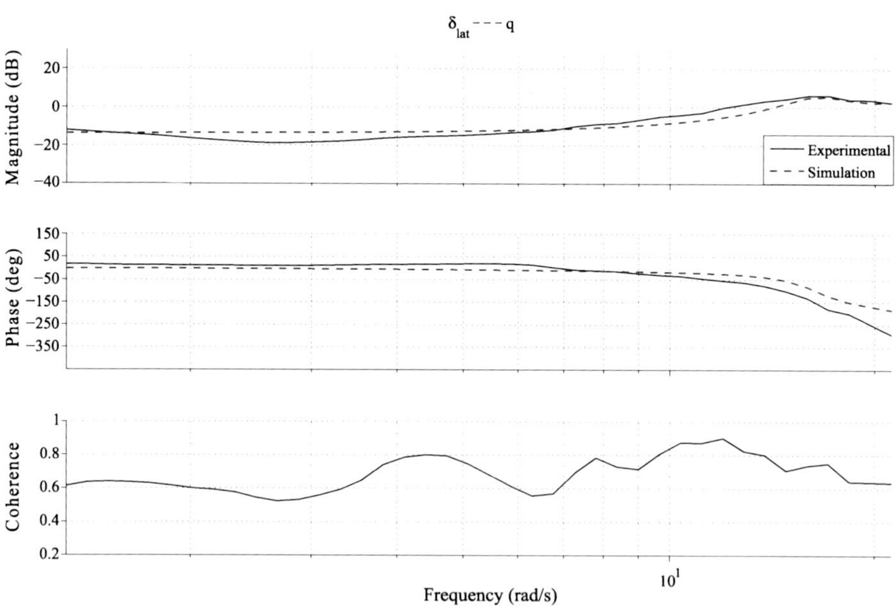

Fig. 6.16 Response comparison using frequency-sweep input (δ_{lat}—q)

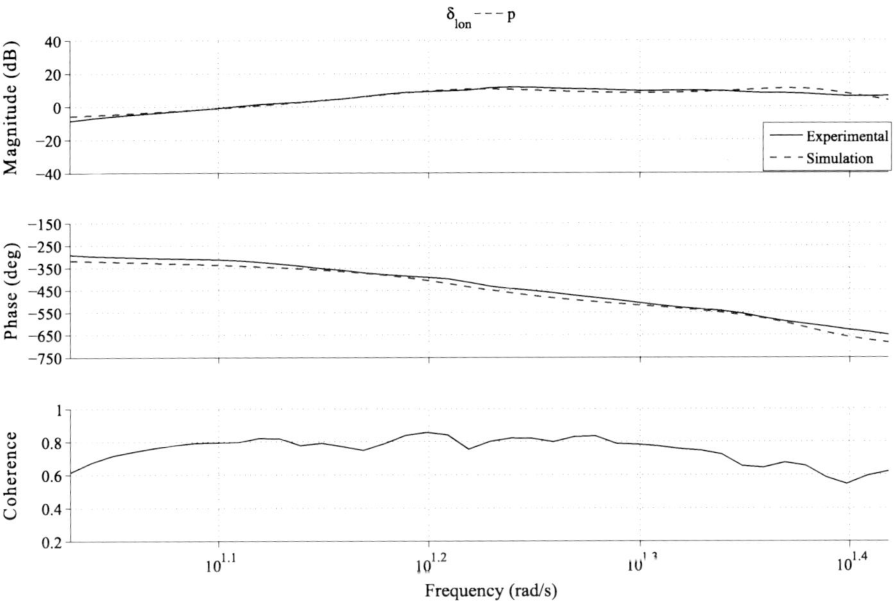

Fig. 6.17 Response comparison using frequency-sweep input (δ_{lon}—p)

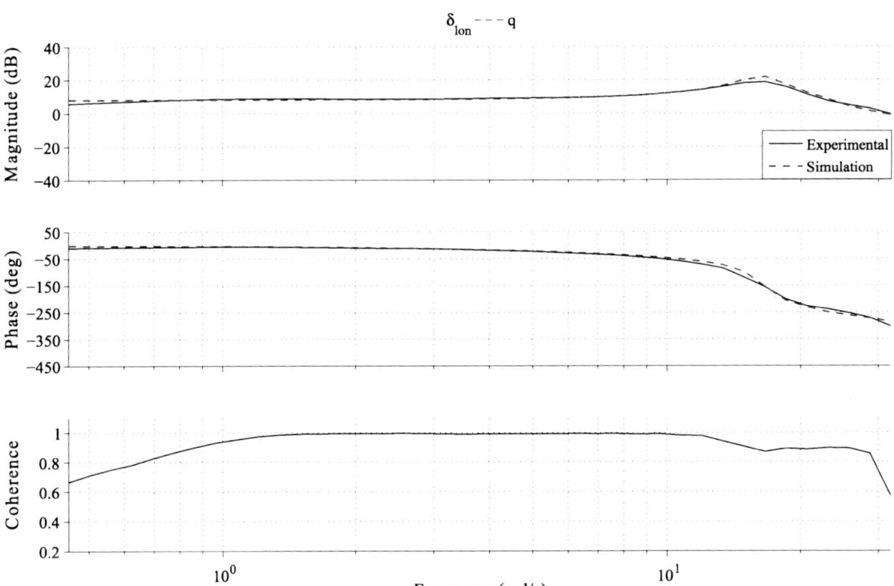

Fig. 6.18 Response comparison using frequency-sweep input (δ_{lon}—q)

Table 6.10 Parameters identified using CIFER system identification toolkit

Parameter	Cramer-Rao bound (%)	Insensitivity (%)	Physical meaning
$L_{b_s} = 583.50\ s^{-2}$	1.88%	0.69%	Lateral rotor spring derivative
$M_{a_s} = 265.30\ s^{-2}$	1.53%	0.64%	Longitudinal rotor spring derivative
$\tau_{mr} + \tau_{sb} = 0.299\ s$	2.68%	1.25%	Effective rotor time constant
$\frac{\tau_{mr} A_{b_s}}{\tau_{mr}+\tau_{sb}} = 2.223\ s^{-1}$	2.51%	5.26%	Coupling effect of rotor flapping dynamics
$\frac{\tau_{mr} B_{a_s}}{\tau_{mr}+\tau_{sb}} = 2.448\ s^{-1}$	5.00%	2.07%	Coupling effect of rotor flapping dynamics
$K_\beta = 114.05$ N·m	NA	NA	Main rotor spring constant

Based on the results obtained, the actual ratios of L_{b_s}/M_{a_s} and J_{yy}/J_{xx} are 2.199 and 2.185, respectively. Such a minor difference (about 0.6%) proves the high validity of the previously determined J_{xx} and J_{yy}.

2. K_β identification: Based on the results listed in Tables 6.4 and 6.10 and using (6.64), we can calculate K_β for the rolling and pitching responses, which yields two numerical values, i.e., 114.76 and 113.33, respectively. We set $K_\beta = 114.05$, the average of the two closely matched results.
3. Initial verification of $C_{l\alpha,mr}$ and $C_{l\alpha,sb}$: The two lift curve slopes, which are estimated previously by the wind-tunnel data, generate another estimation for $\tau_{mr} + \tau_{sb}$. Following from (6.46) and (6.50), we obtain an estimate of $\tau_{mr} + \tau_{sb}$ to be 0.275. The small deviation between the CIFER system identification result (see Table 6.10) and the wind-tunnel-based estimation indicates the good validity of $C_{l\alpha,mr0}$ and $C_{l\alpha,sb0}$. However, since the system identification method only gives us the total sum of τ_{mr} and τ_{sb}, the specific value for $C_{l\alpha,mr}$ and $C_{l\alpha,sb}$ cannot be determined at this stage.
4. Analysis of coupling effect in rotor flapping: The validity of the identified coupling derivatives in Table 6.10 is verified by the close frequency matching, low Cramer-Rao bound (%), and low insensitivity (%). It, however, cannot be predicted by the theoretical calculation resulting from (6.54). This deficiency partially results from the simplified model structure.

6.3.5 Fine Tuning

We aim in this last step of the parameter identification process to determine the three lift curve slopes $C_{l\alpha,mr}$, $C_{l\alpha,sb}$, and $C_{l\alpha,tr}$, as well as A_{b_s} and B_{a_s}. The general idea is to examine the balanced relationship in the ideal hovering condition. The procedure involves solving and/or determining

1. ten equations including the 6-DOF rigid-body dynamics in (6.7) and (6.8), the main-rotor thrust calculation based on (6.12) and (6.13), and the tail-rotor thrust calculation using (6.27) and (6.28);

Table 6.11 Numerical results for the ideal hovering condition

$\phi_{\rm trim}$	$\theta_{\rm trim}$	$a_{\rm s,trim}$	$b_{\rm s,trim}$	$T_{\rm mr,trim}$	$T_{\rm tr,trim}$	$v_{\rm i,mr,trim}$	$v_{\rm i,tr,trim}$	$C_{{\rm l}\alpha,{\rm mr}}$	$C_{{\rm l}\alpha,{\rm tr}}$
0.039	0.001	−0.001	0.005	96.766	4.188	4.90	5.62	5.52	2.82

Table 6.12 Parameters determined after fine tuning

Parameter	Physical meaning
$A_{b_{\rm s}} = 9.720\ {\rm s}^{-1}$	Coupling effect of the bare main rotor flapping
$B_{a_{\rm s}} = 10.704\ {\rm s}^{-1}$	Coupling effect of the bare main rotor flapping
$C_{{\rm l}\alpha,{\rm mr}} = 5.52\ {\rm rad}^{-1}$	Main rotor blade lift curve slope
$C_{{\rm l}\alpha,{\rm sb}} = 2.72\ {\rm rad}^{-1}$	Stabilizer bar lift curve slope
$C_{{\rm l}\alpha,{\rm tr}} = 2.82\ {\rm rad}^{-1}$	Tail rotor blade lift curve slope

2. the trimmed values for $\delta_{\rm col}$ and $\delta_{\rm ped}$ (-0.1746 and 0 for SheLion), which can be easily obtained through a manual experiment and which are necessary to form the balanced relationship; and
3. ten parameters, $\phi_{\rm trim}$, $\theta_{\rm trim}$, $a_{\rm s,trim}$, $b_{\rm s,trim}$, $T_{\rm mr,trim}$, $T_{\rm tr,trim}$, $v_{\rm i,mr,trim}$, $v_{\rm i,tr,trim}$, $C_{{\rm l}\alpha,{\rm mr}}$, and $C_{{\rm l}\alpha,{\rm tr}}$. Among these parameters, the first eight are associated with the trimmed values at the hover flight condition.

With a proper initialization (e.g., utilizing the approximation of $T_{\rm mr} = mg$ to generate the associated $v_{\rm i,mr}$, $T_{\rm tr}$, and $v_{\rm i,tr}$) and using an appropriate numerical searching algorithm (in our case, we adopt the trust-region dogleg method integrated in MATLAB®), the nonlinear equations can quickly converge to an expected result, which is shown in Table 6.11. With the value of $C_{{\rm l}\alpha,{\rm mr}}$, we can calculate the two rotor time constants $\tau_{\rm mr}$ and $\tau_{\rm sb}$ using (6.50). Using the identification result given in Table 6.10, we can completely determine $C_{{\rm l}\alpha,{\rm sb}}$, $A_{b_{\rm s}}$, and $B_{a_{\rm s}}$. The results obtained are listed in Table 6.12.

6.4 Model Validation

In this section, we carry out a comprehensive evaluation on the fidelity of the obtained flight dynamics model. Three manual flight experiments have been conducted for such a purpose, which include the following:

1. A fast forward flight for which SheLion is commanded to start with stable hovering and then gradually accelerate to a pre-set fast forward speed of 14 m/s.
2. A sideslip flight for which SheLion starts with a stable hover. The pilot then accelerates the chopper to a moderate lateral speed of about 7 m/s and finally decelerates it to the near hovering condition.
3. A heave flight test under which SheLion starts again from stable hovering before ascending until its heave speed reaches 2 m/s. It is then commanded to decelerate to hovering at a higher point.

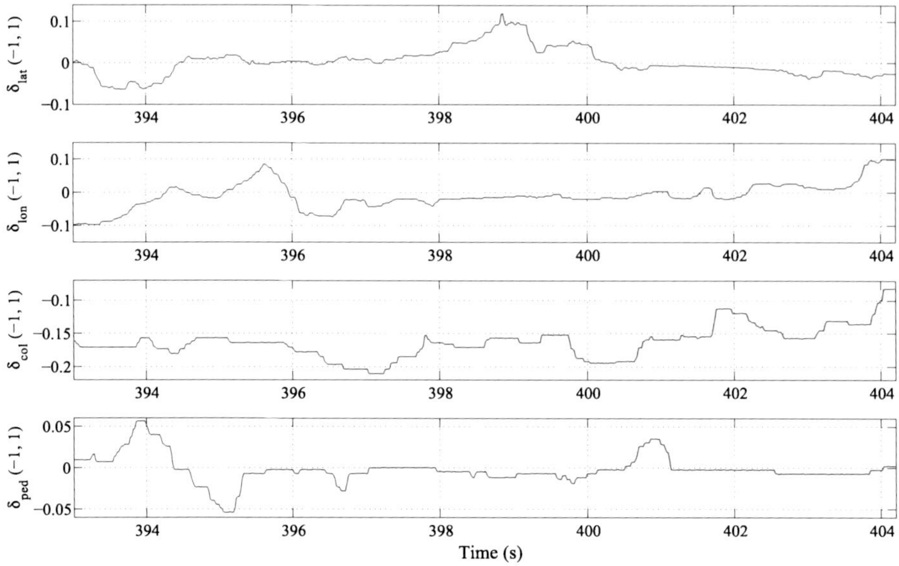

Fig. 6.19 Forward flight test—inputs

The validation procedure is conducted in the time domain. Based on the same inputs, we compare the experimental and simulation responses in terms of the body-frame velocities, Euler angles, and angular velocities. From the obtained results shown in Figs. 6.19 to 6.30, it is clear that the matching between the actual and simulation responses is very close, which persuasively proves that the flight dynamics model is able to capture the flight dynamics of SheLion in a fairly wide envelope. We also note that some small deviations can be observed in velocities, which are mainly caused by wind gusts, unmodeled high frequency dynamics, and inaccuracy of measurement. These problems can be overcome by a properly designed automatic flight control system, which is to be addressed in the coming chapters.

6.5 Flight Envelope Determination

In this section, we proceed to determine the flight envelope of the obtained flight dynamics model for actual flight experiments. Flight envelope determination is essential before proceeding to conduct flight control law design, and flight experiment or trajectory design. We are to determine the flight envelope of SheLion based on its body-frame velocities and yaw rate. More specifically, we consider the following:

1. X-axis velocity (u): The longitudinal velocity envelope is the most important element in determining the flight envelope as the majority of normal helicopter

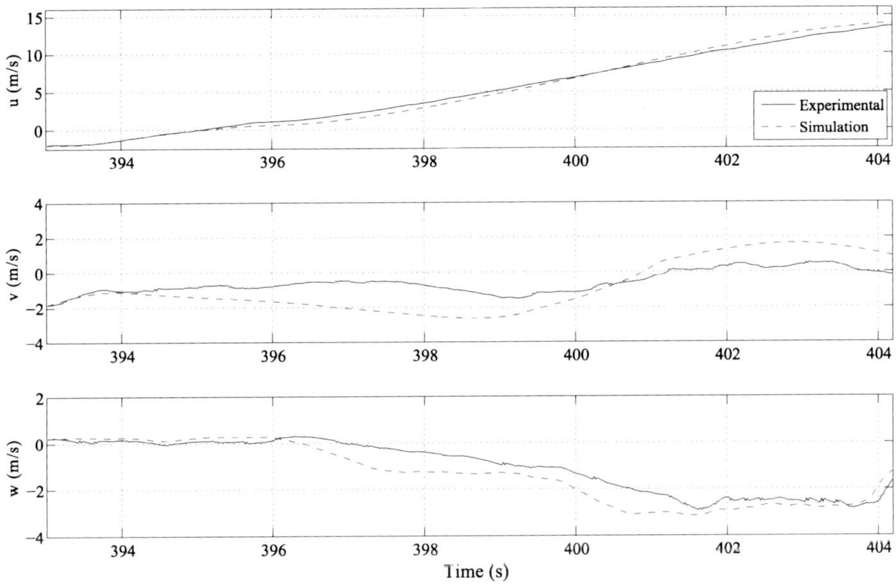

Fig. 6.20 Forward flight test—velocities

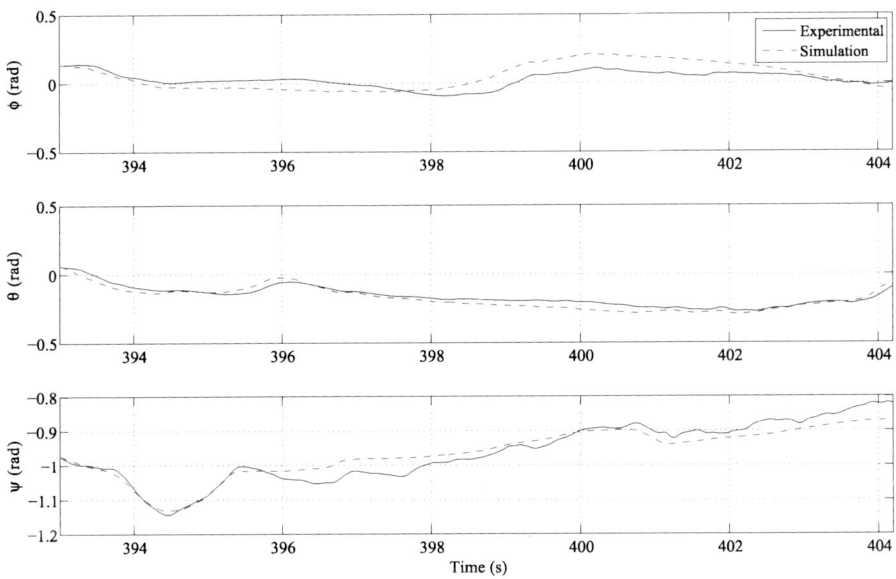

Fig. 6.21 Forward flight test—Euler angles

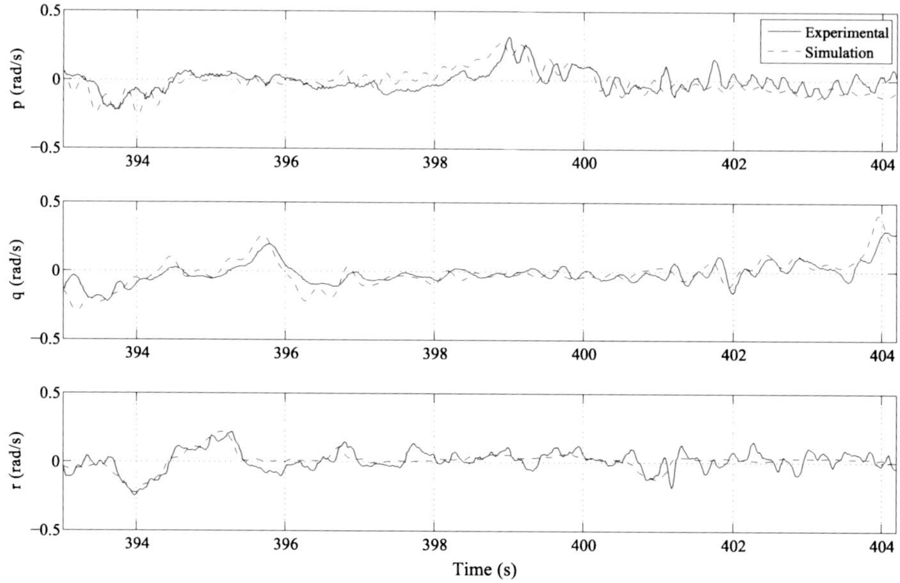

Fig. 6.22 Forward flight test—angular rates

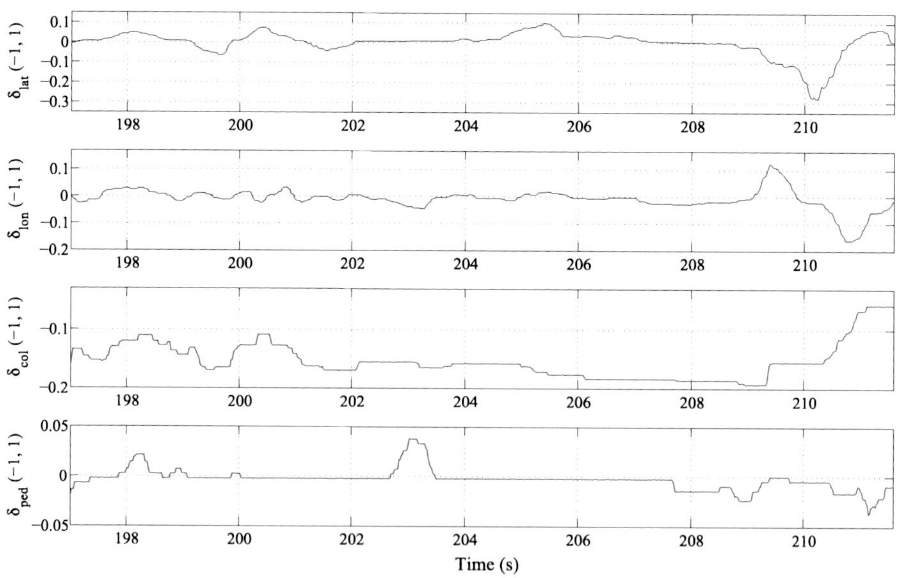

Fig. 6.23 Sideslip flight test—inputs

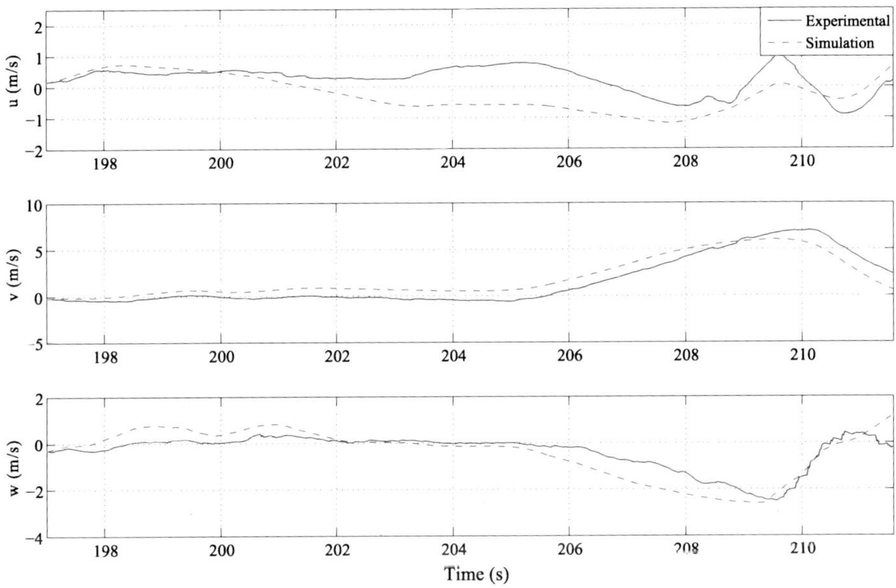

Fig. 6.24 Sideslip flight test—velocities

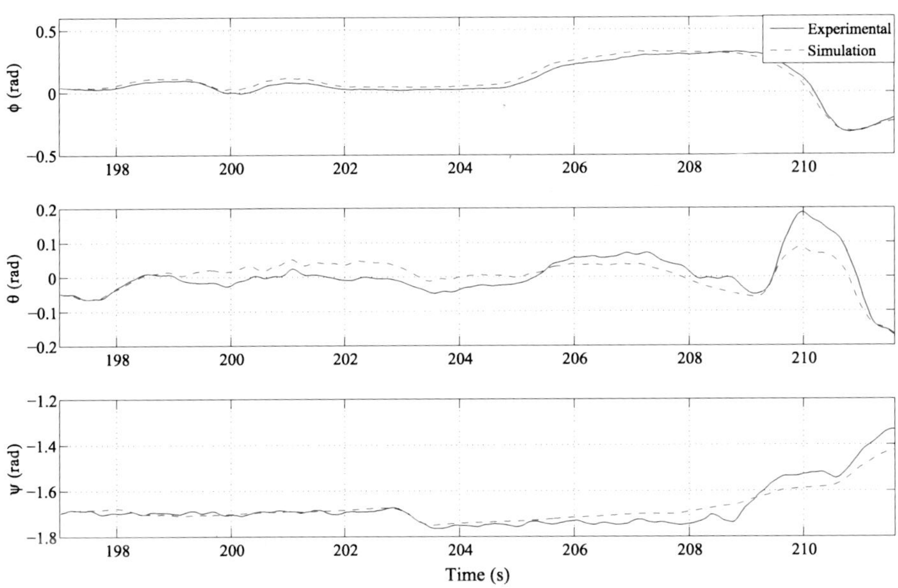

Fig. 6.25 Sideslip flight test—Euler angles

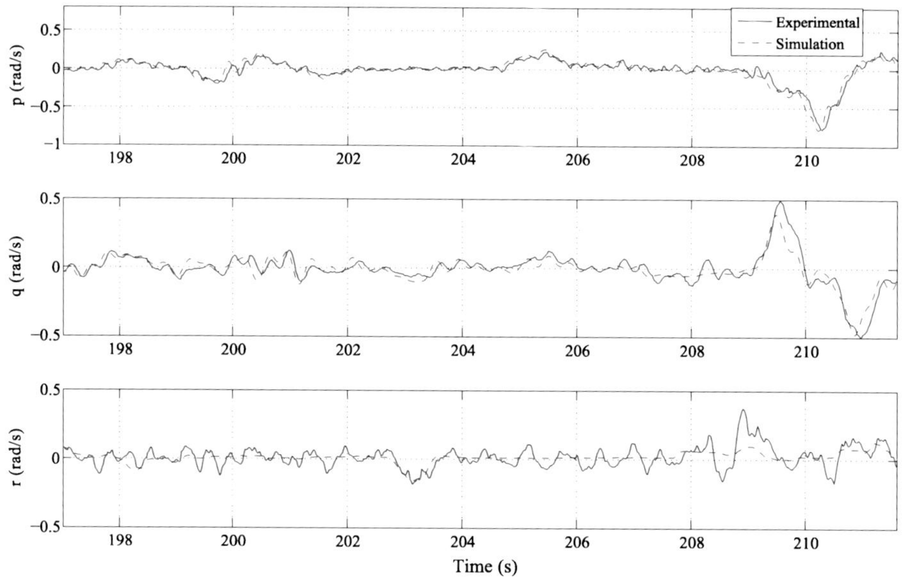

Fig. 6.26 Sideslip flight test—angular rates

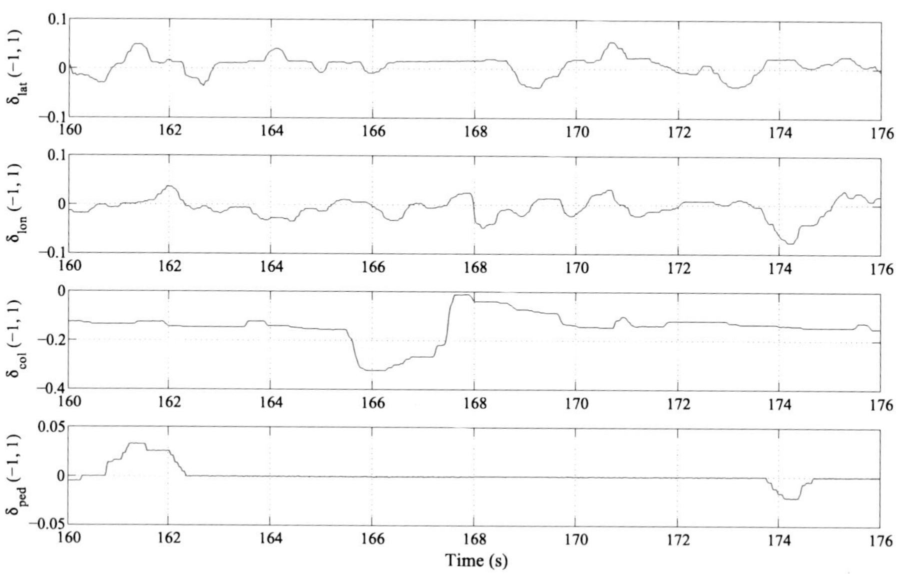

Fig. 6.27 Heave flight test—inputs

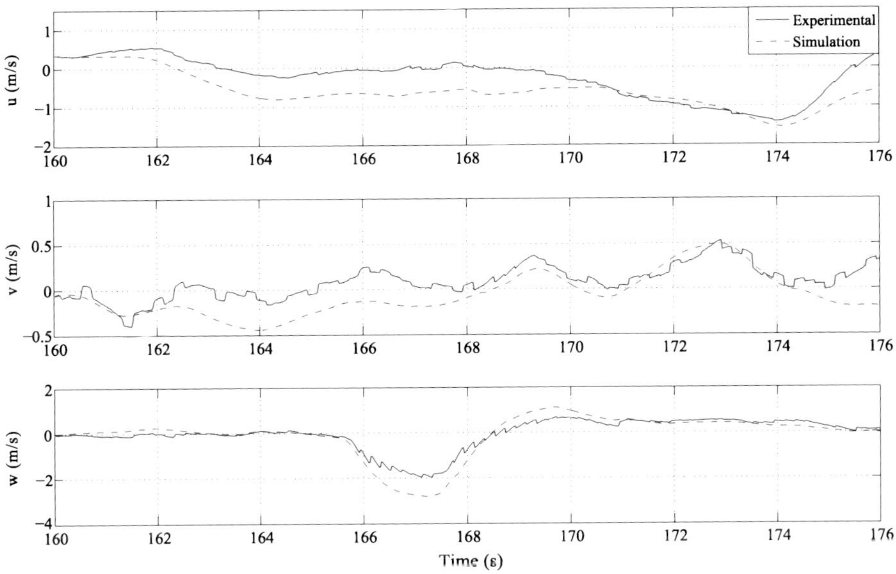

Fig. 6.28 Heave flight test—velocities

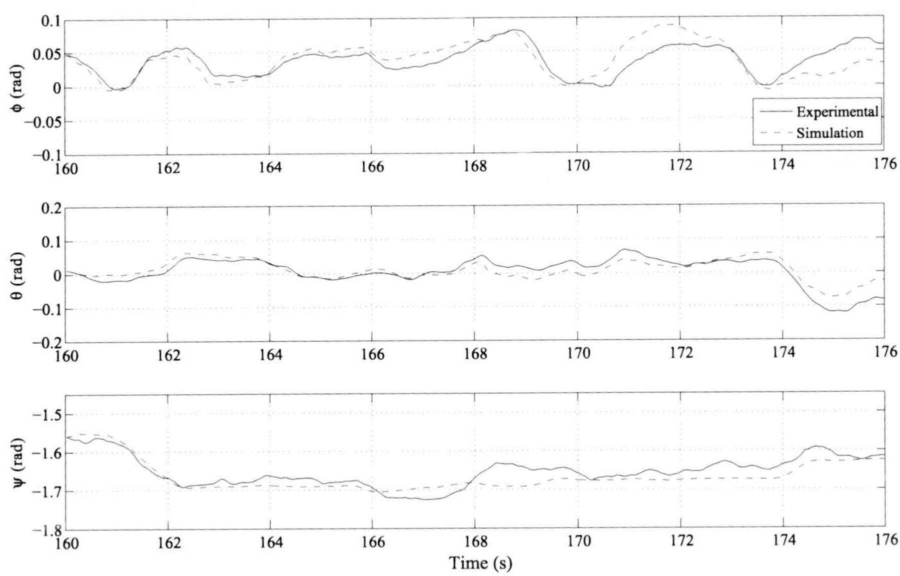

Fig. 6.29 Heave flight test—Euler angles

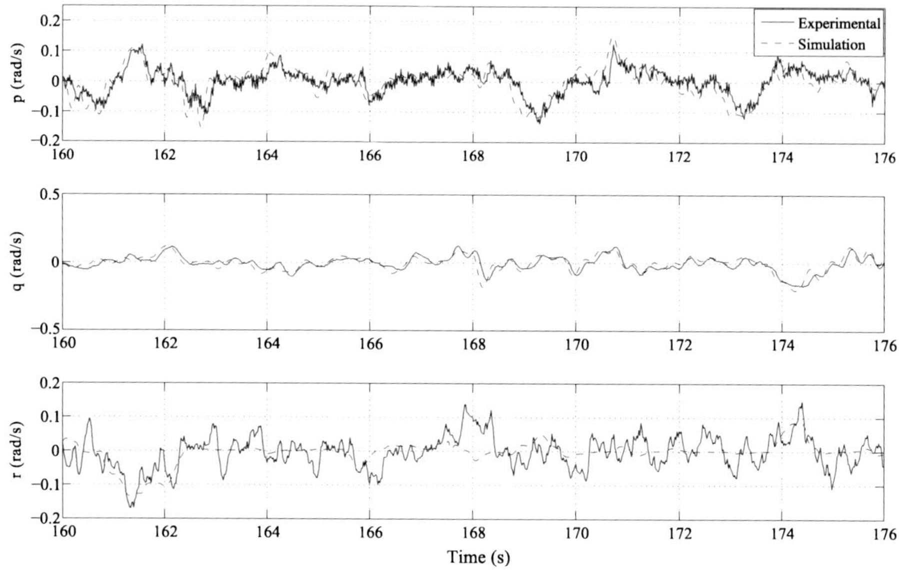

Fig. 6.30 Heave flight test—angular rates

maneuvers exist in hover and forward regimes. We adopt a simple method proposed in [91] to estimate the maximum forward speed, which is given by

$$u_{\max} = \sqrt{\frac{mg}{2\rho\pi R_{\mathrm{mr}}^2}} \left(\frac{4\pi R_{\mathrm{mr}}^2}{S_{\mathrm{fx}}} \right)^{1/3}. \tag{6.67}$$

Based on the parameters identified earlier (see Table 6.2), it can be shown for SheLion, theoretically, $u_{\max} = 19.1$ m/s. It should be noted that the theoretical result neglects the variation of the induced and profile power change with respect to speed. Furthermore, it assumes that the power requirement in the wide envelope is about the same as that at hovering condition [91]. As such, for practical implementation with all of the complexity involved, we need to avoid pushing ourselves to this theoretical limit. In fact, we have conducted a series of manual flight tests and found that the practical speed limit for SheLion is about 15 to 17 m/s. For safety, we set the maximum forward speed of SheLion to 12 m/s for fully autonomous flight. For backward flight tests, we set the maximum speed to −4 m/s as generally there is no point to practically flying at a fast backward speed.

Another key issue associated with the longitudinal direction is to determine respectively the speed ranges for the near hover, low speed, and high speed flight conditions as design specifications and considerations for these regimes (to be discussed in detail later in Chap. 7) are generally different. In ADS-33D-PRF [1], the following boundaries are adopted for military full-scale choppers, such as Bell UH-1H helicopters:

$$\left.\begin{array}{ll}\text{Near hover:} & 0 \sim 7.72 \text{ m/s},\\ \text{Low speed:} & 7.72 \sim 23.15 \text{ m/s},\\ \text{High speed:} & > 23.15 \text{ m/s}.\end{array}\right\} \tag{6.68}$$

We follow the scaling method reported in [121] to scale down the ranges in (6.68) for our miniature UAV, SheLion. The key idea is to maintain the same advance ratio, based on the Froude-scaling hypothesis. Bell UH-1H and Raptor 90 SE helicopters share similar mechanical structures in the rotor head and stabilizer bar, and the ratio of the main rotor diameters of the Bell UH-1H and Raptor 90 SE (SheLion), N_{scale}, is about 10.54. As a result, the speed envelope in (6.68) is to be scaled down by a factor of $\sqrt{N_{\text{scale}}}$ for SheLion, which yields the following:

$$\left.\begin{array}{ll}\text{Near hover:} & 0 \sim 2.38 \text{ m/s},\\ \text{Low speed:} & 2.38 \sim 7.13 \text{ m/s},\\ \text{High speed:} & > 7.13 \text{ m/s}.\end{array}\right\} \tag{6.69}$$

2. Y-axis velocity (v): For the lateral velocity, the theoretical limitation for SheLion is $v_{\max} = 9.27$ m/s. Manual flight tests show that the safe top lateral speed is about 6 m/s.
3. Z-axis velocity (w): In normal helicopter maneuvers, heave motion with high speed rarely happens. As such, we set the speed envelope as $w_{\max} = 3$ m/s for both the up and down motions.
4. Yaw rate (r): The yaw rate limitation is determined based on the suggestion given in ADS-33D-PRF [1]. We follow that in [1] and set the rapid yaw rate change (for turn-to-target flight operations) for SheLion to be 36 deg/s.

The practical flight envelope of SheLion can thus be summarized as the following:

$$\left.\begin{array}{ll}u: & -4 \sim 12 \text{ m/s},\\ v: & -6 \sim 6 \text{ m/s},\\ w: & -3 \sim 3 \text{ m/s},\\ r: & -36 \sim 36 \text{ deg/s}.\end{array}\right\} \tag{6.70}$$

Chapter 7
Inner-Loop Flight Control

7.1 Introduction

As mentioned earlier in the introductory chapter, the automatic flight control system is essential for a UAV to carry out flight missions with minimal or even without interference from human pilots. The classical feedback control method (i.e., PD or PID control) is one of the most common choices because of its simplicity in structure with less requirement on the accuracy of the dynamical model of the UAV. However, there is no guarantee that the simple controllers such as PD or PID can realize the full potential of the unmanned system. As such, there are many attempts that have been reported in the literature to implement flight control systems using various advanced control techniques, such as the neural network approach [50], the differential geometry method [85], the robust and H_∞ control approach [64, 66, 198], the composite nonlinear feedback control with decoupling approach [142], and the model predictive approach [163], to name a few. However, we note that many of the works reported focus merely on the basic autonomy. More specifically, they generally lack evaluation using professional design specifications such as aircraft handling qualities.

We propose a three-layer automatic flight control system (see Fig. 1.9) for our unmanned vehicles based on the time scales of the state variables of the helicopter. The detailed structure of the inner-loop layer and the outer-loop layer of our automatic flight control system is depicted in Fig. 7.1, in which

1. the inner loop stabilizes the dynamics of the helicopter associated with its Euler angles ϕ, θ, and ψ, angular velocities p, q, and r, TPP flapping angles of the main rotor a_s and b_s, and the intermediate state of the built-in yaw rate feedback controller $\delta_{ped,int}$; and
2. the outer loop controls the local-NED-based positions x_n, y_n, and z_n, and their respective velocities u_n, v_n, and w_n. Generally, the dynamics associated with the outer-loop layer are much slower compared to those in the inner loop.

Such a structure for the automatic flight control system is proven to be very effective for miniature unmanned rotorcraft systems. The results of the actual flight tests of

G. Cai et al., *Unmanned Rotorcraft Systems*, Advances in Industrial Control,
DOI 10.1007/978-0-85729-635-1_7,

our HeLion and SheLion given in Chaps. 9 and 10 show that our designs are highly efficient and very successful.

In this chapter, we present the design of the inner-loop control law using the H_∞ control technique. Because they are light in weight and small in size, the small-scale UAV helicopters, such as HeLion and SheLion, possess more agility and maneuverability but are more vulnerable to environmental disturbances such as wind gusts. As such, the H_∞ control method, a technique developed to attenuate external disturbances while maintaining the closed-loop stability, is a natural choice for the inner control loop to realize both internal stabilization and disturbance rejection. More specifically, we focus on issues related to (i) problem formulation, (ii) design specification selection, (iii) control law design, and (iv) performance evaluation. Particularly, the design specifications for military rotorcraft defined by US army aviation are adopted throughout our design process to guarantee a top level performance.

7.2 H_∞ Control Technique

The ultimate goal of a control system designer is to build a system that works in a real environment. Since the real environment may change and the operating conditions may vary from time to time, the control system must be able to withstand these variations. Even if the environment does not change, other factors of life such as model uncertainties and noises need to be taken into consideration. Any mathematical representation of a system often involves simplifying assumptions. Nonlinearities either are unknown, and hence unmodeled, or are modeled and later ignored in order to simplify analysis. High-frequency dynamics are often ignored at the design stage as well. As a consequence, control systems designed based on simplified models may not work on real plants in real environments. The particular property that a control system must possess for it to operate properly in realistic situations is commonly called *robustness*. Mathematically, this means that the controller must perform satisfactorily not just for one plant but for a family of plants. If a controller can be designed such that the whole system to be controlled remains stable when its parameters vary within certain expected limits, the system is said to possess robust stability. In addition, if it can satisfy performance specifications such as steady state tracking, disturbance rejection, and speed of response requirements, it is said to possess robust performance. The problem of designing controllers that satisfy both robust stability and performance requirements is called robust control. H_∞ control theory is one of the cornerstones of modern control theory and was developed in an attempt to solve such a problem. Many robust control problems (such as the robust stability problem of unstructurally perturbed systems, the mixed-sensitivity problem, robust stabilization with additive and multiplicative perturbations, to name a few) can be cast into a general H_∞ control problem (see, e.g., [27]).

Since the original formulation of the H_∞ control problem by Zames [222], a great deal of work has been done on finding the solution to this problem (see,

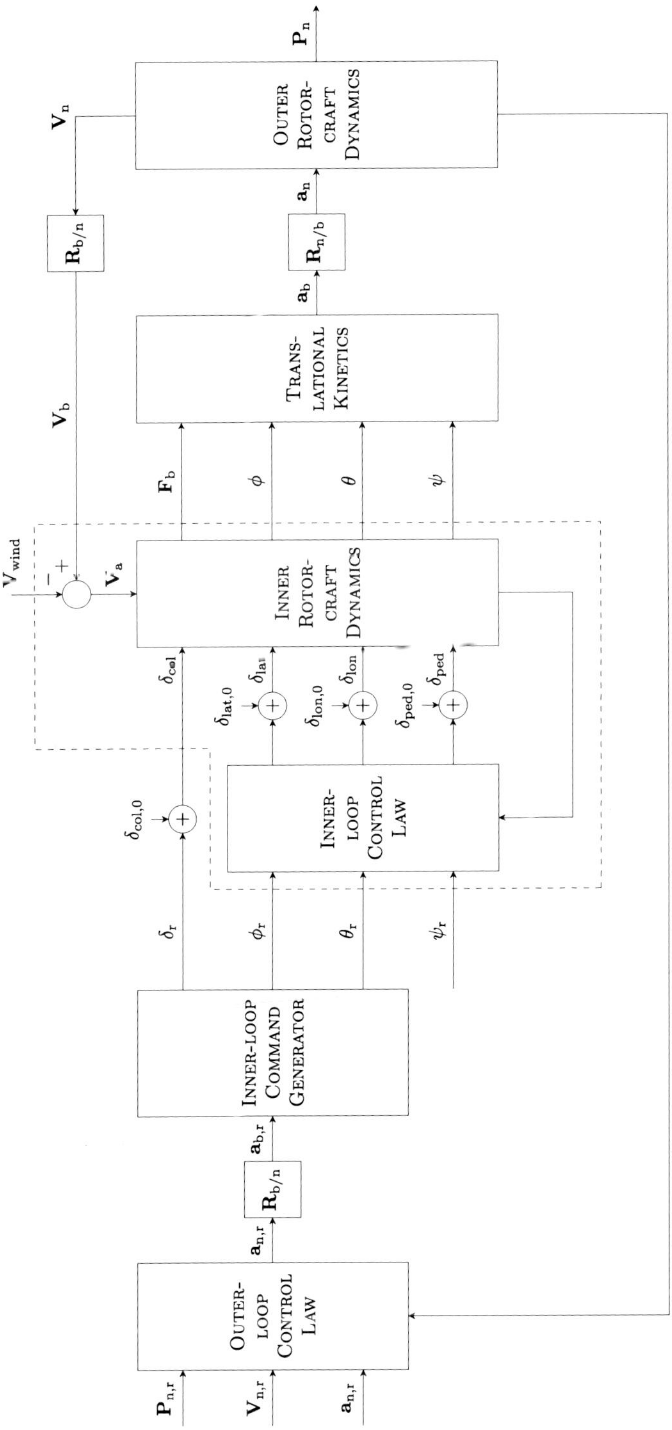

Fig. 7.1 Detailed structure of the inner-loop and outer-loop layers of the flight control system

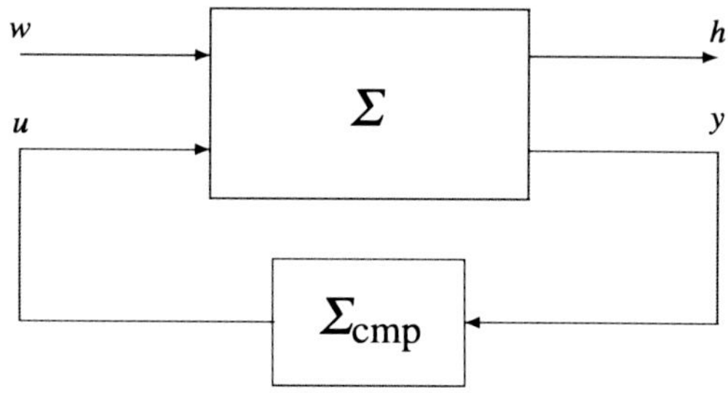

Fig. 7.2 The typical control configuration in state-space setting

e.g., [6, 27, 46, 47, 61, 69, 99, 104, 111, 223] and references cited therein). To be more specific, we consider a generalized system Σ with a state-space description,

$$\Sigma: \quad \begin{cases} \dot{\boldsymbol{x}} = A\ \boldsymbol{x} + B\ \boldsymbol{u} + E\ \boldsymbol{w}, \\ \boldsymbol{y} = C_1\ \boldsymbol{x} + D_1\ \boldsymbol{w}, \\ \boldsymbol{h} = C_2\ \boldsymbol{x} + D_2\ \boldsymbol{u}, \end{cases} \tag{7.1}$$

where $\boldsymbol{x} \in \mathbb{R}^n$ is the state, $\boldsymbol{u} \in \mathbb{R}^m$ is the control input, $\boldsymbol{w} \in \mathbb{R}^{\mathrm{q}}$ is the external disturbance input, $\boldsymbol{y} \in \mathbb{R}^{\mathrm{p}}$ is the measurement output, and $\boldsymbol{h} \in \mathbb{R}^{\ell}$ is the controlled output of Σ. We assume that (A, B) is stabilizable and (A, C_1) is detectable.

The H_∞ control problem is to find an internally stabilizing proper measurement feedback control law,

$$\Sigma_{\mathrm{cmp}}: \quad \begin{cases} \dot{\boldsymbol{v}} = A_{\mathrm{cmp}}\ \boldsymbol{v} + B_{\mathrm{cmp}}\ \boldsymbol{y}, \\ \boldsymbol{u} = C_{\mathrm{cmp}}\ \boldsymbol{v} + D_{\mathrm{cmp}}\ \boldsymbol{y} \end{cases} \tag{7.2}$$

such that the H_∞-norm of the overall closed-loop transfer matrix function from $\boldsymbol{w}$ to $\boldsymbol{h}$ is minimized (see also Fig. 7.2). To be more specific, we will say that the control law Σ_{cmp} of (7.2) is internally stabilizing when applied to the system Σ of (7.1), if the following matrix is asymptotically stable:

$$A_{\mathrm{cl}} := \begin{bmatrix} A + BD_{\mathrm{cmp}}C_1 & BC_{\mathrm{cmp}} \\ B_{\mathrm{cmp}}C_1 & A_{\mathrm{cmp}} \end{bmatrix}, \tag{7.3}$$

i.e., all of its eigenvalues lie in the open left-half complex plane. It is straightforward to verify that the closed-loop transfer matrix from the disturbance $\boldsymbol{w}$ to the controlled output $\boldsymbol{h}$ is given by

$$T_{\boldsymbol{hw}}(s) = C_{\mathrm{e}}(sI - A_{\mathrm{e}})^{-1}B_{\mathrm{e}} + D_{\mathrm{e}}, \tag{7.4}$$

where

$$\left.\begin{aligned} A_{\mathrm{e}} &:= \begin{bmatrix} A + BD_{\mathrm{cmp}}C_1 & BC_{\mathrm{cmp}} \\ B_{\mathrm{cmp}}C_1 & A_{\mathrm{cmp}} \end{bmatrix}, \\ B_{\mathrm{e}} &:= \begin{bmatrix} E + BD_{\mathrm{cmp}}D_1 \\ B_{\mathrm{cmp}}D_1 \end{bmatrix}, \\ C_{\mathrm{e}} &:= [\, C_2 + D_2D_{\mathrm{cmp}}C_1 \quad D_2C_{\mathrm{cmp}} \,], \\ D_{\mathrm{e}} &:= D_2D_{\mathrm{cmp}}D_1. \end{aligned}\right\} \tag{7.5}$$

It is simple to note that if $\Sigma_{\rm cmp}$ is a static state feedback law, i.e., $\boldsymbol{u} = F\boldsymbol{x}$, then the closed-loop transfer matrix from $\boldsymbol{w}$ to $\boldsymbol{h}$ is given by

$$T_{\boldsymbol{hw}}(s) = (C_2 + D_2F)(sI - A - BF)^{-1}E. \tag{7.6}$$

The H_∞-norm of a stable continuous-time transfer matrix, e.g., $T_{\boldsymbol{hw}}(s)$, is defined as follows:

$$\|T_{\boldsymbol{hw}}\|_\infty := \sup_{\omega\in[0,\infty)} \sigma_{\max}[T_{\boldsymbol{hw}}(j\omega)] = \sup_{\|\boldsymbol{w}\|_2=1} \frac{\|\boldsymbol{h}\|_2}{\|\boldsymbol{w}\|_2}, \tag{7.7}$$

where $\boldsymbol{w}$ and $\boldsymbol{h}$ are, respectively, the input and output of $T_{\boldsymbol{hw}}(s)$, $\sigma_{\max}[\cdot]$ denotes the maximal singular value of the matrix, and $\|\cdot\|_2$ is the L_2-norm of the corresponding signal. It is clear that the H_∞-norm of $T_{\boldsymbol{hw}}(s)$ corresponds to the worst case gain from its input to its output. For future use, we define

$$\gamma^* := \inf\big\{\|T_{\boldsymbol{hw}}(\Sigma \times \Sigma_{\rm cmp})\|_\infty \mid \Sigma_{\rm cmp} \text{ internally stabilizes } \Sigma\big\}. \tag{7.8}$$

We note that the determination of this γ^* is rather tedious. For a fairly large class of systems, γ^* can be exactly computed using some numerically stable algorithms. In general, an iterative scheme is required to determine γ^*. We refer interested readers to the work of Chen [27] for a detailed treatment of this particular issue. For simplicity, we assume throughout this section that γ^* has been determined and hence it is known.

It transpires that, for H_∞ control, it is almost impossible to find a control law with a finite gain to achieve the optimal performance, i.e., γ^*. As such, we focus on designing H_∞ suboptimal controllers instead. To be more specific, given a scalar $\gamma > \gamma^*$, we focus on finding a control law that yields $\|T_{\boldsymbol{hw}}\|_\infty < \gamma$, where $T_{\boldsymbol{hw}}(s)$ is the corresponding closed-loop transfer matrix. Hereafter, we call a control law that possesses such a property an H_∞ γ-suboptimal controller.

Next, we proceed to construct a solution to the regular problem, i.e., for the problem with the given system satisfying the following conditions:

1. (A, B, C_2, D_2) has no invariant zeros on the imaginary axis and D_2 is of maximal column rank.
2. (A, E, C_1, D_1) has no invariant zeros on the imaginary axis and D_1 is of maximal row rank.

Given a scalar $\gamma > \gamma^*$, we solve for positive semi-definite stabilizing solutions $P \geq 0$ and $Q \geq 0$, respectively, to the following Riccati equations:

$$\begin{aligned} &A^{\rm T}P + PA + C_2^{\rm T}C_2 + \gamma^{-2}PEE^{\rm T}P \\ &\quad - (PB + C_2^{\rm T}D_2)(D_2^{\rm T}D_2)^{-1}(B^{\rm T}P + D_2^{\rm T}C_2) = 0 \end{aligned} \tag{7.9}$$

and

$$\begin{aligned} &AQ + QA^{\rm T} + EE^{\rm T} + \gamma^{-2}QC_2^{\rm T}C_2Q \\ &\quad - (QC_1^{\rm T} + ED_1^{\rm T})(D_1D_1^{\rm T})^{-1}(C_1Q + D_1E^{\rm T}) = 0. \end{aligned} \tag{7.10}$$

The H_∞ γ-suboptimal control law is given by (see also [47]),

$$\Sigma_{\rm cmp}: \quad \begin{cases} \dot{\boldsymbol{v}} = A_{\rm cmp}\, \boldsymbol{v} + B_{\rm cmp}\, \boldsymbol{y}, \\ \boldsymbol{u} = C_{\rm cmp}\, \boldsymbol{v} \end{cases} \tag{7.11}$$

where

$$\begin{aligned} A_{\rm cmp} &= A + \gamma^{-2} E E^{\rm T} P + B F \\ &\qquad + \left(I - \gamma^{-2} Q P\right)^{-1} K \left(C_1 + \gamma^{-2} D_1 E^{\rm T} P\right), \end{aligned} \tag{7.12}$$

$$B_{\rm cmp} = -\left(I - \gamma^{-2} Q P\right)^{-1} K, \tag{7.13}$$

$$C_{\rm cmp} = F, \tag{7.14}$$

and where

$$F = -(D_2^{\rm T} D_2)^{-1}(D_2^{\rm T} C_2 + B^{\rm T} P), \qquad K = -(Q C_1^{\rm T} + E D_1^{\rm T})(D_1 D_1^{\rm T})^{-1}. \tag{7.15}$$

Note that, for the state feedback case, the H_∞ γ-suboptimal control law is given by $\boldsymbol{u} = F\boldsymbol{x}$ with F being given as in (7.15).

For the singular case, i.e., the given problem does not satisfy the conditions for the regular problem, the following perturbation method can be utilized. For $\gamma > \gamma^*$ and a positive scalar $\varepsilon > 0$, define $\tilde{E}$, $\tilde{D}_1$, $\tilde{C}_2$, and $\tilde{D}_2$ as

$$\tilde{E} := [\, E \quad \varepsilon I \quad 0 \,], \qquad \tilde{D}_1 := [\, D_1 \quad 0 \quad \varepsilon I \,] \tag{7.16}$$

and

$$\tilde{C}_2 := \begin{bmatrix} C_2 \\ \varepsilon I \\ 0 \end{bmatrix}, \qquad \tilde{D}_2 := \begin{bmatrix} D_2 \\ 0 \\ \varepsilon I \end{bmatrix} \tag{7.17}$$

and solve the following Riccati equations:

$$\begin{aligned} & A^{\rm T} \tilde{P} + \tilde{P} A + \tilde{C}_2^{\rm T} \tilde{C}_2 + \gamma^{-2} \tilde{P} \tilde{E} \tilde{E}^{\rm T} \tilde{P} \\ & \quad - (\tilde{P} B + \tilde{C}_2^{\rm T} \tilde{D}_2)(\tilde{D}_2^{\rm T} \tilde{D}_2)^{-1} (B^{\rm T} \tilde{P} + \tilde{D}_2^{\rm T} \tilde{C}_2) = 0 \end{aligned} \tag{7.18}$$

and

$$\begin{aligned} & A \tilde{Q} + \tilde{Q} A^{\rm T} + \tilde{E} \tilde{E}^{\rm T} + \gamma^{-2} \tilde{Q} \tilde{C}_2^{\rm T} \tilde{C}_2 \tilde{Q} \\ & \quad - (\tilde{Q} C_1^{\rm T} + \tilde{E} \tilde{D}_1^{\rm T})(\tilde{D}_1 \tilde{D}_1^{\rm T})^{-1} (C_1 \tilde{Q} + \tilde{D}_1 \tilde{E}^{\rm T}) = 0 \end{aligned} \tag{7.19}$$

for $\tilde{P} > 0$ and $\tilde{Q} > 0$. Then, it can be shown that there exists an $\varepsilon^* > 0$ such that for all $\varepsilon \in (0, \varepsilon^*]$, the following control law is an H_∞ γ-suboptimal for the given system:

$$\tilde{\Sigma}_{\rm cmp}: \quad \begin{cases} \dot{\boldsymbol{v}} = \tilde{A}_{\rm cmp}\, \boldsymbol{v} + \tilde{B}_{\rm cmp}\, \boldsymbol{y}, \\ \boldsymbol{u} = \tilde{C}_{\rm cmp}\, \boldsymbol{v} \end{cases} \tag{7.20}$$

where

$$\tilde{A}_{\text{cmp}} = A + \gamma^{-2} E E^{\mathrm{T}} \tilde{P} + B\tilde{F} + \left(I - \gamma^{-2}\tilde{Q}\tilde{P}\right)^{-1} \tilde{K}\left(C_1 + \gamma^{-2} D_1 E^{\mathrm{T}} \tilde{P}\right), \tag{7.21}$$

$$\tilde{B}_{\text{cmp}} = -\left(I - \gamma^{-2}\tilde{Q}\tilde{P}\right)^{-1} \tilde{K}, \tag{7.22}$$

$$\tilde{C}_{\text{cmp}} = \tilde{F}, \tag{7.23}$$

and where

$$\tilde{F} = -(\tilde{D}_2^{\mathrm{T}}\tilde{D}_2)^{-1}(\tilde{D}_2^{\mathrm{T}}\tilde{C}_2 + B^{\mathrm{T}}\tilde{P}), \tag{7.24}$$

$$\tilde{K} = -(\tilde{Q}C_1^{\mathrm{T}} + \tilde{E}\tilde{D}_1^{\mathrm{T}})(\tilde{D}_1\tilde{D}_1^{\mathrm{T}})^{-1}. \tag{7.25}$$

Note that for the state feedback case, the H_∞ γ-suboptimal control law is given by $\boldsymbol{u} = \tilde{F}\boldsymbol{x}$ with $\tilde{F}$ being given as in (7.24).

Alternatively, the singular H_∞ control problem can also be solved in a more systematic approach given in [27]. It can also be shown (see, e.g., [27]) that for the case when the subsystem (A, E, C_1, D_1) is left invertible and of minimum phase, the optimal achievable H_∞ control performance, i.e., γ^*, under the state feedback and the measurement feedback are identical. Furthermore, in such a situation, we can follow the procedure given below to obtain a reduced-order measurement feedback control law that can recover the performance of the state feedback law.

First, without loss of generality and for simplicity of presentation, we assume that the matrices C_1 and D_1 are already in the form

$$C_1 = \begin{bmatrix} 0 & C_{1,02} \\ I_k & 0 \end{bmatrix} \quad \text{and} \quad D_1 = \begin{bmatrix} D_{1,0} \\ 0 \end{bmatrix}, \tag{7.26}$$

where $k = \ell - \operatorname{rank}(D_1)$ and $D_{1,0}$ is of full rank. Then, the given system in (7.1) can be written as

$$\begin{cases} \begin{pmatrix} \dot{\boldsymbol{x}}_1 \\ \dot{\boldsymbol{x}}_2 \end{pmatrix} = \begin{bmatrix} A_{11} & A_{12} \\ A_{21} & A_{22} \end{bmatrix} \begin{pmatrix} \boldsymbol{x}_1 \\ \boldsymbol{x}_2 \end{pmatrix} + \begin{bmatrix} B_1 \\ B_2 \end{bmatrix} \boldsymbol{u} + \begin{bmatrix} E_1 \\ E_2 \end{bmatrix} \boldsymbol{w}, \\ \begin{pmatrix} \boldsymbol{y}_0 \\ \boldsymbol{y}_1 \end{pmatrix} = \begin{bmatrix} 0 & C_{1,02} \\ I_k & 0 \end{bmatrix} \begin{pmatrix} \boldsymbol{x}_1 \\ \boldsymbol{x}_2 \end{pmatrix} + \begin{bmatrix} D_{1,0} \\ 0 \end{bmatrix} \boldsymbol{w}, \\ \boldsymbol{h} = [\, C_{2,1} \quad C_{2,2} \,] \begin{pmatrix} \boldsymbol{x}_1 \\ \boldsymbol{x}_2 \end{pmatrix} + D_2 \boldsymbol{u}, \end{cases} \tag{7.27}$$

where the original state $\boldsymbol{x}$ is partitioned into two parts, $\boldsymbol{x}_1$ and $\boldsymbol{x}_2$; and $\boldsymbol{y}$ is partitioned into $\boldsymbol{y}_0$ and $\boldsymbol{y}_1$ with $\boldsymbol{y}_1 \equiv \boldsymbol{x}_1$. Thus, one needs to estimate only the state $\boldsymbol{x}_2$ in the reduced-order controller design. Next, define an auxiliary subsystem Σ_{QR} characterized by a matrix quadruple $(A_{\text{R}}, E_{\text{R}}, C_{\text{R}}, D_{\text{R}})$, where

$$(A_{\text{R}}, E_{\text{R}}, C_{\text{R}}, D_{\text{R}}) = \left(A_{22}, E_2, \begin{bmatrix} C_{1,02} \\ A_{12} \end{bmatrix}, \begin{bmatrix} D_{1,0} \\ E_1 \end{bmatrix} \right). \tag{7.28}$$

The following is a step-by-step algorithm that constructs the reduced-order output feedback controller for the general H_∞ optimization.

STEP 1 (Construction of the gain matrix F) Define an auxiliary system

$$\begin{cases} \dot{\boldsymbol{x}} = A\ \boldsymbol{x} + B\ \boldsymbol{u} + E\ \boldsymbol{w}, \\ \boldsymbol{y} = \quad \boldsymbol{x}, \\ \boldsymbol{h} = C_2\, \boldsymbol{x} + D_2\, \boldsymbol{u}. \end{cases} \tag{7.29}$$

Given a $\gamma > \gamma^*$, compute its corresponding H_∞ γ-suboptimal state feedback gain matrix F as given either in (7.15) if it is a regular problem or in (7.24) otherwise.

STEP 2 (Construction of the gain matrix K) Define another auxiliary system

$$\begin{cases} \dot{\boldsymbol{x}} = A_{\mathrm{R}}^{\mathrm{T}}\, \boldsymbol{x} + C_{\mathrm{R}}^{\mathrm{T}}\, \boldsymbol{u} + C_{2,2}^{\mathrm{T}}\, \boldsymbol{w}, \\ \boldsymbol{y} = \quad \boldsymbol{x}, \\ \boldsymbol{h} = E_{\mathrm{R}}^{\mathrm{T}}\, \boldsymbol{x} + D_{\mathrm{R}}^{\mathrm{T}}\, \boldsymbol{u}. \end{cases} \tag{7.30}$$

Given a sufficient small $\gamma > 0$, compute its corresponding H_∞ γ-suboptimal state feedback gain matrix F_{R} as given either in (7.15) or in (7.24). We let $K_{\mathrm{R}} = F_{\mathrm{R}}^{\mathrm{T}}$.

STEP 3 (Construction of the reduced-order controller Σ_{RC}) Let us partition F and K_{R} as

$$F = [\,F_1 \quad F_2\,], \qquad K_{\mathrm{R}} = [\,K_{\mathrm{R}0} \quad K_{\mathrm{R}1}\,] \tag{7.31}$$

in conformity with the partitions of $\boldsymbol{x} = \binom{\boldsymbol{x}_1}{\boldsymbol{x}_2}$ and $\boldsymbol{y} = \binom{\boldsymbol{y}_0}{\boldsymbol{y}_1}$ of (7.27), respectively. Then define

$$G_{\mathrm{R}} = [\,-K_{\mathrm{R}0}, \quad A_{21} + K_{\mathrm{R}1}A_{11} - (A_{\mathrm{R}} + K_{\mathrm{R}}C_{\mathrm{R}})K_{\mathrm{R}1}\,].$$

Finally, the reduced-order output feedback controller is given by

$$\Sigma_{\mathrm{RC}}: \quad \begin{cases} \dot{\boldsymbol{v}} = A_{\mathrm{cmp}}\, \boldsymbol{v} + B_{\mathrm{cmp}}\, \boldsymbol{y}, \\ \boldsymbol{u} = C_{\mathrm{cmp}}\, \boldsymbol{v} + D_{\mathrm{cmp}}\, \boldsymbol{y}, \end{cases} \tag{7.32}$$

where

$$\left.\begin{aligned} A_{\mathrm{cmp}} &= A_{\mathrm{R}} + B_2F_2 + K_{\mathrm{R}}C_{\mathrm{R}} + K_{\mathrm{R}1}B_1F_2, \\ B_{\mathrm{cmp}} &= G_{\mathrm{R}} + (B_2 + K_{\mathrm{R}1}B_1)[\,0, \quad F_1 - F_2K_{\mathrm{R}1}\,], \\ C_{\mathrm{cmp}} &= F_2, \\ D_{\mathrm{cmp}} &= [\,0, \quad F_1 - F_2K_{\mathrm{R}1}\,]. \end{aligned}\right\} \tag{7.33}$$

This concludes the procedure for constructing the H_∞ γ-suboptimal reduced-order output feedback controller for the overall given system.

7.3 Inner-Loop Control System Design

In this section, the above H_∞ control technique is employed to design an inner-loop control law for HeLion and SheLion. More specifically, in our design formulation of the inner-loop controller, we formulate the wind gusts as an external disturbance input to inner-loop dynamics. The H_∞ optimization technique is thus utilized to attenuate such a disturbance and minimize the effects of wind gusts. Our design procedure consists of six steps, i.e., (i) dynamics model linearization, (ii) problem formulation, (iii) specification selection, (iv) H_∞ state feedback control law design, (v) reduced-order observer design, and (vi) performance evaluation.

We note that our approach is rather different from those under the H_∞ control framework reported in the literature. For example, in Weilenmann et al. [198], the H_∞ control technique was employed in two decoupled subsystems (one for the translational, pitch and roll motions, and the other for the heave and yaw motions) without actual experimental tests. Gadewadikar et al. [66] formulated the inner-loop control of an unmanned helicopter as a static measurement output feedback H_∞ control problem. Likewise, the design is presented without experimental verification on the actual platform. In Fujiwara et al. [64], an H_∞ automatic path tracking control law for a small-scale UAV helicopter was proposed and realized. It was, however, utilized to control the horizontal velocity.

7.3.1 Model Linearization

In order to utilize the H_∞ control technique to design an efficient automatic flight control system, we need to first linearize the nonlinear dynamics model at the operating conditions of interest. We have the following linearized model for the inner loop,

$$\dot{\mathbf{x}} = A\mathbf{x} + B\mathbf{u} + B_{\text{col}}\left(\delta_{\text{col}} - \delta_{\text{col},0}\right), \tag{7.34}$$

where $\mathbf{x} = \mathbf{x}_{\text{act}} - \mathbf{x}_{\text{trim}}$ is the difference between the actual state variables and their trimmed values, and similarly, $\mathbf{u} = \mathbf{u}_{\text{act}} - \mathbf{u}_{\text{trim}}$ is the difference between the actual input variables and their trimmed values, $\delta_{\text{col},0}$ is the trim value of δ_{col}, and where $\mathbf{x}_{\text{act}}$ and $\mathbf{u}_{\text{act}}$ are respectively given as

$$\mathbf{x}_{\text{act}} = [\,\phi \quad \theta \quad p \quad q \quad a_{\text{s}} \quad b_{\text{s}} \quad r \quad \delta_{\text{ped,int}} \quad \psi\,]^{\text{T}} \tag{7.35}$$

and

$$\mathbf{u}_{\text{act}} = [\,\delta_{\text{lat}} \quad \delta_{\text{lon}} \quad \delta_{\text{ped}}\,]^{\text{T}}. \tag{7.36}$$

In the proposed three-layer flight control structure, it turns out that matrices A and B do not change much with respect to the flight velocities of interest. The nominal values of A and B are respectively taken as

$$A = \begin{bmatrix} 0 & 0 & 1 & 0 & 0 & 0 & 0.0009 & 0 & 0 \\ 0 & 0 & 0 & 0.9992 & 0 & 0 & -0.0389 & 0 & 0 \\ 0 & 0 & -0.0302 & -0.0056 & -0.0003 & 585.1165 & 11.4448 & -59.529 & 0 \\ 0 & 0 & 0 & -0.0707 & 267.7499 & -0.0003 & 0 & 0 & 0 \\ 0 & 0 & 0 & -1.0000 & -3.3607 & 2.2223 & 0 & 0 & 0 \\ 0 & 0 & -1 & 0 & 2.4483 & -3.3607 & 0 & 0 & 0 \\ 0 & 0 & 0.0579 & 0.0108 & 0.0049 & 0.0037 & -21.9557 & 114.2 & 0 \\ 0 & 0 & 0 & 0 & 0 & 0 & -1 & 0 & 0 \\ 0 & 0 & 0 & 0.0389 & 0 & 0 & 0.9992 & 0 & 0 \end{bmatrix}$$

Table 7.1 Open-loop modes and their physical interpretations

Eigenvalues	Physical interpretations
$-1.6590 \pm j23.9114$	Short-period rolling angular dynamics
$-1.7462 \pm j16.4222$	Short-period pitching angular dynamics
-8.4617	Short-period yawing angular dynamics
-13.5059	Short-period yawing angular dynamics
0, 0, 0	Euler kinematics

and

$$B = \begin{bmatrix} 0 & 0 & 0 \\ 0 & 0 & 0 \\ 0 & 0 & 43.3635 \\ 0 & 0 & 0 \\ 0.2026 & 2.5878 & 0 \\ 2.5878 & -0.0663 & 0 \\ 0 & 0 & -83.1883 \\ 0 & 0 & -3.8500 \\ 0 & 0 & 0 \end{bmatrix}.$$

For completeness, we list in Table 7.1 the open-loop modes of the corresponding system, i.e., the eigenvalues of A and their associated physical interpretations. Obviously, a control law is required to stabilize the inner-loop dynamics and improve its overall performance.

Finally, the trim values of the state and input variables for hovering are respectively listed in Table 7.2. We note that for other flight conditions, the trim values are slightly different for some state and input variables.

7.3.2 Problem Formulation

Based on the above linearized model, we formulate the inner-loop controller design into the framework of an H_∞ control problem. Since the wind gust $\boldsymbol{V}_{\text{wind}}$, which is defined in (6.5), affects the aerodynamic force generation and further the angular velocity dynamics, it is considered as the disturbance input of the inner-loop dynamics. We note that in our design, the control input δ_{col} is not utilized to control the

Table 7.2 Trim values of the state and input variables for hovering

ϕ	θ	p	q	a_s	b_s	r	$\delta_{\text{ped,int}}$	ψ	δ_{col}	δ_{lat}	δ_{lon}	δ_{ped}
0.0389	0.0009	0	0	−0.0009	0.0049	0	0	$0^\dagger$	−0.1746	0.0072	−0.0054	0

† ψ can be arbitrarily selected within $(-\pi, \pi]$

inner-loop dynamics. As such, the linearized model of (7.34) can then be modified as

$$\dot{\mathbf{x}} = A\,\mathbf{x} + B\,\mathbf{u} + E\,\mathbf{w}, \tag{7.37}$$

where $\mathbf{x}$, $\mathbf{u}$, A, and B are as defined in the previous section, and

$$E = \begin{bmatrix} 0 & 0 & 0 \\ 0 & 0 & 0 \\ -0.0001 & 0.1756 & -0.0395 \\ 0.0000 & 0.0003 & 0.0338 \\ 0 & 0 & 0 \\ 0 & 0 & 0 \\ -0.0002 & -0.3396 & 0.6424 \\ 0 & 0 & 0 \\ 0 & 0 & 0 \end{bmatrix},$$

which is obtained by linearizing the whole flight dynamics of the unmanned system with an injection of V_{wind} as a disturbance input to its respective channels. It turns out that the wind gust disturbance only directly affects the part associated with p, q, and r in the inner loop.

The measurement output is given by

$$\mathbf{y} = [\,\phi \quad \theta \quad p \quad q \quad r \quad \psi\,]^{\mathrm{T}} - \mathbf{y}_{\text{trim}} := C_1\mathbf{x}, \tag{7.38}$$

where $\mathbf{y}_{\text{trim}}$ is the trim value of the corresponding measurable state variables, and C_1 can be defined in an obvious fashion. As mentioned earlier, the primary task of the inner-loop control system is to internally stabilize the attitude dynamics and at the same time to yield a good attitude response. Thus, the primary output to be controlled is selected as

$$\mathbf{h}_{\text{out}} := [\,\phi \quad \theta \quad \psi\,]^{\mathrm{T}} - \mathbf{h}_{\text{out,trim}} := C_{\text{out}}\mathbf{x}, \tag{7.39}$$

where $\mathbf{h}_{\text{out,trim}}$ is the trim value of the corresponding $\mathbf{h}_{\text{out}}$, and C_{out} is the corresponding constant matrix. In order to handle the input constraints and constraints on other state variables, we adopt the following controlled output in the design process,

$$\mathbf{h}_{\text{in}} = C_2\mathbf{x} + D_2\mathbf{u} \tag{7.40}$$

with

$$C_2 = \left[\begin{array}{ccccccccc} & & & & \mathbf{0}_{3\times 9} & & & & \\ \hline b_1 & 0 & 0 & 0 & 0 & 0 & 0 & 0 & 0 \\ 0 & b_2 & 0 & 0 & 0 & 0 & 0 & 0 & 0 \\ 0 & 0 & b_3 & 0 & 0 & 0 & 0 & 0 & 0 \\ 0 & 0 & 0 & b_4 & 0 & 0 & 0 & 0 & 0 \\ 0 & 0 & 0 & 0 & 0 & 0 & b_5 & 0 & 0 \\ 0 & 0 & 0 & 0 & 0 & 0 & 0 & 0 & b_6 \end{array}\right], \qquad D_2 = \left[\begin{array}{ccc} a_1 & 0 & 0 \\ 0 & a_2 & 0 \\ 0 & 0 & a_3 \\ \hline & \mathbf{0}_{6\times 3} & \end{array}\right], \tag{7.41}$$

where a_i's and b_i's are weighting parameters to be determined later. We can now proceed to design a measurement feedback H_∞ control law (either a full order or

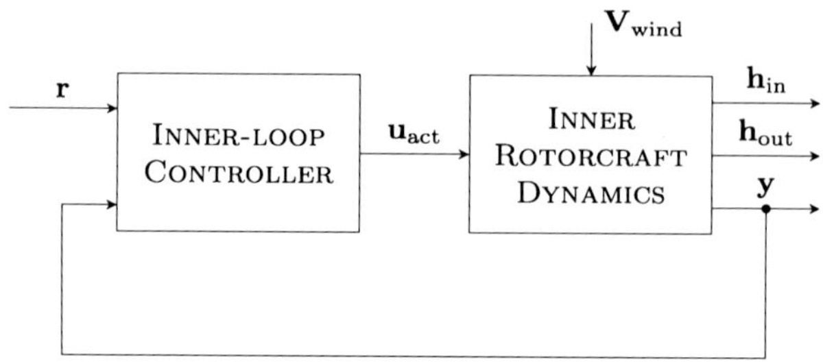

Fig. 7.3 System setup for the inner-loop controller design

reduced order) that would minimize the effect of the wind gust disturbance, i.e., to minimize the H_∞ norm of the closed-loop transfer matrix from the disturbance $\mathbf{w}$ to the controlled output $\mathbf{h}_{\rm in}$ or $\mathbf{h}_{\rm out}$ or both. The overall system setup for the inner-loop controller design is depicted in Fig. 7.3. It can be verified that the subsystem characterized by the matrix triple (A, E, C_1) is left invertible and of minimum phase. As mentioned in Sect. 7.2, it can be shown that the H_∞ optimization performance of the problem with measurement feedback is equivalent to that of the problem with state feedback.

7.3.3 Selection of Design Specifications

We follow the guideline given in [176] to select a series of design specifications to guarantee good performance in various categories, such as handling qualities, disturbance rejection, stability, and control usage. These specifications were originally defined in the military rotorcraft standards (see, e.g., ADS-33D-PRF [1] and SAE-AS94900 [153]). For our design, the specifications are selected as follows:

1. LOCATIONS OF EIGENVALUES: All of the eigenvalues are required to be located at the left-half-plane to guarantee system stability.
2. BANDWIDTH OF PITCH AND ROLL ATTITUDE RESPONSES: This specification has requirements on both bandwidth $\omega_{\rm BW}$ and phase delay $\tau_{\rm p}$, which are defined by

$$\left.\begin{aligned} \omega_{\rm BW} &= \min(\omega_{\rm BW,gain}, \omega_{\rm BW,phase}), \\ \tau_{\rm p} &= \frac{\Delta\Phi_{2\omega_{180}}}{57.3(2\omega_{180})}, \end{aligned}\right\} \tag{7.42}$$

   where ω_{180} is the frequency point where the phase crosses 180 degrees; $\omega_{\rm BW,gain}$ is the lowest frequency point where the corresponding gain is 6 dB larger than the ω_{180} gain value; $\omega_{\rm BW,phase}$ is the lowest frequency where the phase crosses 135 degrees; and $\Delta\Phi_{2\omega_{180}}$ is the phase difference between ω_{180} and $2\omega_{180}$.
3. COUPLING EFFECT BETWEEN ROLL AND PITCH RESPONSES: For this specification, the step input signal is injected in $\delta_{\rm lat}$ (or $\delta_{\rm lon}$). The resulting off- and on-axis attitude responses are compared. An upper limitation is set for $k_{\rm att}$, the ratio of the off- to on-axis peak attitudes during the transient period.

4. CROSSOVER FREQUENCY: The crossover frequency ($\omega_{\rm CF}$) is defined as the frequency where the magnitude curve crosses 0 dB [36]. Following [36], broken points are set at the input channels $\delta_{\rm lat}$ and $\delta_{\rm lon}$, and the frequency response of the resulting broken-loop input and output is examined. An upper limitation for the crossover frequency is set in this specification.
5. DISTURBANCE REJECTION BANDWIDTH FOR ATTITUDE CONTROL: The disturbance rejection bandwidth, $\omega_{\rm dst}$, is defined as the lowest frequency where the magnitude curve of attitude response to disturbance crosses -3 dB [36]. The attitude disturbance signal is required to be added to the bare attitude output generated by the helicopter dynamics. The main purpose of this specification is to evaluate the hold capability of the system in the presence of attitude disturbance [36].
6. QUICKNESS OF PITCH, ROLL, AND YAW RESPONSES: For each of the three cases, the spike signal is adopted as the input. The ratio of peak rate to the corresponding peak angle, $k_{\rm qik}$, measuring the quickness of the response, is required to be larger than the defined lower limitation, which varies with respect to the minimum angle response.
7. ATTITUDE HOLD FOR SPIKE DISTURBANCE INPUT: This specification evaluates the time-domain attitude hold capacity for the short-period spike input. Upper limitation is provided for the settling time, $t_{\rm set}$, when the attitude response returns to within 10% of the peak attitude response.

Among them, the first specification is for the stability and is necessary for any control system. The performance in terms of the remaining specifications can be categorized into three performance levels with Level 1 being the best. In our design, we aim to achieve top level performance in all categories as set in [1]. The detailed specifications for the Level 1 requirements are to be given later together with the performance evaluation of our design.

7.3.4 H_∞ Control Law

Based on the problem formulation and analysis given in the previous subsections, we follow the design procedure outlined in Sect. 7.2 to complete the state feedback control law design.

1. First, we need to determine the weighting parameters a_i's for D_2 and b_i's for C_2, respectively. Once these parameters are fixed, we compute $\gamma_{\rm in}^*$, which is the optimal H_∞ performance for the closed-loop system from the disturbance input $\mathbf{w}$ (wind gust) to the controlled output $\mathbf{h}_{\rm in}$ over all the possible internally stabilizing controllers.
2. It can be verified that matrix D_2 is of full column rank and the matrix quadruple (A, B, C_2, D_2) is left invertible and is free of invariant zeros. It is a regular

problem. Thus, for any given $\gamma > \gamma_{\rm in}^*$, its corresponding H_∞ γ-suboptimal state feedback law can be obtained as follows:

$$\mathbf{u} = F\mathbf{x} + G(\mathbf{r} - \mathbf{h}_{\rm out,trim}) = -(D_2^{\rm T}D_2)^{-1}(D_2^{\rm T}C_2 + B^{\rm T}P)\mathbf{x} + G(\mathbf{r} - \mathbf{h}_{\rm out,trim}), \tag{7.43}$$

where $\mathbf{r} = [\,\phi_{\rm r} \quad \theta_{\rm r} \quad \psi_{\rm r}\,]^{\rm T}$ is the reference signal vector generated by a command generator linked to the outer-loop control law,

$$F = -(D_2^{\rm T}D_2)^{-1}(D_2^{\rm T}C_2 + B^{\rm T}P), \tag{7.44}$$

with P being the positive semi-definite stabilizing solution of the following H_∞ algebraic Riccati equation

$$\begin{aligned} A^{\rm T}P + PA + C_2^{\rm T}C_2 + \gamma^{-2}PEE^{\rm T}P \\ - (PB + C_2^{\rm T}D_2)(D_2^{\rm T}D_2)^{-1}(D_2^{\rm T}C_2 + B^{\rm T}P) = 0, \end{aligned}$$

and lastly, the reference feed forward matrix G is given by

$$G = -[C_{\rm out}(A + BF)^{-1}B]^{-1}. \tag{7.45}$$

After a few trials, the final selections for the weighting parameters are

$$a_1 = 13, \qquad a_2 = 12, \qquad a_3 = 30$$

and

$$b_1 = 13, \qquad b_2 = 12, \qquad b_3 = 1, \qquad b_4 = 1, \qquad b_5 = 1, \qquad b_6 = 6$$

and the corresponding $\gamma_{\rm in}^* = 0.0476$. We then select $\gamma = 0.065 > \gamma_{\rm in}^*$ and obtain the following γ-suboptimal state feedback gain matrix,

$$F = \left[\begin{matrix} -1.0368 & -0.0604 & -0.0230 & -0.0083 & -0.2857 & -2.6165 \\ 0.0760 & -0.9970 & 0.0174 & -0.0378 & -1.8340 & -0.1130 \\ -0.0002 & -0.0185 & -0.0066 & 0.0004 & 0.0353 & 0.0990 \end{matrix}\right.$$
$$\left.\begin{matrix} -0.0312 & 0.0499 & -0.0746 \\ 0.0026 & 0.0024 & -0.0169 \\ 0.0044 & 0.2295 & 0.2441 \end{matrix}\right]. \tag{7.46}$$

The reference feed forward matrix G is then given as

$$G = \begin{bmatrix} 1.0368 & 0.0604 & 0.0746 \\ -0.0760 & 0.9970 & 0.0169 \\ 0.0002 & 0.0185 & -0.2441 \end{bmatrix}. \tag{7.47}$$

As there are three variables, i.e., $a_{\rm s}$, $b_{\rm s}$ and $\delta_{\rm ped,int}$, that cannot be directly measured, we follow the procedure given in Sect. 7.2 to design a reduced-order measurement feedback control law that recovers the performance obtained under the state feedback law. The following is the reduced-order observer

$$\dot{\mathbf{x}}_{\rm in,cmp} = A_{\rm in,cmp}\,\mathbf{x}_{\rm in,cmp} + B_{\rm in,cmp}\,\mathbf{y} + H_{\rm in,cmp}\,\mathbf{u}, \tag{7.48}$$

where

$$A_{\mathrm{in,cmp}} = \begin{bmatrix} -10 & 0 & 0 \\ 0 & -10 & 0 \\ 0 & 0 & -12 \end{bmatrix},$$

$$H_{\mathrm{in,cmp}} = \begin{bmatrix} 0.2026 & 2.5878 & 0 \\ 2.5878 & -0.0663 & 0 \\ 0 & 0 & 4.8913 \end{bmatrix}, \tag{7.49}$$

$$B_{\mathrm{in,cmp}} = 10^{-2} \times \begin{bmatrix} 0 & 0 & -3.7980 & -124.6213 & -1.9798 & 0 \\ 0 & 0 & -111.3469 & -9.0793 & -5.9148 & 0 \\ 0 & 0 & -0.6076 & -0.1112 & 4.6136 & 0 \end{bmatrix}, \tag{7.50}$$

and

$$\begin{pmatrix} \hat{a}_{\mathrm{s}} \\ \hat{b}_{\mathrm{s}} \\ \hat{\delta}_{\mathrm{ped,int}} \end{pmatrix} = \mathbf{x}_{\mathrm{in,cmp}} + K_{\mathrm{in,cmp}}\, \mathbf{y} \tag{7.51}$$

and where

$$K_{\mathrm{in,cmp}} = 10^{-3} \times \begin{bmatrix} 0 & 0 & 3.7980 & 24.7966 & 1.9798 & 0 \\ 0 & 0 & 11.3469 & 9.1439 & 5.9148 & 0 \\ 0 & 0 & 0 & 0 & 105.0784 & 0 \end{bmatrix}. \tag{7.52}$$

The control law of (7.43) is implemented by replacing the unmeasured state variables with their estimates given in (7.51), and it is used for the wide flight envelope throughout.

7.3.5 Performance Evaluation

In this section, we carry out a comprehensive performance evaluation on the above H_∞ inner-loop control law. CONDUIT toolkit [175, 176], a software package developed by the NASA Ames Research Center for assisting the flight control law design of military rotorcraft and aircraft, has been adopted for the straightforward graphical display of the selected design specifications (as shown in Fig. 7.17). We have the following evaluation results:

1. WIND GUST DISTURBANCE ATTENUATION: Although the controlled output $\mathbf{h}_{\mathrm{in}}$ is used in the control law design, the disturbance rejection capacity of the controlled output $\mathbf{h}_{\mathrm{out}}$ is practically more important as it is directly related to the Euler angles of the unmanned system. As such, we evaluate the frequency domain responses of the closed-loop transfer matrix from $\mathbf{w}$ to $\mathbf{h}_{\mathrm{in}}$ and that from $\mathbf{w}$ to $\mathbf{h}_{\mathrm{out}}$. The singular values of the closed-loop systems shown in Fig. 7.4 clearly show that the effect of wind gust can be almost completely attenuated by our design. In the worst situation, the wind gust effect (in terms of the worst case L_2-gain) to $\mathbf{h}_{\mathrm{out}}$ can be reduced by more than 99%.

   Next, we verify the wind gust attenuation of the overall closed-loop system in the time domain. The wind gust input and the corresponding output responses are shown in Fig. 7.5. In the simulation process, the 20-second-long "$1-\cos(\cdot)$"-type wind gust disturbance as suggested in [63] has been sequentially injected

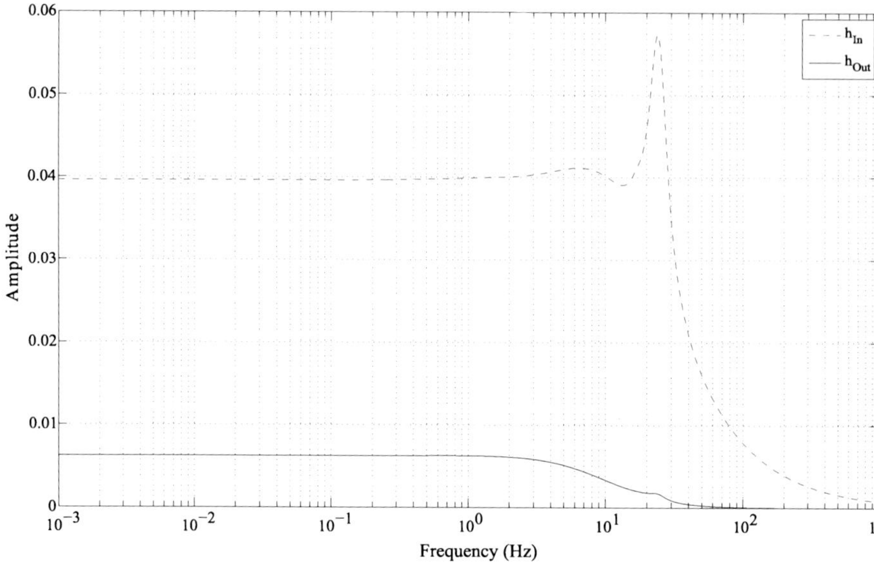

Fig. 7.4 Maximum singular values of the closed-loop transfer matrices

to the X-, Y-, and Z-directions of the body frame, with the peaking amplitude of 5 m/s, 5 m/s, and 2 m/s, respectively. It can be clearly observed that the wind gust effect has been effectively attenuated by the H_∞ control law. The largest deviations for the Euler angles and angular rate are only -0.0139 rad and 0.0022 rad/s, respectively.

2. LOCATIONS OF EIGENVALUES: The eigenvalues of the closed-loop system are depicted in Fig. 7.6. All of them are placed at the proper locations.
3. BANDWIDTH OF PITCH AND ROLL ATTITUDE RESPONSES: The frequency responses of pitch and roll attitudes are shown in Figs. 7.7 and 7.8, respectively, along with the resulting values of ω_{180}, $\omega_{\mathrm{BW,gain}}$, $\omega_{\mathrm{BW,phase}}$, and $\Delta\Phi_{2\omega_{180}}$. We have $\omega_{\mathrm{BW}} = 2.03$ rad and $\tau_{\mathrm{p}} = 0.037$ s for the pitch attitude response, and $\omega_{\mathrm{BW}} = 3.22$ rad and $\tau_{\mathrm{p}} = 0.026$ s for the roll attitude response. In accordance with the standards set in [1], it is depicted in Fig. 7.17 that both channels achieve the top level performance.
4. COUPLING EFFECT BETWEEN ROLL AND PITCH RESPONSES: To evaluate the performance in terms of this specification, a step input signal with the amplitude of 0.15 is adopted. Figures 7.9 and 7.10 demonstrate the attitude responses when the step input is injected in δ_{lon} and δ_{lat}, respectively. The resulting ratios k_{att} are 0.064 and 0.077, which are both less than 0.25, the Level 1 performance requirement, as summarized in Fig. 7.17.
5. CROSSOVER FREQUENCIES: The examination result for the crossover frequencies is shown in Fig. 7.11. The crossover frequencies for the aileron and elevator channels are 2.65 rad/s and 2.54 rad/s, respectively. Both are less than the minimum requirement for the Level 1 performance, which is 10 rad/s.

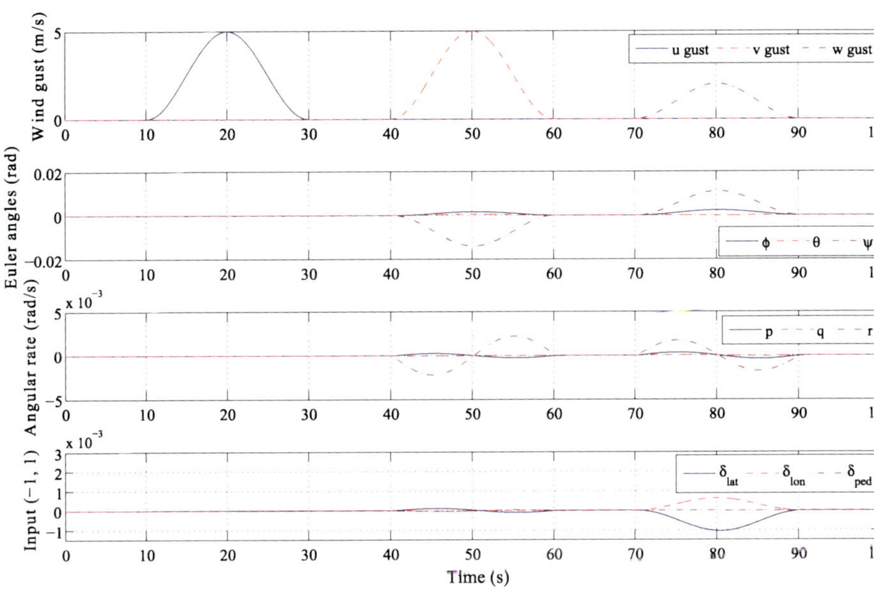

Fig. 7.5 Simulation results—wind gust attenuation of the closed-loop system

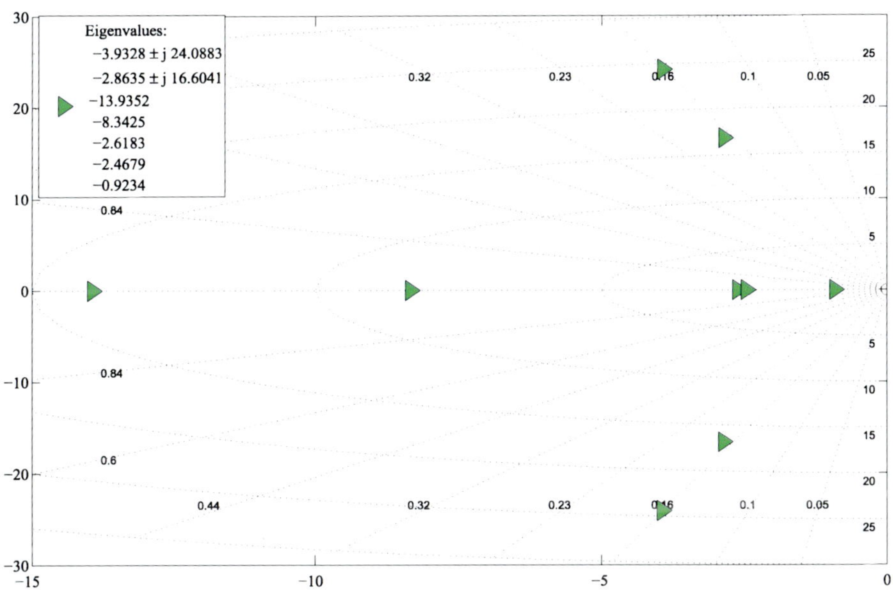

Fig. 7.6 Eigenvalues of the closed-loop system

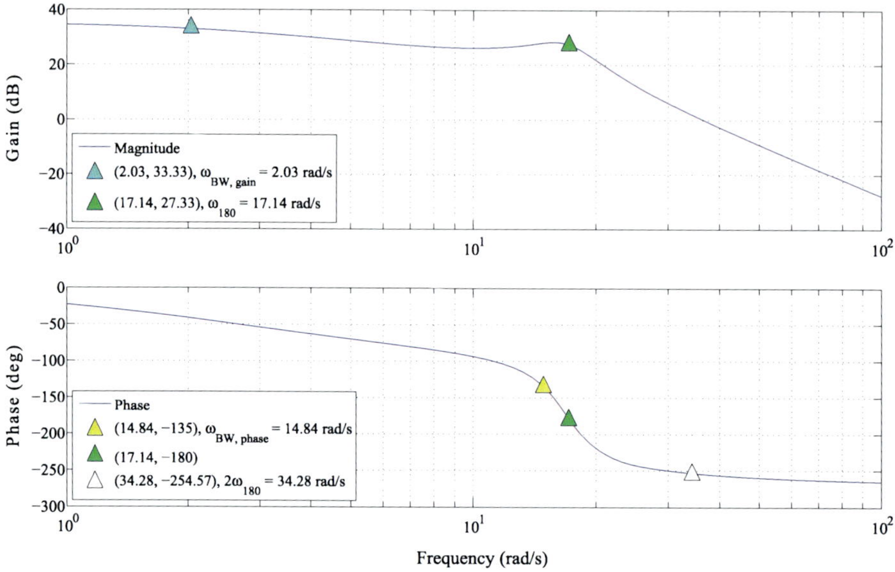

Fig. 7.7 Pitch attitude frequency response

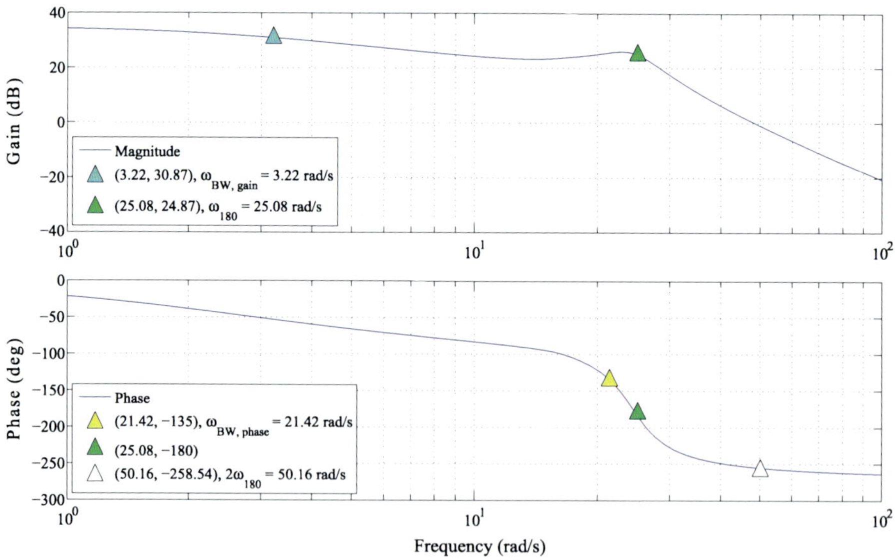

Fig. 7.8 Roll attitude frequency response

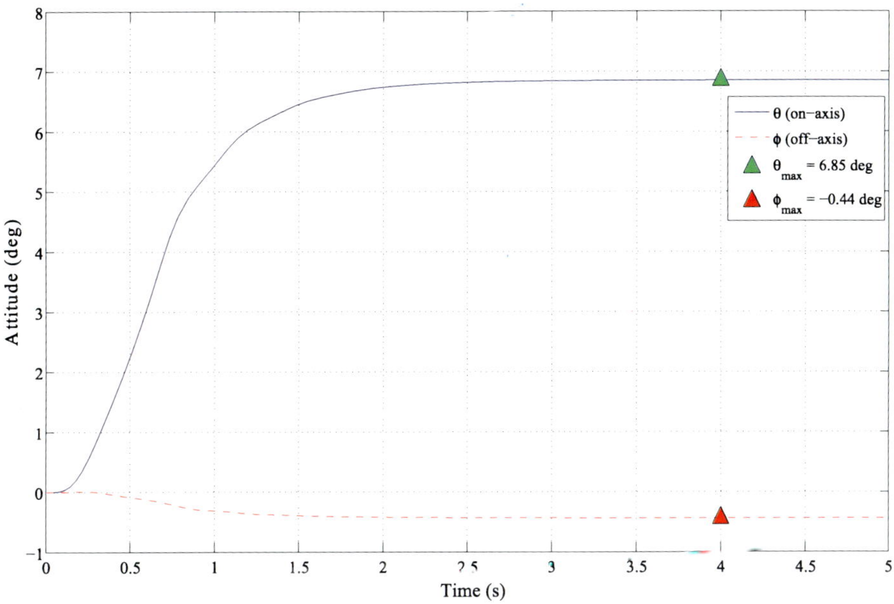

Fig. 7.9 Coupling effect with a step input in δ_{lon}

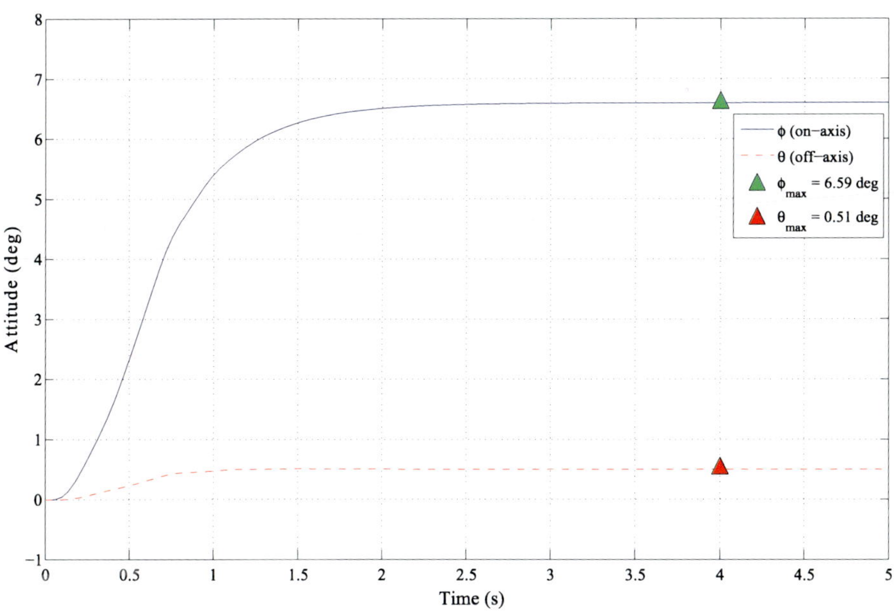

Fig. 7.10 Coupling effect with a step input in δ_{lat}

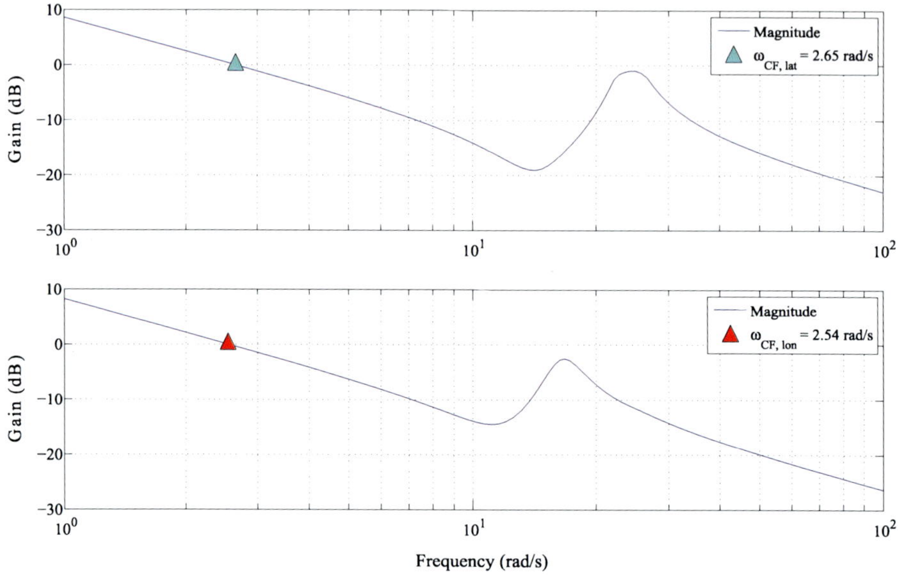

Fig. 7.11 Crossover frequencies

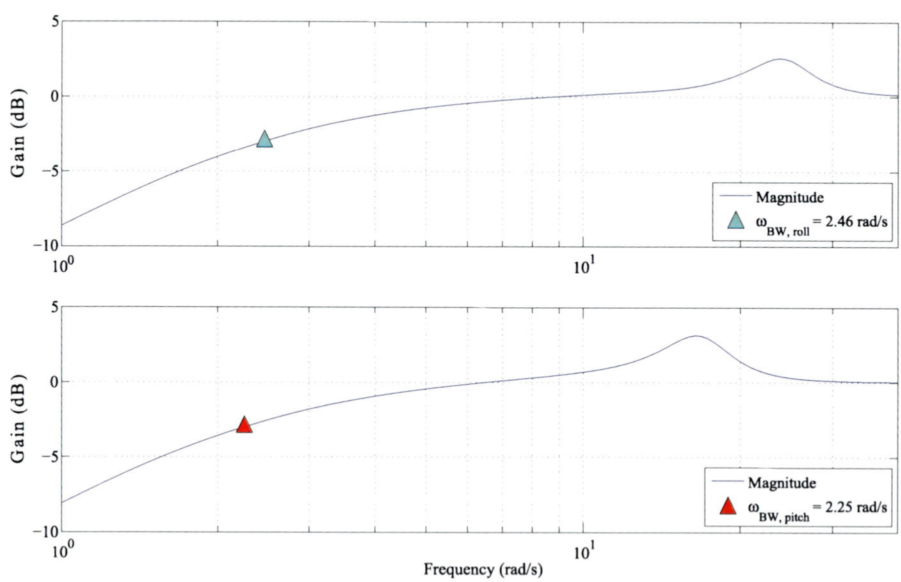

Fig. 7.12 Disturbance rejection examination for attitude control

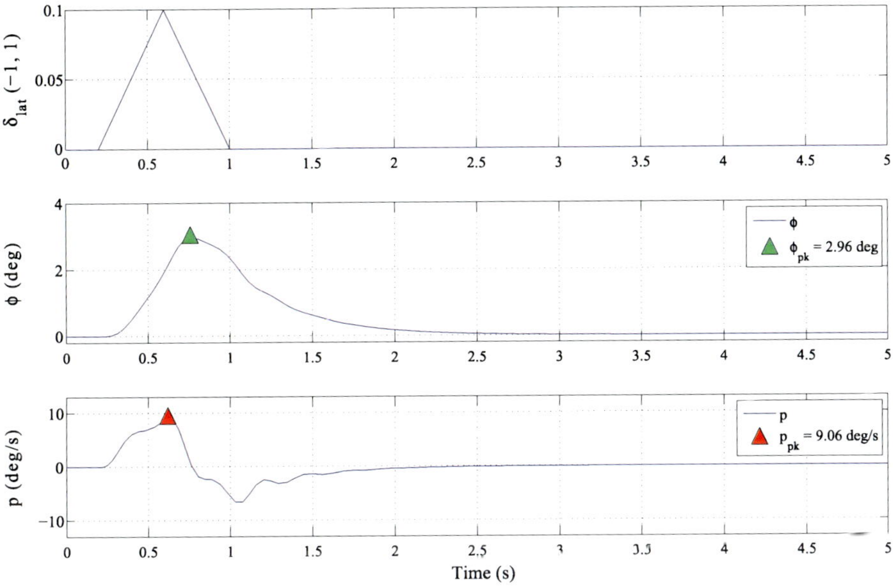

Fig. 7.13 Quickness evaluation for the roll response

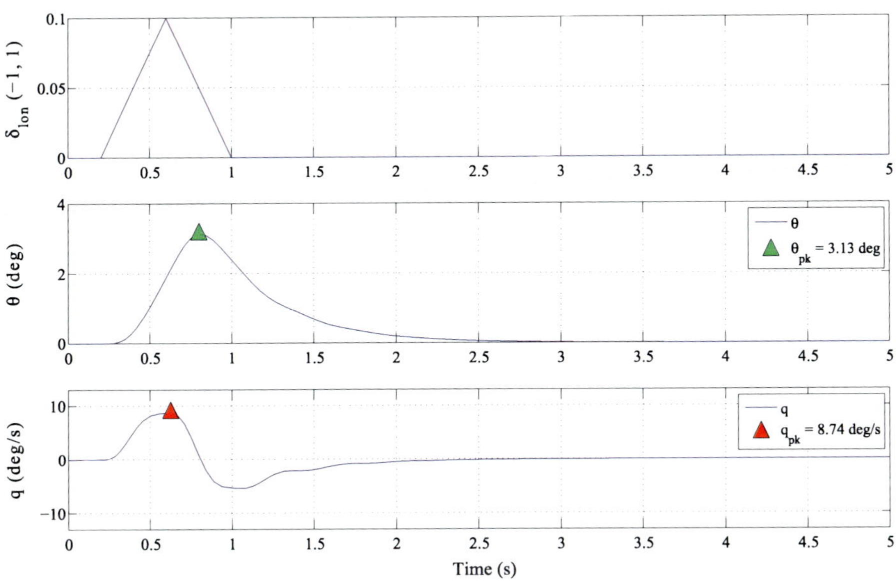

Fig. 7.14 Quickness evaluation for the pitch response

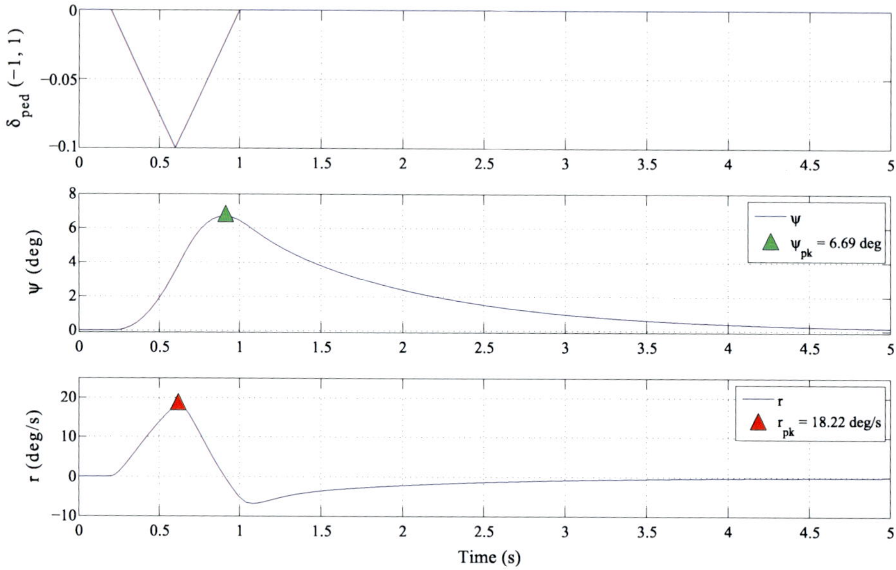

Fig. 7.15 Quickness evaluation for the yaw response

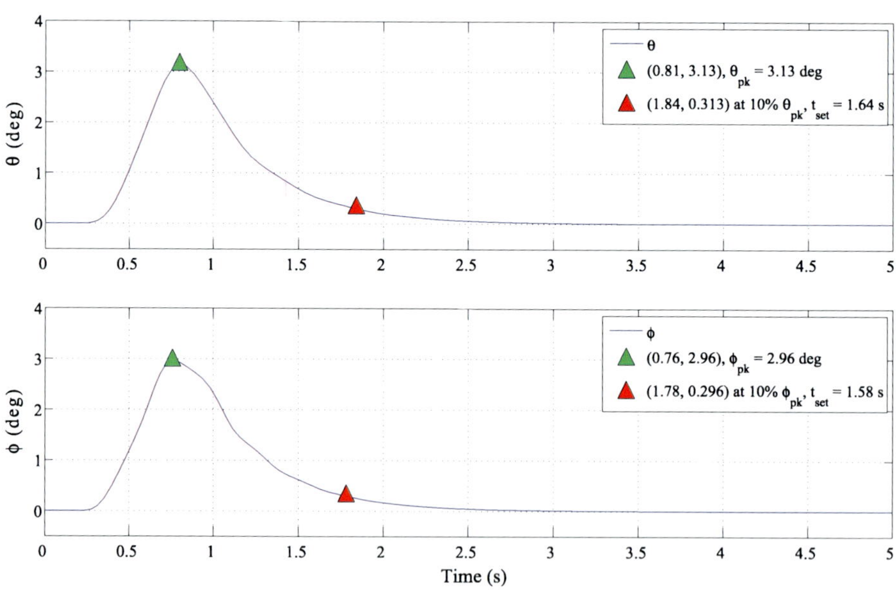

Fig. 7.16 Attitude hold examination for spike disturbance input

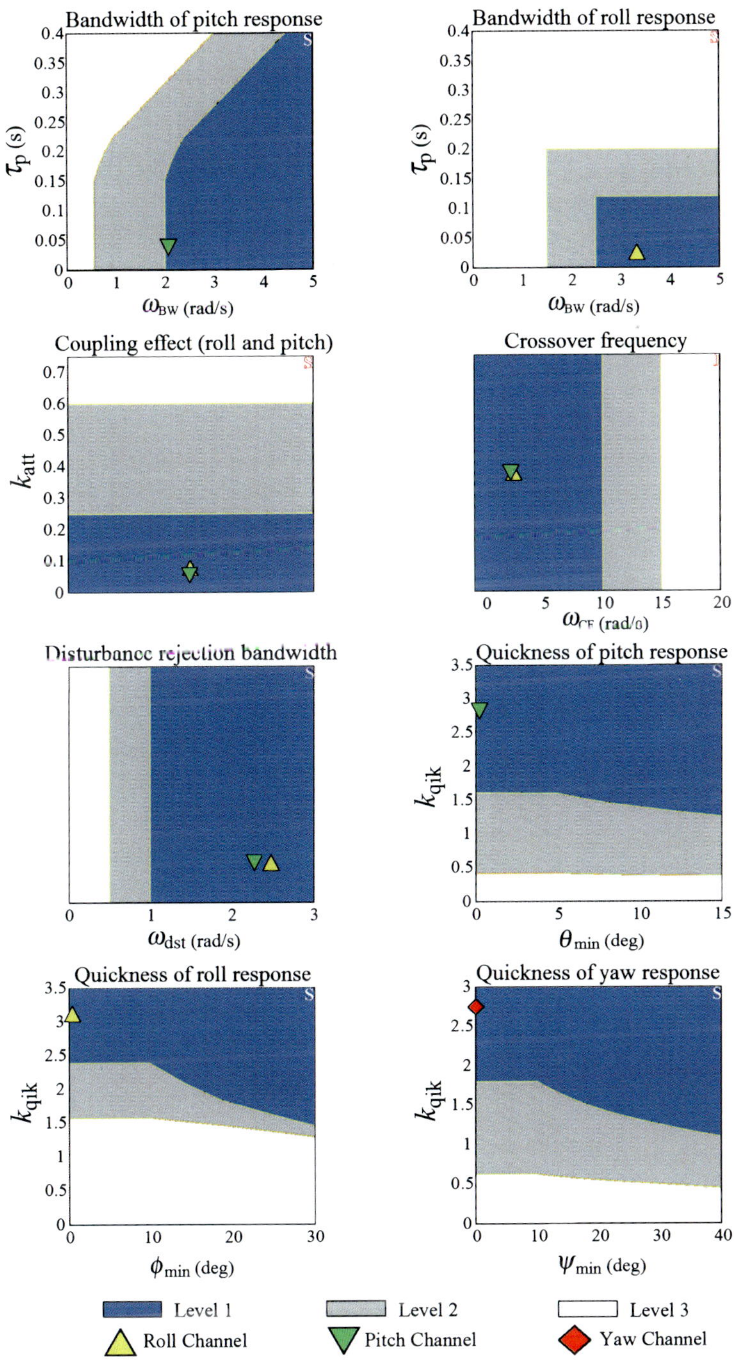

Fig. 7.17 Summary of evaluation results for the inner-loop control system

6. DISTURBANCE REJECTION BANDWIDTH FOR ATTITUDE CONTROL: The frequency responses from attitude disturbances to attitude output are shown in Fig. 7.12. The disturbance bandwidths for the roll and pitch angles are 2.46 rad/s and 2.25 rad/s, respectively. Both of them exceed the Level 1 performance requirements (1 rad/s), as summarized in Fig. 7.17.
7. QUICKNESS OF PITCH, ROLL, AND YAW RESPONSES: Figs. 7.13, 7.14, 7.15 demonstrate the evaluation results of quickness for the pitch, roll, and yaw channels. The amplitude, starting time, and time interval of the spike input injected in $\delta_{\rm lon}$, $\delta_{\rm lat}$, or $\delta_{\rm ped}$ are, respectively, set to be 0.1, 0.2 s, and 0.8 s for all channels. The quickness ratios $k_{\rm qik}$ for the pitch, roll, and yaw channels are respectively 2.79, 3.06, and 2.72, which satisfy the Level 1 performance requirement (see Fig. 7.17).
8. ATTITUDE HOLD FOR SPIKE DISTURBANCE INPUT: Lastly, we examine the attitude hold response for a spike disturbance input injected in $\delta_{\rm lon}$ or $\delta_{\rm lat}$. The amplitude, starting time, and time interval for the spike input are also set to 0.1, 0.2 s, and 0.8 s, respectively. The corresponding attitude response results are depicted in Fig. 7.16. We note that the settling times $t_{\rm set}$ for the pitch and roll responses are 1.64 s and 1.58 s, respectively, which are less than 10 s, the Level 1 performance requirement.

In summary, the H_∞ control law that we have obtained achieves the top level performance in all the categories under examination. We summarize the overall evaluation results in Fig. 7.17 for easy reference.

Chapter 8
Outer-Loop Flight Control

8.1 Introduction

The outer loop of our proposed automatic flight control system is for controlling the position of the unmanned system in the local NED frame. Traditionally, the outer-loop layer can be controlled by simple controllers, such as PID or even proportional control laws (see, e.g., [14]). However, the flight control system with simple outer-loop controllers can only provide reasonable performance for position and heading control. When it comes to situations in which complicated maneuvers are required, it generally results in poor performance. We propose in this chapter the design of the outer-loop controllers for our unmanned systems using the so-called robust and perfect tracking (RPT) control technique developed by Chen and his co-workers (see, e.g., [27, 28, 112]). Given a system that satisfies certain conditions, the RPT control technique is for designing a controller such that the resulting closed-loop system is asymptotically stable and the controlled output almost perfectly tracks a given reference signal in the presence of any initial conditions and external disturbances. Almost perfect tracking means the ability of a controller to track a given reference signal with an arbitrarily fast settling time in the face of external disturbances and initial conditions. Of course, in real life, a certain tradeoff has to be made in order to design a physically implementable control law.

We should highlight that one of the most interesting features in the RPT control method is its capability of utilizing all possible information available in its controller structure. More specifically, given a reference, if its derivatives are also available, all of them can be fed into the RPT controller to yield a better performance. Such a feature is highly desirable for flight missions involving complicated maneuvers, in which not only the position reference is useful, but also its velocity and even acceleration information are important or even necessary to be used in order to achieve a good overall performance. It is soon to be seen in Chap. 10 that the RPT control renders the flight formation of multiple UAVs a trivial task. In what follows, we first recall the basic theory behind the RPT technique. Interested readers are referred to [27] for the rigorous treatment of the RPT control theory.

G. Cai et al., *Unmanned Rotorcraft Systems*, Advances in Industrial Control,
DOI 10.1007/978-0-85729-635-1_8,

8.2 Robust and Perfect Tracking Control

Consider the following continuous-time system:

$$\Sigma: \quad \begin{cases} \dot{x} = A\ x + B\ \ u + E\ \ \ w, \quad x(0) = x_0, \\ y = C_1\ x \qquad\qquad + D_1\ \ w, \\ h = C_2\ x + D_2\ u + D_{22}\ w, \end{cases} \tag{8.1}$$

where $\boldsymbol{x} \in \mathbb{R}^n$ is the state, $\boldsymbol{u} \in \mathbb{R}^m$ is the control input, $\boldsymbol{w} \in \mathbb{R}^q$ is the external disturbance, $\boldsymbol{y} \in \mathbb{R}^p$ is the measurement output, and $\boldsymbol{h} \in \mathbb{R}^\ell$ is the output to be controlled. Given the external disturbance $\boldsymbol{w} \in L_p$, $p \in [1, \infty)$, and any reference signal vector $\boldsymbol{r} \in \mathbb{R}^\ell$ with $r, \dot{\boldsymbol{r}}, \ldots, \boldsymbol{r}^{(\kappa-1)}$, $\kappa \geq 1$, being available, and $\boldsymbol{r}^{(\kappa)}$ being either a vector of delta functions or in L_p, the RPT problem for the system in (8.1) is to find a parameterized dynamic measurement control law of the following form:

$$\begin{cases} \dot{\boldsymbol{v}} = A_{\text{cmp}}(\varepsilon)\boldsymbol{v} + B_{\text{cmp}}(\varepsilon)\boldsymbol{y} + G_0(\varepsilon)\boldsymbol{r} + \cdots + G_{\kappa-1}(\varepsilon)\boldsymbol{r}^{(\kappa-1)}, \\ \boldsymbol{u} = C_{\text{cmp}}(\varepsilon)\boldsymbol{v} + D_{\text{cmp}}(\varepsilon)\boldsymbol{y} + H_0(\varepsilon)\boldsymbol{r} + \cdots + H_{\kappa-1}(\varepsilon)\boldsymbol{r}^{(\kappa-1)} \end{cases} \tag{8.2}$$

such that when the controller of (8.2) is applied to the system of (8.1), we have the following

1. There exists an $\varepsilon^* > 0$ such that the resulting closed-loop system with $\boldsymbol{r} = 0$ and $\boldsymbol{w} = 0$ is asymptotically stable for all $\varepsilon \in (0, \varepsilon^*]$.
2. Let $\boldsymbol{h}(t, \varepsilon)$ be the closed-loop controlled output response and let $\boldsymbol{e}(t, \varepsilon)$ be the resulting tracking error, i.e., $\boldsymbol{e}(t, \varepsilon) := \boldsymbol{h}(t, \varepsilon) - \boldsymbol{r}(t)$. Then, for any initial condition of the state, $\boldsymbol{x}_0 \in \mathbb{R}^n$,

$$\|\boldsymbol{e}\|_p = \left(\int_0^\infty |\boldsymbol{e}(t)|^p \, \mathrm{d}t \right)^{1/p} \to 0 \quad \text{as } \varepsilon \to 0. \tag{8.3}$$

We introduce in the above formulation some additional information besides the reference signal $\boldsymbol{r}$, i.e., $\dot{\boldsymbol{r}}, \ddot{\boldsymbol{r}}, \ldots, \boldsymbol{r}^{(\kappa-1)}$, as additional controller inputs. Note that, in general, these additional signals can easily be generated without any extra costs. For example, if $\boldsymbol{r}(t) = t^2$, then one can easily obtain its first-order derivative $\dot{\boldsymbol{r}}(t) = 2t$ and its second-order derivative $\ddot{\boldsymbol{r}}(t) = 2$. In flight control systems, taking $\boldsymbol{r}$ as a position reference, generally, its associated velocity, $\dot{\boldsymbol{r}}$, and acceleration, $\ddot{\boldsymbol{r}}$, are readily available. These $\dot{\boldsymbol{r}}(t)$ and $\ddot{\boldsymbol{r}}(t)$ can be used to improve the overall tracking performance. We also note that the above formulation covers all possible reference signals that have the form $\boldsymbol{r}(t) = t^k$, $0 \leq k < \infty$. Thus, it can be applied to track approximately those reference signals that have a Taylor series expansion at $t = 0$. This can be done by truncating the higher-order terms of the Taylor series of the given signal.

It is shown (see [27]) that the RPT problem for the system in (8.1) is solvable if and only if the following conditions hold:

1. (A, B) is stabilizable and (A, C_1) is detectable.
2. $D_{22} + D_2 S D_1 = 0$, where $S = -(D_2^{\mathrm{T}} D_2)^{\dagger} D_2^{\mathrm{T}} D_{22} D_1^{\mathrm{T}} (D_1 D_1^{\mathrm{T}})^{\dagger}$.
3. (A, B, C_2, D_2) is right invertible and of minimum phase.
4. $\mathrm{Ker}(C_2 + D_2 S C_1) \supset C_1^{-1}\{\mathrm{Im}(D_1)\}$.

Here, we note that $X^\dagger$ denotes the Moore-Penrose (pseudo) inverse of a constant matrix X, $\mathrm{Im}(X)$ and $\mathrm{Ker}(X)$ are respectively the range and null spaces of X, and lastly, $C^{-1}\{\mathcal{X}\} := \{x \mid Cx \in \mathcal{X}\}$, where $\mathcal{X}$ is a subspace and C is a constant matrix. We also note that for the case when $D_1 = 0$, then the above solvability conditions can be simplified as follows:

1. (A, B) is stabilizable and (A, C_1) is detectable.
2. $D_{22} = 0$.
3. (A, B, C_2, D_2) is right invertible and of minimum phase.
4. $\mathrm{Ker}(C_2) \supset \mathrm{Ker}(C_1)$.

The last condition is automatically satisfied if the controlled output $\boldsymbol{h}$ of the given system is part of its measurement output $\boldsymbol{y}$.

We assume throughout the rest of this section that the above conditions are satisfied, and we move on to construct solutions to the RPT problem. As usual, we focus on the following three cases: (i) the state feedback case, (ii) the full-order measurement feedback case, and (iii) the reduced-order measurement feedback case.

i. State Feedback Case When all states of the plant are measured for feedback, the problem can be solved by a static control law. We construct a parameterized state feedback control law,

$$\boldsymbol{u} = F(\varepsilon)\boldsymbol{x} + H_0(\varepsilon)\boldsymbol{r} + \cdots + H_{\kappa-1}(\varepsilon)\boldsymbol{r}^{(\kappa-1)}, \tag{8.4}$$

that solves the RPT problem for the system in (8.1). It is simple to note that we can rewrite the given reference in the following form:

$$\frac{\mathrm{d}}{\mathrm{d}t}\begin{pmatrix} \boldsymbol{r} \\ \vdots \\ \boldsymbol{r}^{(\kappa-2)} \\ \boldsymbol{r}^{(\kappa-1)} \end{pmatrix} = \begin{bmatrix} 0 & I_\ell & \cdots & 0 \\ \vdots & \vdots & \ddots & \vdots \\ 0 & 0 & \cdots & I_\ell \\ 0 & 0 & \cdots & 0 \end{bmatrix}\begin{pmatrix} \boldsymbol{r} \\ \vdots \\ \boldsymbol{r}^{(\kappa-2)} \\ \boldsymbol{r}^{(\kappa-1)} \end{pmatrix} + \begin{bmatrix} 0 \\ \vdots \\ 0 \\ I_\ell \end{bmatrix}\boldsymbol{r}^{(\kappa)}. \tag{8.5}$$

Combining (8.5) with the given system, we obtain the following augmented system:

$$\Sigma_{\mathrm{AUG}}: \quad \begin{cases} \dot{\mathbf{x}} = \boldsymbol{A}\ \mathbf{x} + \boldsymbol{B}\ \boldsymbol{u} + \boldsymbol{E}\ \mathbf{w}, \\ \mathbf{y} = \mathbf{x}, \\ \boldsymbol{e} = \boldsymbol{C}_2\ \mathbf{x} + \boldsymbol{D}_2\ \boldsymbol{u}, \end{cases} \tag{8.6}$$

where

$$\mathbf{x} := \begin{pmatrix} \boldsymbol{r} \\ \vdots \\ \boldsymbol{r}^{(\kappa-2)} \\ \boldsymbol{r}^{(\kappa-1)} \\ \boldsymbol{x} \end{pmatrix}, \qquad \mathbf{w} := \begin{pmatrix} \boldsymbol{w} \\ \boldsymbol{r}^{(\kappa)} \end{pmatrix}, \tag{8.7}$$

$$\boldsymbol{A} = \begin{bmatrix} 0 & I_\ell & \cdots & 0 & 0 \\ \vdots & \vdots & \ddots & \vdots & \vdots \\ 0 & 0 & \cdots & I_\ell & 0 \\ 0 & 0 & \cdots & 0 & 0 \\ 0 & 0 & \cdots & 0 & A \end{bmatrix}, \qquad \boldsymbol{B} = \begin{bmatrix} 0 \\ \vdots \\ 0 \\ 0 \\ B \end{bmatrix}, \qquad \boldsymbol{E} = \begin{bmatrix} 0 & 0 \\ \vdots & \vdots \\ 0 & 0 \\ 0 & I_\ell \\ E & 0 \end{bmatrix}, \tag{8.8}$$

and

$$\boldsymbol{C}_2=[\,-I_\ell \quad 0 \quad 0 \quad \cdots \quad 0 \quad C_2\,], \qquad \boldsymbol{D}_2=D_2. \tag{8.9}$$

It is then straightforward to show that the subsystem from $\boldsymbol{u}$ to $\boldsymbol{e}$ in the augmented system of (8.6), i.e., the quadruple $(\boldsymbol{A}, \boldsymbol{B}, \boldsymbol{C}_2, \boldsymbol{D}_2)$, is right invertible and has the same infinite zero structure as that of (A, B, C_2, D_2). Furthermore, its invariant zeros contain those of (A, B, C_2, D_2) and $\ell \times \kappa$ extra ones at $s=0$.

Next, we define

$$\tilde{\boldsymbol{C}}_2=\begin{bmatrix} \boldsymbol{C}_2 \\ \varepsilon I_{\kappa\ell+n} \\ 0 \end{bmatrix}, \qquad \tilde{\boldsymbol{D}}_2=\begin{bmatrix} \boldsymbol{D}_2 \\ 0 \\ \varepsilon I_m \end{bmatrix}, \tag{8.10}$$

$$\tilde{\boldsymbol{A}}=\begin{bmatrix} \tilde{A}_0 & 0 \\ 0 & A \end{bmatrix}, \qquad \tilde{A}_0=-\varepsilon_0 I_{\kappa\ell}+\begin{bmatrix} 0 & I_\ell & \cdots & 0 \\ \vdots & \vdots & \ddots & \vdots \\ 0 & 0 & \cdots & I_\ell \\ 0 & 0 & \cdots & 0 \end{bmatrix}, \tag{8.11}$$

where ε_0 is a sufficiently small scalar (introduced to fool the Riccati equation), and solve the following Riccati equation:

$$\boldsymbol{P}\tilde{\boldsymbol{A}}+\tilde{\boldsymbol{A}}^{\mathrm{T}}\boldsymbol{P}+\tilde{\boldsymbol{C}}_2^{\mathrm{T}}\tilde{\boldsymbol{C}}_2-\big(\boldsymbol{P}\boldsymbol{B}+\tilde{\boldsymbol{C}}_2^{\mathrm{T}}\tilde{\boldsymbol{D}}_2\big)\big(\tilde{\boldsymbol{D}}_2^{\mathrm{T}}\tilde{\boldsymbol{D}}_2\big)^{-1}\big(\boldsymbol{P}\boldsymbol{B}+\tilde{\boldsymbol{C}}_2^{\mathrm{T}}\tilde{\boldsymbol{D}}_2\big)^{\mathrm{T}}=0 \tag{8.12}$$

for a positive-definite solution $\boldsymbol{P}>0$. The required state feedback gain matrix that solves the RPT problem for the given system is then given by

$$\begin{aligned}\tilde{\boldsymbol{F}}(\varepsilon)&=-(\tilde{\boldsymbol{D}}_2^{\mathrm{T}}\tilde{\boldsymbol{D}}_2)^{-1}\big(\boldsymbol{P}\boldsymbol{B}+\tilde{\boldsymbol{C}}_2^{\mathrm{T}}\tilde{\boldsymbol{D}}_2\big)^{\mathrm{T}} \\ &=[\,H_0(\varepsilon) \quad \cdots \quad H_{\kappa-1}(\varepsilon) \quad F(\varepsilon)\,],\end{aligned} \tag{8.13}$$

where $H_i(\varepsilon)\in\mathbb{R}^{m\times\ell}$ and $F(\varepsilon)\in\mathbb{R}^{m\times n}$.

Finally, we note that solutions to the Riccati equation in (8.12) might have severe numerical problems as ε becomes smaller and smaller. Alternatively, one can solve the RPT control problem using a structural decomposition approach, which can be found in Chen [27].

ii. Full-Order Measurement Feedback Case The full-order measurement output feedback controller can be constructed as follows: First, we compute an appropriate state feedback gain matrix as in (8.13). Then, we solve the following Riccati equation:

$$AQ+QA^{\mathrm{T}}+(EE^{\mathrm{T}}+I)-(QC_1^{\mathrm{T}}+ED_1^{\mathrm{T}})(D_1D_1^{\mathrm{T}}+\varepsilon I)^{-1}(C_1Q+D_1E^{\mathrm{T}})=0 \tag{8.14}$$

for a positive-definite solution $Q>0$, and calculate an observer gain matrix

$$K(\varepsilon)=-(QC_1^{\mathrm{T}}+ED_1^{\mathrm{T}})(D_1D_1^{\mathrm{T}}+\varepsilon I)^{-1}. \tag{8.15}$$

The full-order measurement output feedback controller that solves the RPT problem for the given system is given by

$$\begin{cases}\dot{\boldsymbol{v}}=A_{\mathrm{cmp}}(\varepsilon)\boldsymbol{v}-K(\varepsilon)\boldsymbol{y}+BH_0(\varepsilon)\boldsymbol{r}+\cdots+BH_{\kappa-1}(\varepsilon)\boldsymbol{r}^{(\kappa-1)}, \\ \boldsymbol{u}=F(\varepsilon)\boldsymbol{v}+H_0(\varepsilon)\boldsymbol{r}+\cdots+H_{\kappa-1}(\varepsilon)\boldsymbol{r}^{(\kappa-1)},\end{cases} \tag{8.16}$$

where $A_{\mathrm{cmp}}(\varepsilon)=A+BF(\varepsilon)+K(\varepsilon)C_1$.

iii. Reduced-Order Measurement Feedback Case We now present solutions to the RPT problem via reduced-order measurement feedback control laws. For simplicity of presentation, we assume that matrices C_1 and D_1 have already been transformed into the following forms:

$$C_1 = \begin{bmatrix} 0 & C_{1,02} \\ I_k & 0 \end{bmatrix} \quad \text{and} \quad D_1 = \begin{bmatrix} D_{1,0} \\ 0 \end{bmatrix}, \tag{8.17}$$

where $D_{1,0}$ is of full row rank. Before we present a step-by-step algorithm to construct a parameterized reduced-order measurement feedback controller, we first partition the following system:

$$\begin{cases} \dot{\boldsymbol{x}} = A\ \boldsymbol{x} + B\,\boldsymbol{u} + [\,E \quad I_n\,]\,\tilde{\boldsymbol{w}}, \\ \boldsymbol{y} = C_1\,\boldsymbol{x} \qquad\quad + [\,D_1 \quad 0\,]\,\tilde{\boldsymbol{w}}, \end{cases} \tag{8.18}$$

in conformity with the structures of C_1 and D_1 in (8.17), i.e.,

$$\begin{cases} \begin{pmatrix} \dot{\boldsymbol{x}}_1 \\ \dot{\boldsymbol{x}}_2 \end{pmatrix} = \begin{bmatrix} A_{11} & A_{12} \\ A_{21} & A_{22} \end{bmatrix} \begin{pmatrix} \boldsymbol{x}_1 \\ \boldsymbol{x}_2 \end{pmatrix} + \begin{bmatrix} B_1 \\ B_2 \end{bmatrix} \boldsymbol{u} + \begin{bmatrix} E_1 & I_k & 0 \\ E_2 & 0 & I_{n-k} \end{bmatrix} \tilde{\boldsymbol{w}}, \\ \begin{pmatrix} \boldsymbol{y}_0 \\ \boldsymbol{y}_1 \end{pmatrix} = \begin{bmatrix} 0 & C_{1,02} \\ I_k & 0 \end{bmatrix} \begin{pmatrix} \boldsymbol{x}_1 \\ \boldsymbol{x}_2 \end{pmatrix} \qquad\qquad + \begin{bmatrix} D_{1,0} & 0 & 0 \\ 0 & 0 & 0 \end{bmatrix} \tilde{\boldsymbol{w}}, \end{cases} \tag{8.19}$$

where

$$\tilde{\boldsymbol{w}} = \begin{pmatrix} \boldsymbol{w} \\ \boldsymbol{x}_0 \cdot \delta(t) \end{pmatrix}, \tag{8.20}$$

and where $\delta(t)$ is a unit impulse function. Obviously, $\boldsymbol{y}_1 = \boldsymbol{x}_1$ is directly available and hence need not be estimated. Next, we define Σ_{QR} to be characterized by

$$(A_{\mathrm{R}}, E_{\mathrm{R}}, C_{\mathrm{R}}, D_{\mathrm{R}}) = \left(A_{22}, [\,E_2 \quad 0 \quad I_{n-k}\,], \begin{bmatrix} C_{1,02} \\ A_{12} \end{bmatrix}, \begin{bmatrix} D_{1,0} & 0 & 0 \\ E_1 & I_k & 0 \end{bmatrix} \right).$$

It is again straightforward to verify that Σ_{QR} is right invertible with no finite and infinite zeros. Moreover, $(A_{\mathrm{R}}, C_{\mathrm{R}})$ is detectable if and only if (A, C_1) is detectable. We are ready to present the following algorithm:

STEP R.1: We first construct an appropriate state feedback gain matrix as in (8.13), i.e.,

$$\boldsymbol{F}(\varepsilon) = [\,H_0(\varepsilon) \quad \cdots \quad H_{\kappa-1}(\varepsilon) \quad F(\varepsilon)\,], \tag{8.21}$$

and then partition $F(\varepsilon)$ in conformity with $\boldsymbol{x}_1$ and $\boldsymbol{x}_2$ of (8.19) as follows:

$$F(\varepsilon) = [\,F_1(\varepsilon) \quad F_2(\varepsilon)\,]. \tag{8.22}$$

STEP R.2: Let K_{R} be an appropriate dimensional constant matrix such that the eigenvalues of

$$A_{\mathrm{R}} + K_{\mathrm{R}} C_{\mathrm{R}} = A_{22} + [\,K_{\mathrm{R}0} \quad K_{\mathrm{R}1}\,] \begin{bmatrix} C_{1,02} \\ A_{12} \end{bmatrix} \tag{8.23}$$

are all in the left half complex plane, i.e., it is a stable matrix. This can be done because $(A_{\mathrm{R}}, C_{\mathrm{R}})$ is detectable.

STEP R.3: Let

$$G_{\mathrm{R}} = [\,-K_{\mathrm{R}0}, \quad A_{21} + K_{\mathrm{R}1}A_{11} - (A_{\mathrm{R}} + K_{\mathrm{R}}C_{\mathrm{R}})K_{\mathrm{R}1}\,] \tag{8.24}$$

and

$$\left.\begin{aligned}
A_{\mathrm{cmp}}(\varepsilon) &= A_{\mathrm{R}} + B_2F_2(\varepsilon) + K_{\mathrm{R}}C_{\mathrm{R}} + K_{\mathrm{R}1}B_1F_2(\varepsilon),\\
B_{\mathrm{cmp}}(\varepsilon) &= G_{\mathrm{R}} + (B_2 + K_{\mathrm{R}1}B_1)\,[\,0, \quad F_1(\varepsilon) - F_2(\varepsilon)K_{\mathrm{R}1}\,],\\
C_{\mathrm{cmp}}(\varepsilon) &= F_2(\varepsilon),\\
D_{\mathrm{cmp}}(\varepsilon) &= [\,0, \quad F_1(\varepsilon) - F_2(\varepsilon)K_{\mathrm{R}1}\,].
\end{aligned}\right\} \tag{8.25}$$

STEP R.4: The reduced-order measurement feedback control law that solves the RPT control problem for the given system is then given by

$$\left\{\begin{aligned}
\dot{\boldsymbol{v}} &= A_{\mathrm{cmp}}(\varepsilon)\boldsymbol{v} + B_{\mathrm{cmp}}(\varepsilon)\boldsymbol{y} + G_0(\varepsilon)\boldsymbol{r} + \cdots + G_{\kappa-1}(\varepsilon)\boldsymbol{r}^{(\kappa-1)},\\
\boldsymbol{u} &= C_{\mathrm{cmp}}(\varepsilon)\boldsymbol{v} + D_{\mathrm{cmp}}(\varepsilon)\boldsymbol{y} + H_0(\varepsilon)\boldsymbol{r} + \cdots + H_{\kappa-1}(\varepsilon)\boldsymbol{r}^{(\kappa-1)},
\end{aligned}\right. \tag{8.26}$$

where $G_i(\varepsilon) = (B_2 + K_{\mathrm{R}1}B_1)H_i(\varepsilon)$, $i = 0, 1, \ldots, \kappa - 1$.

8.3 Outer-Loop Control System Design

As depicted in Fig. 8.1, the outer loop of the flight control system is for controlling the position of the unmanned rotorcraft, i.e., $\mathbf{P}_{\mathrm{n}}$. In Fig. 8.1, the inner-loop command generator is computed as the following:

$$\begin{pmatrix}\delta_{\mathrm{r}}\\ \phi_{\mathrm{r}}\\ \theta_{\mathrm{r}}\end{pmatrix} = \left(G_{\mathrm{in,cl},0}^{-1}\right)\mathbf{a}_{\mathrm{b,r}}, \tag{8.27}$$

where $G_{\mathrm{in,cl},0}$ is the DC gain of the inner closed-loop system with its input variables being δ_{r}, ϕ_{r}, and θ_{r} and its output variables being a_{x}, a_{y}, and a_{z}, respectively. For SheLion and HeLion with the H_∞ inner-loop controller as given in Chap. 7, the resulting DC gain is given by

$$G_{\mathrm{in,cl},0}^{-1} = \begin{bmatrix} -0.0001 & 0.0019 & 0.0478\\ 0.0022 & -0.1031 & -0.0048\\ 0.1022 & 0 & 0.0002 \end{bmatrix}. \tag{8.28}$$

For the outer-loop control system design, we treat the closed inner-loop and the inner-loop command generator, i.e., the portion inside the dashed box in Fig. 8.1, as a *virtual actuator* (such a design idea is illustrated in Fig. 8.2 for easy reference). Our design will work properly, if $\mathbf{a}_{\mathrm{n,r}}$ with frequencies in the working range of the outer loop is able to freely pass through the virtual actuator. It indeed turns out to be the case.

Shown in Figs. 8.3 and 8.4 are the frequency responses of the linearized model of the virtual actuator, which clearly indicate that all its three channels are almost perfectly decoupled. Moreover, the characteristics of both the X- and Y-channels are of low-pass systems with cutoff frequencies around 1 rad/s, whereas the Z-channel

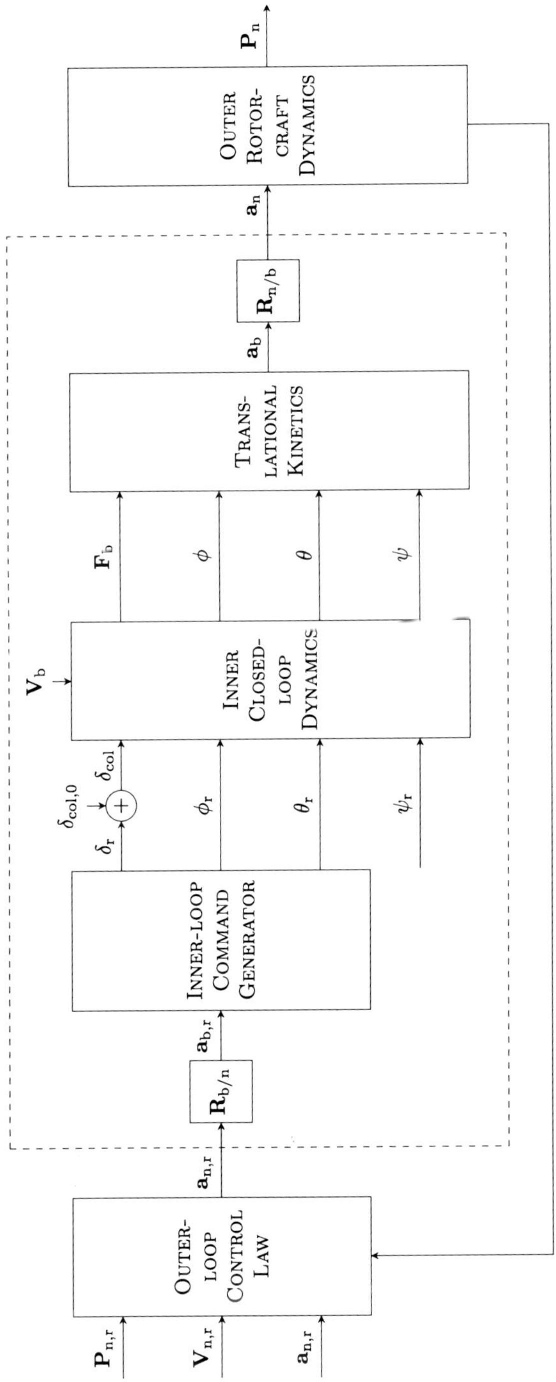

Fig. 8.1 Outer loop of the rotorcraft flight control system

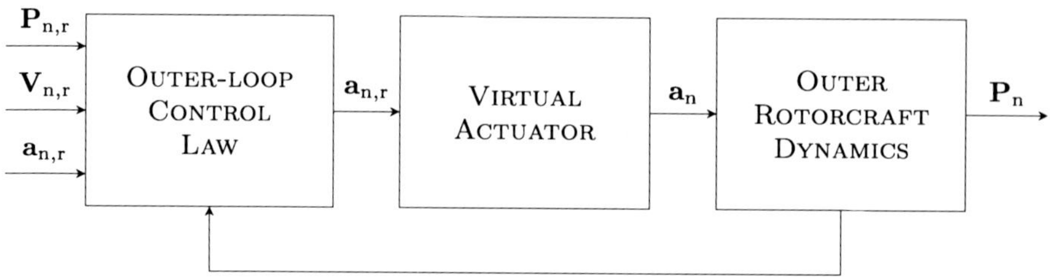

Fig. 8.2 Reconfiguration of the outer-loop flight control system

is an all-pass system. As such, it is pretty safe for us to separate the outer rotorcraft dynamics into three decoupled channels, respectively, in the X-, Y-, and Z-axes of the local NED frame, with each channel being characterized by a double integrator, provided that the actual working frequency of the outer loop is kept within the bandwidth of the virtual actuator, i.e., 1 rad/s. More specifically, in such a situation, the dynamical equation for the X-axis can be expressed as

$$\begin{pmatrix} \dot{x}_n \\ \dot{u}_n \end{pmatrix} = \begin{bmatrix} 0 & 1 \\ 0 & 0 \end{bmatrix} \begin{pmatrix} x_n \\ u_n \end{pmatrix} + \begin{bmatrix} 0 \\ 1 \end{bmatrix} a_{x,n}, \tag{8.29}$$

where x_n is the X-axis position of the UAV in the local NED frame, and u_n and $a_{x,n}$ are respectively the local NED velocity and acceleration projected onto the X-axis. Similarly, the dynamical equations for the Y- and Z-axes are given by

$$\begin{pmatrix} \dot{y}_n \\ \dot{v}_n \end{pmatrix} = \begin{bmatrix} 0 & 1 \\ 0 & 0 \end{bmatrix} \begin{pmatrix} y_n \\ v_n \end{pmatrix} + \begin{bmatrix} 0 \\ 1 \end{bmatrix} a_{y,n} \tag{8.30}$$

and

$$\begin{pmatrix} \dot{z}_n \\ \dot{w}_n \end{pmatrix} = \begin{bmatrix} 0 & 1 \\ 0 & 0 \end{bmatrix} \begin{pmatrix} z_n \\ w_n \end{pmatrix} + \begin{bmatrix} 0 \\ 1 \end{bmatrix} a_{z,n}, \tag{8.31}$$

respectively, with all of its state and control variables defined in the same fashion as those in (8.29). We should note that in the unmanned rotorcraft system, its position, velocity, and acceleration are all measurable and available for feedback control. As all three channels share the same dynamical structure, we proceed in what follows to focus on the design of the outer-loop controller for the X-axis only using the RPT control technique introduced in the previous section. The controllers for the Y-axis and the heave direction can be carried out with the same procedure.

To control the position of the UAV, we defined the controlled output associated with (8.29) as

$$h_x = x_n = [\,1 \quad 0\,] \begin{pmatrix} x_n \\ u_n \end{pmatrix}. \tag{8.32}$$

It is straightforward to verify that the RPT control problem for the given system comprising (8.29) and (8.32) is solvable under state feedback. Since the position

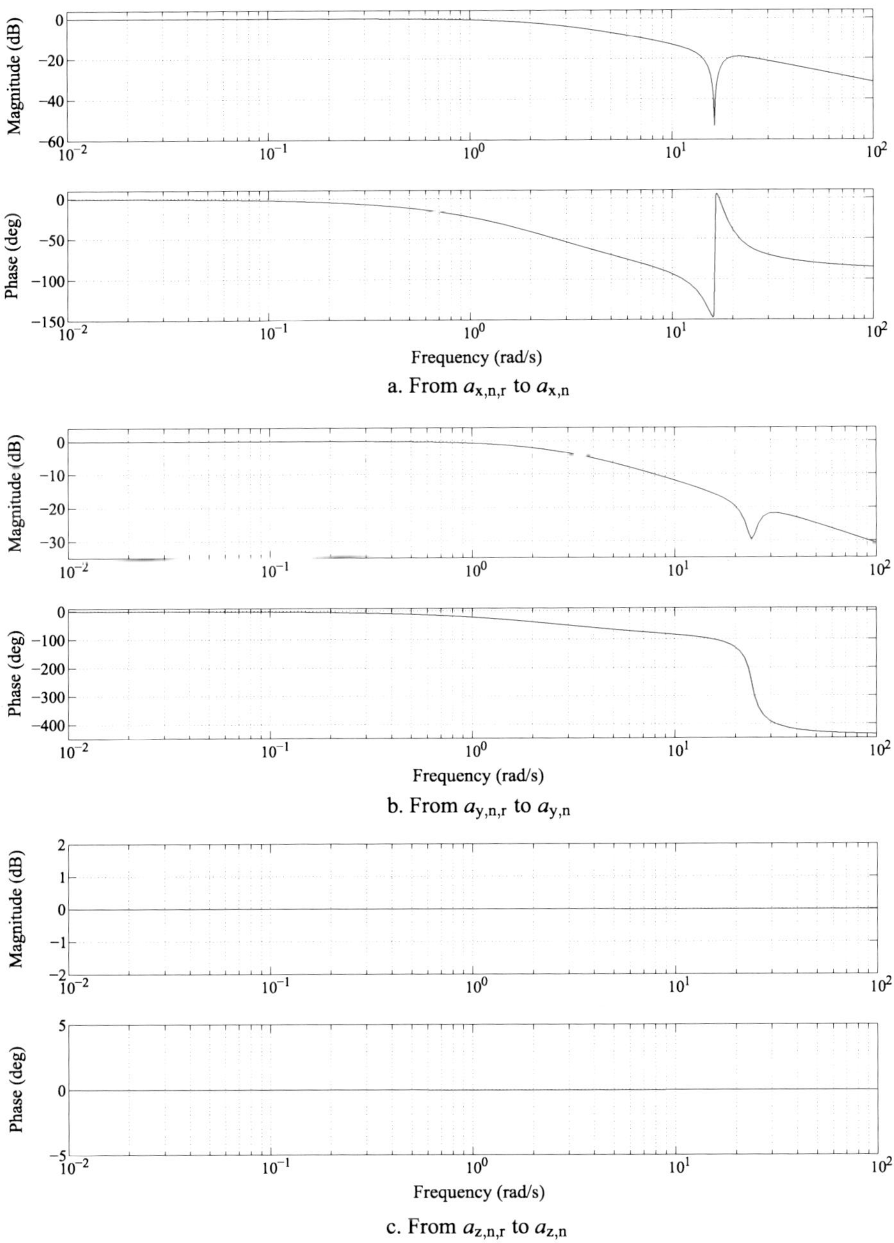

Fig. 8.3 Frequency responses of the main channels of the virtual actuator

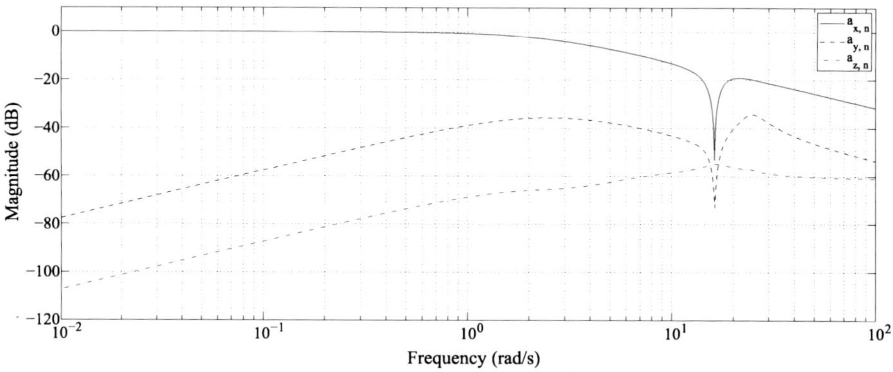

a. From $a_{\mathrm{x,n,r}}$ to all output channels

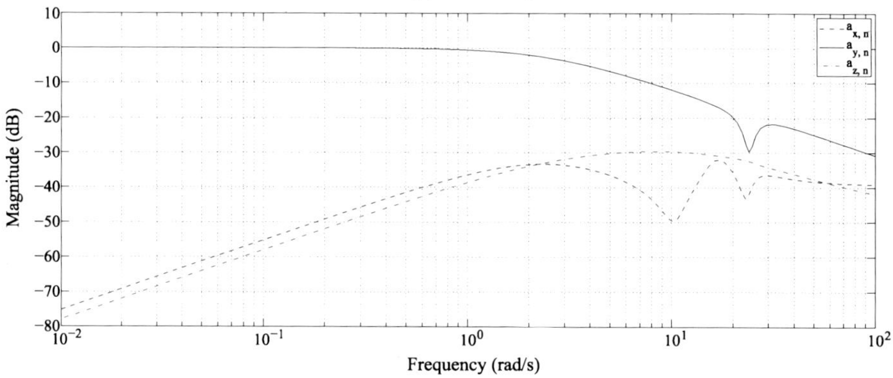

b. From $a_{\mathrm{y,n,r}}$ to all output channels

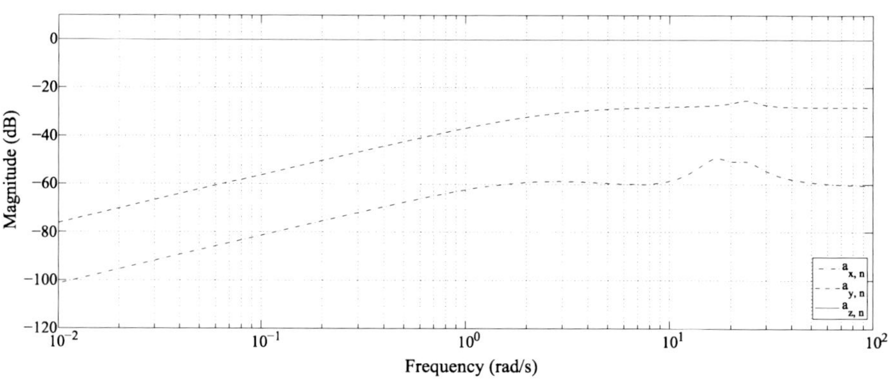

c. From $a_{\mathrm{z,n,r}}$ to all output channels

Fig. 8.4 Magnitude responses of all channels of the virtual actuator

reference $x_{\mathrm{n,r}}$ and its associated velocity, $u_{\mathrm{n,r}}$, and acceleration, $a_{\mathrm{x,n,r}}$, are all available, we formulate the problem into the RPT design framework by defining

$$\frac{\mathrm{d}}{\mathrm{d}t}\begin{pmatrix} x_{\mathrm{n,r}} \\ u_{\mathrm{n,r}} \\ a_{\mathrm{x,n,r}} \end{pmatrix} = \begin{bmatrix} 0 & 1 & 0 \\ 0 & 0 & 1 \\ 0 & 0 & 0 \end{bmatrix}\begin{pmatrix} x_{\mathrm{n,r}} \\ u_{\mathrm{n,r}} \\ a_{\mathrm{x,n,r}} \end{pmatrix} + \begin{bmatrix} 0 \\ 0 \\ 1 \end{bmatrix}\dot{a}_{\mathrm{x,n,r}}. \tag{8.33}$$

We obtain an augmented system of the following form:

$$\Sigma_{\mathrm{AUG}}: \quad \begin{cases} \dot{\boldsymbol{x}}_{\mathrm{x}} = \boldsymbol{A}_{\mathrm{x}}\ \boldsymbol{x}_{\mathrm{x}} + \boldsymbol{B}_{\mathrm{x}}\, a_{\mathrm{x,n}} + \boldsymbol{E}_{\mathrm{x}}\, \boldsymbol{w}_{\mathrm{x}}, \\ \boldsymbol{y}_{\mathrm{x}} = \boldsymbol{x}_{\mathrm{x}}, \\ \boldsymbol{e}_{\mathrm{x}} = \boldsymbol{C}_{2,\mathrm{x}}\, \boldsymbol{x}_{\mathrm{x}} \end{cases} \tag{8.34}$$

where

$$\boldsymbol{x}_{\mathrm{x}} := \begin{pmatrix} x_{\mathrm{n,r}} \\ u_{\mathrm{n,r}} \\ a_{\mathrm{x,n,r}} \\ x_{\mathrm{n}} \\ u_{\mathrm{n}} \end{pmatrix}, \qquad \boldsymbol{w}_{\mathrm{x}} := \dot{a}_{\mathrm{x,n,r}}, \tag{8.35}$$

$$\boldsymbol{A}_{\mathrm{x}} = \begin{bmatrix} 0 & 1 & 0 & 0 & 0 \\ 0 & 0 & 1 & 0 & 0 \\ 0 & 0 & 0 & 0 & 0 \\ 0 & 0 & 0 & 0 & 1 \\ 0 & 0 & 0 & 0 & 0 \end{bmatrix}, \qquad \boldsymbol{B}_{\mathrm{x}} = \begin{bmatrix} 0 \\ 0 \\ 0 \\ 0 \\ 1 \end{bmatrix}, \qquad \boldsymbol{E}_{\mathrm{x}} = \begin{bmatrix} 0 \\ 0 \\ 1 \\ 0 \\ 0 \end{bmatrix}, \tag{8.36}$$

and

$$\boldsymbol{C}_{2,\mathrm{x}} = [\,-1 \quad 0 \quad 0 \quad 1 \quad 0\,]. \tag{8.37}$$

Using the procedure given in [27], we are able to obtain a closed-form solution for the state feedback gain for the system of (8.34) that solves the RPT control problem. The closed-form solution is given by

$$a_{\mathrm{x,n}} = \boldsymbol{F}_{\mathrm{x}}\boldsymbol{x}_{\mathrm{x}} \tag{8.38}$$

with

$$\boldsymbol{F}_{\mathrm{x}} = \left[\, -\frac{\omega_{\mathrm{n,x}}^2}{\varepsilon_{\mathrm{x}}^2} \quad -\frac{2\zeta_{\mathrm{x}}\omega_{\mathrm{n,x}}}{\varepsilon_{\mathrm{x}}} \quad \frac{\omega_{\mathrm{n,x}}^2}{\varepsilon_{\mathrm{x}}^2} \quad \frac{2\zeta_{\mathrm{x}}\omega_{\mathrm{n,x}}}{\varepsilon_{\mathrm{x}}} \quad 1 \,\right]. \tag{8.39}$$

Equivalently, we have

$$\begin{aligned} a_{\mathrm{x,n}} &= F_{\mathrm{x}}(\varepsilon_{\mathrm{x}})\begin{pmatrix} x_{\mathrm{n}} \\ u_{\mathrm{n}} \end{pmatrix} + H_{\mathrm{x},0}(\varepsilon_{\mathrm{x}})x_{\mathrm{n,r}} + H_{\mathrm{x},1}(\varepsilon_{\mathrm{x}})u_{\mathrm{n,r}} + H_{\mathrm{x},2}(\varepsilon_{\mathrm{x}})a_{\mathrm{x,n,r}} \\ &= -\left[\, \frac{\omega_{\mathrm{n,x}}^2}{\varepsilon_{\mathrm{x}}^2} \quad \frac{2\zeta_{\mathrm{x}}\omega_{\mathrm{n,x}}}{\varepsilon_{\mathrm{x}}} \,\right]\begin{pmatrix} x_{\mathrm{n}} \\ u_{\mathrm{n}} \end{pmatrix} + \left(\frac{\omega_{\mathrm{n,x}}^2}{\varepsilon_{\mathrm{x}}^2}\right)x_{\mathrm{n,r}} + \left(\frac{2\zeta_{\mathrm{x}}\omega_{\mathrm{n,x}}}{\varepsilon_{\mathrm{x}}}\right)u_{\mathrm{n,r}} + a_{\mathrm{x,n,r}}, \end{aligned} \tag{8.40}$$

where ε_{x} is the tuning parameter, and $\omega_{\mathrm{n,x}}$ and ζ_{x} are respectively the nominal natural frequency and damping ratio associated with the closed-loop system of the

X-axis dynamics. More specifically, the closed-loop eigenvalues of the X-axis dynamical system under the state feedback control are given by

$$-\frac{\zeta_x \omega_{n,x}}{\varepsilon_x} \pm j\frac{\omega_{n,x}\sqrt{1-\zeta_x^2}}{\varepsilon_x}. \tag{8.41}$$

Similarly, following the same procedure, we can obtain the controllers for the Y-axis dynamics and the heave dynamics respectively as

$$a_{y,n} = -\begin{bmatrix} \frac{\omega_{n,y}^2}{\varepsilon_y^2} & \frac{2\zeta_y \omega_{n,y}}{\varepsilon_y} \end{bmatrix} \begin{pmatrix} y_n \\ v_n \end{pmatrix} + \left(\frac{\omega_{n,y}^2}{\varepsilon_y^2}\right) y_{n,r} + \left(\frac{2\zeta_y \omega_{n,y}}{\varepsilon_y}\right) v_{n,r} + a_{y,n,r} \tag{8.42}$$

and

$$a_{z,n} = -\begin{bmatrix} \frac{\omega_{n,z}^2}{\varepsilon_z^2} & \frac{2\zeta_z \omega_{n,z}}{\varepsilon_z} \end{bmatrix} \begin{pmatrix} z_n \\ w_n \end{pmatrix} + \left(\frac{\omega_{n,z}^2}{\varepsilon_z^2}\right) z_{n,r} + \left(\frac{2\zeta_z \omega_{n,z}}{\varepsilon_z}\right) w_{n,r} + a_{z,n,r}. \tag{8.43}$$

We note that, in principle, the RPT controllers above are capable of achieving an arbitrarily fast response if the tuning parameters are chosen to be sufficiently small. However, due to the limitations of the physical system and the constraints of the inner-loop dynamics, the response of the outer-loop system is required to be slower than the bandwidth of the virtual actuator, i.e., 1 rad/s. Based on these guidelines and observation, we select the following outer-loop controller parameters for HeLion and SheLion:

$$\omega_{n,x} = 0.54, \qquad \omega_{n,y} = 0.62, \qquad \omega_{n,z} = 0.78, \tag{8.44}$$

$$\varepsilon_x = \varepsilon_y = \varepsilon_z = 1, \tag{8.45}$$

and

$$\zeta_x = 1, \qquad \zeta_y = 1, \qquad \zeta_z = 1.1. \tag{8.46}$$

We note that in order to minimize overshoots in time-domain responses, the damping ratios for all three channels are selected to be greater than or equal to unity.

8.4 Performance Evaluation

In this section, we evaluate the performance of the RPT outer-loop controllers based on some simulation results. In order to verify the robustness of the outer-loop flight control system, we examine the frequency response of each individual channel of the outer-loop system, respectively shown in Figs. 8.5, 8.6 and 8.7. It is clear that all of the channels have an infinite gain margin and a phase margin greater than 75 degrees. The robustness of the outer-loop control system is excellent.

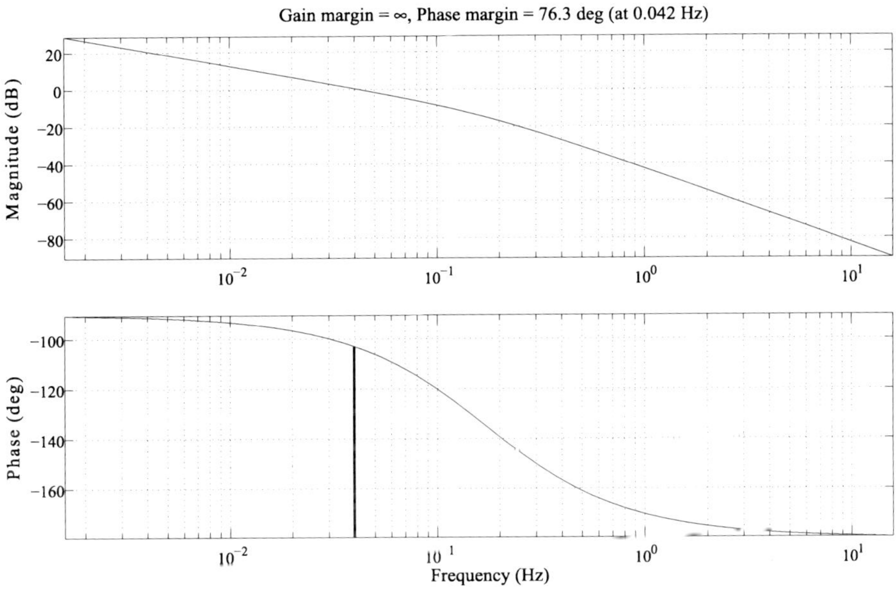

Fig. 8.5 Gain and phase margins of local NED X-axis position control

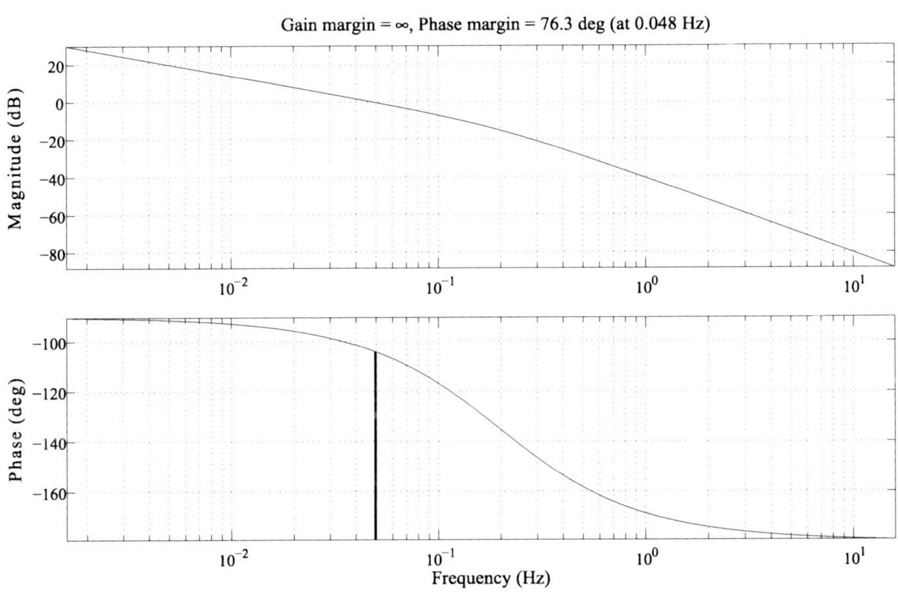

Fig. 8.6 Gain and phase margins of local NED Y-axis position control

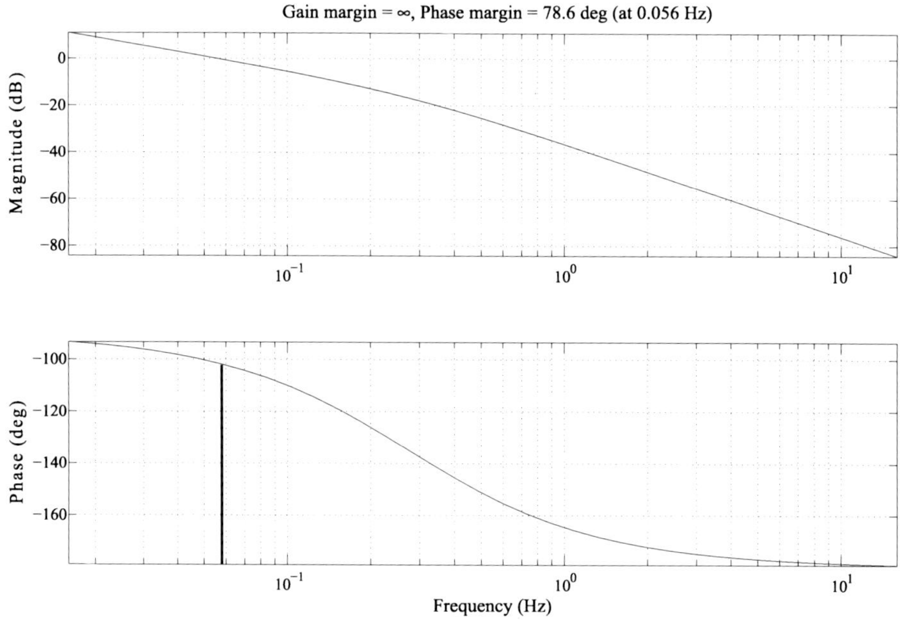

Fig. 8.7 Gain and phase margins of local NED Z-axis position control

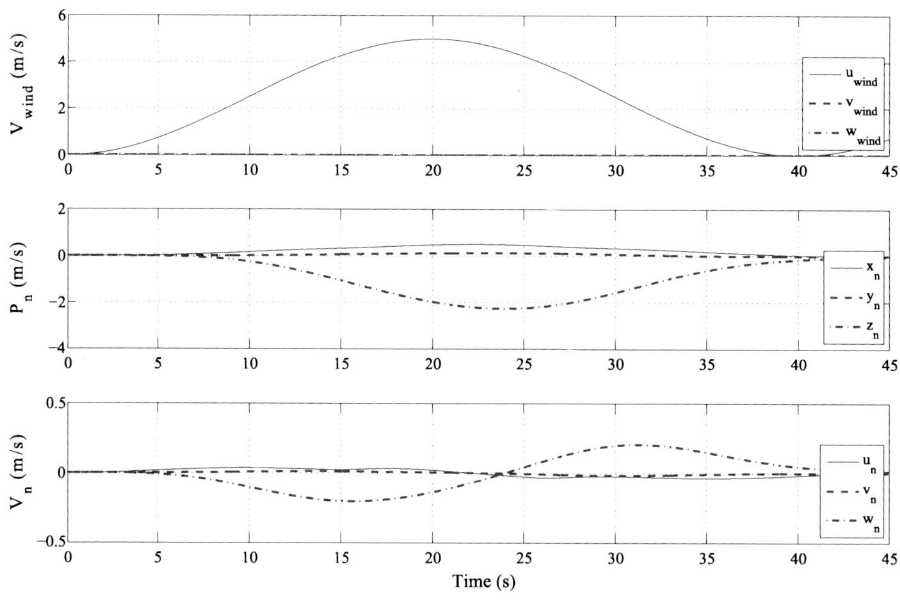

Fig. 8.8 Wind gust disturbance attenuation along local NED X-axis direction

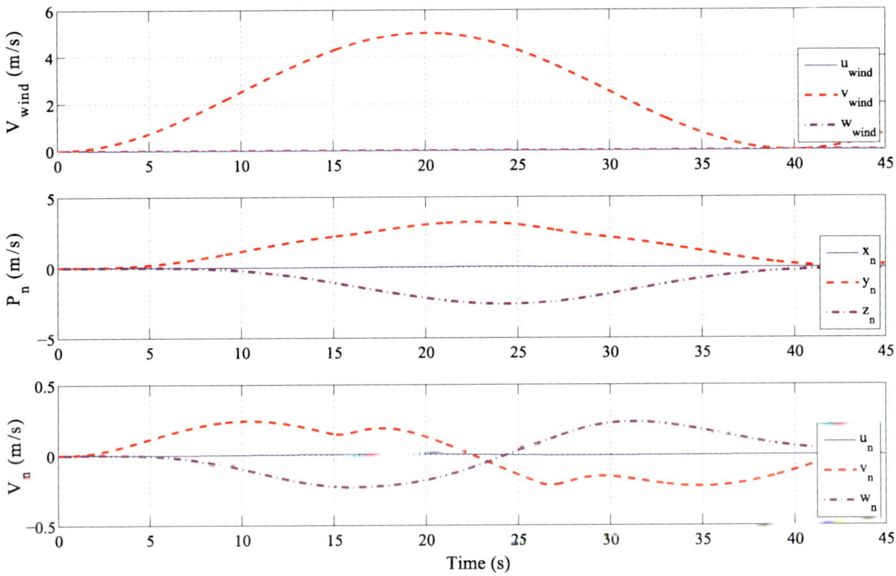

Fig. 8.9 Wind gust disturbance attenuation along local NED Y-axis direction

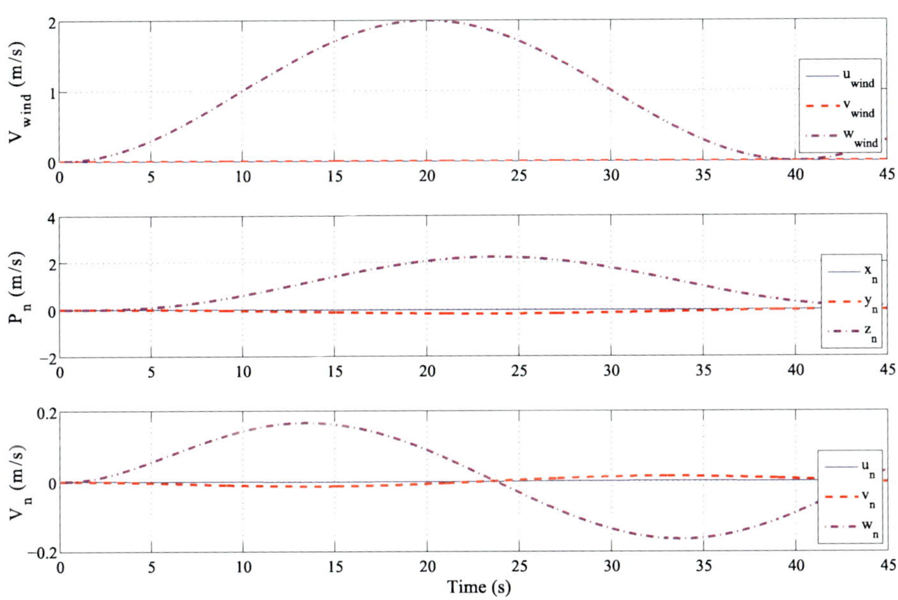

Fig. 8.10 Wind gust disturbance attenuation along local NED Z-axis direction

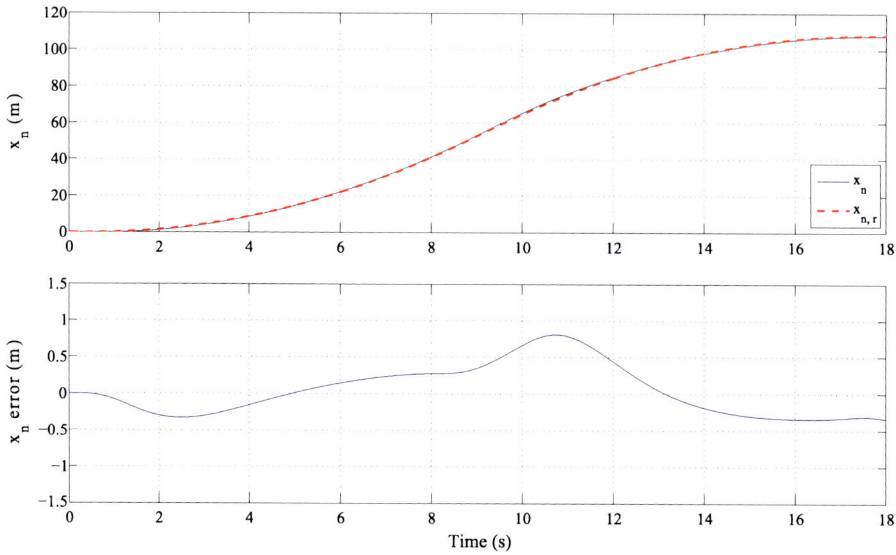

a. Position tracking along local NED X-axis direction

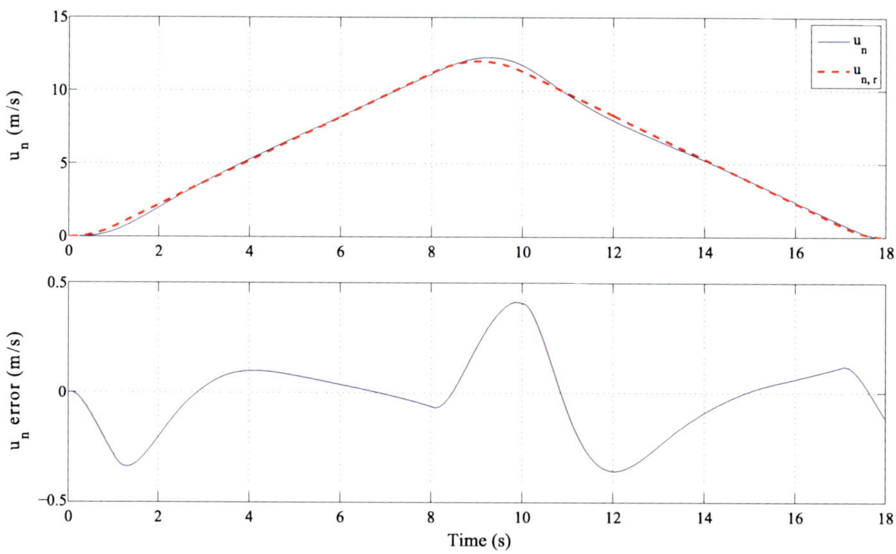

b. Velocity tracking along local NED X-axis direction

Fig. 8.11 Tracking performance evaluation along local NED X-axis direction

For testing wind gust disturbance attenuation and tracking performance of the flight control system, we include the inner-loop controller of Chap. 7. Figures 8.8, 8.9 and 8.10 show the position and velocity hold performance with the overall system due to a wind gust. For tracking performance, we examine three flight motions of our unmanned system. In the first maneuver, the UAV starts with a stable hover

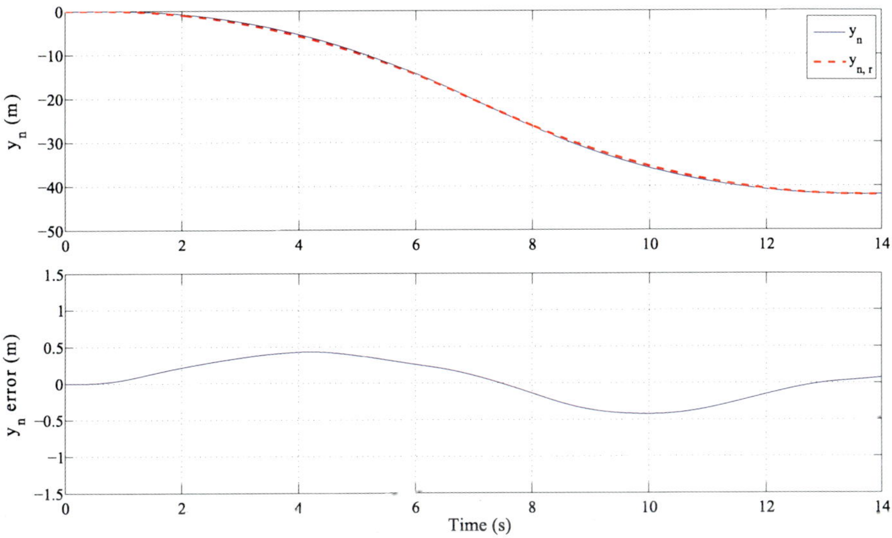

a. Position tracking along local NED Y-axis direction

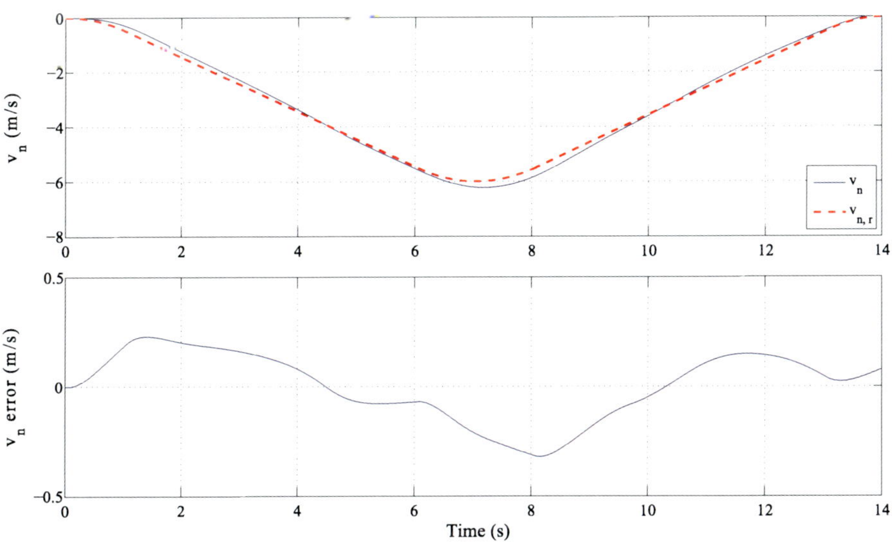

b. Velocity tracking along local NED Y-axis direction

Fig. 8.12 Tracking performance evaluation along local NED Y-axis direction

with heading to the north direction, then conducts a forward acceleration to 12 m/s, and finally decelerates to another stable hover. The other two maneuvers are similar to the first one, but with the acceleration/deceleration directions being changed to the east and upward and with the top sideslip and heave speed being 6 m/s and 2.5 m/s, respectively. Figures 8.11, 8.12 and 8.13 demonstrate the evaluation re-

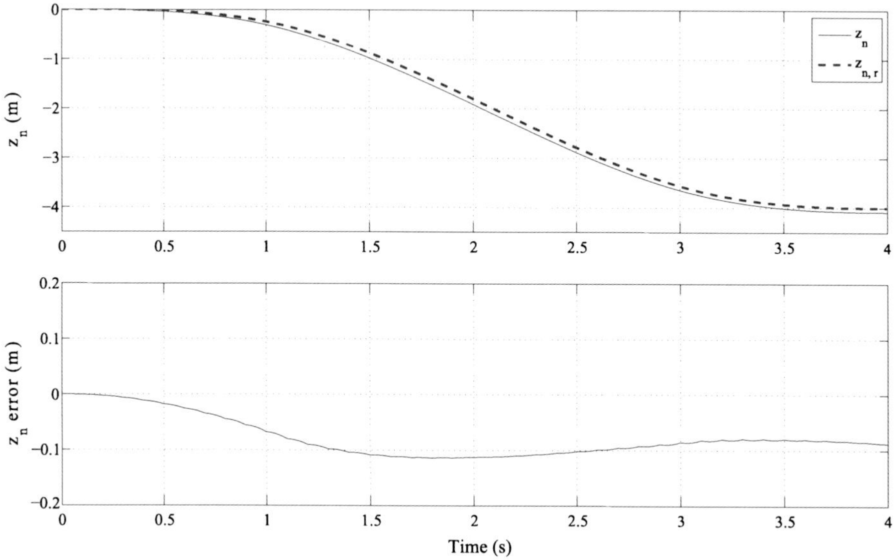

a. Position tracking along local NED Z-axis direction

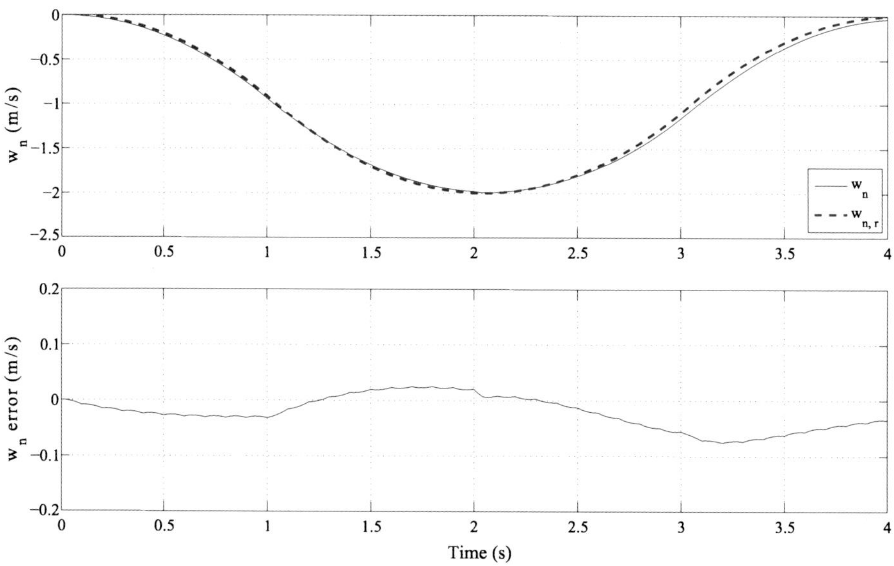

b. Velocity tracking along local NED Z-axis direction

Fig. 8.13 Tracking performance evaluation along local NED Z-axis direction

sults. It is clear that the overall performance is very satisfactory in simulation. More completed tests on the actual flight implementation will be presented in the next chapter.

Chapter 9
Flight Simulation and Experiment

9.1 Introduction

In this chapter, we aim to present the results of a series of simulations and actual flight test experiments to evaluate the performance of the unmanned hardware and software systems constructed earlier together with the automatic flight control laws given in Chaps. 7 and 8. We aim to evaluate the performance and robustness of our unmanned systems by a careful selection of mission-task-elements (MTEs) adopted from ADS-33D-PRF [1], which was originally set for military rotorcraft by US army aviation and used in military-based applications (see, e.g., [175, 176]). Based on the selected MTEs, we generate corresponding appropriate flight trajectories for the outmost layer, i.e., the flight scheduling layer, in our proposed flight control structure (see Fig. 7.1). The actual flight tests are then conducted after intensive simulations executed on our hardware-in-the-loop (HITL) simulation system. The results obtained clearly indicate that our design is very successful. The unmanned rotorcraft system is capable of achieving the desired performance in accordance with the military standard under examination.

We should highlight that the real-time HITL simulation is one of the most effective methods for the verification of the overall performance and safety of the unmanned systems before conducting actual flight tests. In our HITL simulation framework, four modules, which include the onboard hardware system, automatic flight control system, ground control station, and software system, are integrated to realize the simulation. Communications between the onboard system and the ground control station are done through the wireless modules and the actual rotorcraft is substituted with the nonlinear dynamics model obtained in Chap. 6. Such a configuration is proven to be highly effective and useful. It helps us avoid some potential crashes in the real flight experiment. We should also note that the HITL simulation system is to be used as well in testing the cooperative control of the flight formation of multiple unmanned rotorcraft systems, which will be presented later in Chap. 10.

G. Cai et al., *Unmanned Rotorcraft Systems*, Advances in Industrial Control,
DOI 10.1007/978-0-85729-635-1_9,

9.2 Flight Scheduling

As mentioned earlier, the flight scheduling layer generates appropriate flight trajectories for the selected MTEs to evaluate the robustness and handling quality of the overall unmanned systems in their wide operational envelope. In ADS-33D-PRF [1], each MTE is defined in terms of a clear objective, description of maneuver, and performance requirement. For the performance requirement, it is categorized into two qualitative levels, i.e., the *desired level* (satisfactory) and the *adequate level* (barely acceptable). After careful consideration, we have selected the following nine MTEs for testing on our UAVs (based on their appearing sequence in the complete flight trajectory):

1. Depart/abort (forward flight)
2. Hover
3. Depart/abort (backward flight)
4. Hovering turn
5. Vertical maneuver
6. Lateral reposition
7. Turn-to-target
8. Slalom
9. Pirouette

9.2.1 Depart/Abort (Forward Flight)

The depart/abort forward flight is a flight operation with moderate aggressiveness.

1. Objectives: For the depart/abort flight, we mainly examine the UAV helicopter in four aspects, which include (i) testing handling quality and control performance of pitch and heave axes in the transient process, (ii) verifying for the existence of any undesirable coupling between the longitudinal and lateral directions, (iii) obtaining the working performance in the condition with the predefined maximum forward speed, and lastly, (iv) verifying the ability to re-establish automatic hover after the forward flight.
2. Maneuver description: The depart/abort forward flight operation starts from a hover condition at an altitude with good eyesight (ranging from 8 to 10 m based on our own practical experience). The flight path is generally a straight line, with its destination point 204 m away from the starting position. The overall procedure consists of the following three stages: (i) performing an accelerating departure until the predefined top forward speed (12 m/s) is achieved in 9 s, (ii) maintaining a dash with the top forward speed (12 m/s) for 7 s, and (iii) aborting the dash and decelerating to hover at the destination point in 9 s. Both the acceleration and deceleration maneuvers should be accomplished smoothly. Overshoot is not allowed.
3. Performance requirements: See Table 9.1.

Table 9.1 Depart/abort (forward flight) performance requirements

Specification	Desired level	Adequate level
Longitudinal position error	≤ 3 m	$3 \sim 6$ m
Lateral position error	≤ 3 m	$3 \sim 6$ m
Altitude error	≤ 3 m	$3 \sim 5$ m
Heading error	≤ 10 deg	$10 \sim 15$ deg
Time to complete the maneuver	≤ 25 s	$25 \sim 30$ s

Table 9.2 Hover performance requirements

Specification	Desired level	Adequate level
Time to decelerate to a stabilized hover	≤ 3 s	$3 \sim 8$ s
Duration in maintaining a stabilized hover	≥ 30 s	≥ 30 s
Longitudinal position maintaining range	≤ 1.5 m	$1.5 \sim 3$ m
Lateral position maintaining range	≤ 1.5 m	$1.5 \sim 3$ m
Altitude maintaining range	≤ 3 m	$3 \sim 5$ m
Heading maintaining range	≤ 5 deg	$5 \sim 10$ deg

9.2.2 Hover

Hover is the most essential and important flight operation for rotorcraft.

1. Objectives: We aim to examine the ability of transition from translating flight to a stabilized hover with sufficient precision and reasonable aggressiveness and to examine the hovering precision in position and heading, in the presence of wind gusts.
2. Maneuver description: Hover maneuver starts with a deceleration. The initial horizontal speed in the NED frame shall be no more than 2.38 m/s (threshold of the near hover envelope), at an altitude with good eyesight. Hovering shall be accomplished at a certain predefined point with wind gusts blowing from the rear side of the helicopter.
3. Performance requirements: See Table 9.2.

9.2.3 Depart/Abort (Backward Flight)

The depart/abort backward flight is for verifying the flight performance in backward motion with moderate aggressiveness.

1. Objectives: For the depart/abort backward flight, we are to examine the handling quality and control performance of the pitch and heave axes in the transient

Table 9.3 Depart/abort (backward flight) performance requirements

Specification	Desired level	Adequate level
Longitudinal position error	≤ 3 m	$3 \sim 6$ m
Lateral position error	≤ 3 m	$3 \sim 6$ m
Altitude error	≤ 3 m	$3 \sim 6$ m
Heading error	≤ 10 deg	$10 \sim 15$ deg
Time to complete the maneuver	≤ 25 s	$25 \sim 30$ s

process, to examine the existence of any undesirable coupling between the longitudinal and lateral directions, to test the heading control in the condition with middle-speed backward speed, and to verify the ability to re-establish automatic hover from the backward flight.

2. Maneuver description: The backward depart/abort maneuver is very similar to its forward counterpart. More specifically, it starts from a stable hover and then travels a straight-line path to a destination point 80 m away from the initial position. It involves performing a backward accelerating departure until the predefined top backward speed (-4 m/s) is achieved at 5 s, maintaining a dash with the top backward speed (-4 m/s) for another 15 s, and aborting the dash and decelerating to hover at the destination point in 5 s. Again, the acceleration and deceleration maneuvers should be sufficiently smooth and no overshoot is allowed.
3. Performance requirements: See Table 9.3.

9.2.4 Hovering Turn

Hovering turn is a flight operation that can often be seen in practical implementations.

1. Objectives: We aim to check the ability of position recovery, to examine inter-axis coupling between the yaw and heave directions, and to identify any undesirable handling qualities.
2. Maneuver description: The hovering turn flight starts from a stable hover at an altitude with good eyesight, followed by a 270-degree heading turn.
3. Performance requirements: See Table 9.4.

9.2.5 Vertical Maneuver

The vertical maneuver is adopted for examining the controllability of the heave axis.

1. Objectives: We aim to examine the heave damping, i.e., the ability to effectively start and stop a vertical motion, and to identify the existence of any undesirable coupling between the heave channel and the roll, pitch, and yaw channels.

Table 9.4 Hovering turn performance requirements

Specification	Desired level	Adequate level
Longitudinal position maintaining range	≤ 1.5 m	$1.5 \sim 3$ m
Lateral position maintaining range	≤ 1.5 m	$1.5 \sim 3$ m
Altitude maintaining range	≤ 3 m	$3 \sim 6$ m
Heading error	≤ 3 deg	$3 \sim 6$ deg
Time to complete the maneuver	≤ 15 s	$15 \sim 22.5$ s

Table 9.5 Vertical maneuver performance requirements

Specification	Desired level	Adequate level
Longitudinal position error	≤ 1.5 m	$1.5 \sim 3$ m
Lateral position error	≤ 1.5 m	$1.5 \sim 3$ m
Altitude error	≤ 2 m	$2 \sim 4$ m
Heading change	≤ 3 deg	$3 \sim 6$ deg
Time to complete the maneuver	≤ 10 s	$10 \sim 15$ s

2. Maneuver description: The vertical maneuver starts from a stable hover at a good-eyesight altitude. The rotorcraft then performs a vertical ascent of 5 m, hovers for 2 s, and finally descends back to the initial hovering position.
3. Performance requirements: See Table 9.5.

9.2.6 Lateral Reposition

Lateral reposition examines the flight performance in sideslip motion with moderate aggressiveness.

1. Objectives: We aim to examine the handling qualities of the roll and heave axes and to verify the existence of any undesirable coupling between the roll and other axes.
2. Maneuver description: The lateral reposition maneuver starts at hovering with a good-eyesight altitude. The trajectory is a straight path of 84 m. Before starting, the X-axis of the rotorcraft is required to be oriented to 90 degrees to the flight path. The rotorcraft is required to first perform an accelerating sideslip until it reaches the predefined top lateral speed (-6 m/s) in 7 s, then to maintain the sideslip with -6 m/s for 7 s, and finally to decelerate to hover at the destination in another 7 s. Similar to the aforementioned depart/abort MTEs, the acceleration and deceleration process should be smooth and no overshoot is allowed.
3. Performance requirements: See Table 9.6.

Table 9.6 Lateral reposition performance requirements

Specification	Desired level	Adequate level
Longitudinal position error	≤ 3 m	$3 \sim 6$ m
Lateral position error	≤ 3 m	$3 \sim 6$ m
Altitude error	≤ 3 m	$3 \sim 6$ m
Heading error	≤ 10 deg	$10 \sim 15$ deg
Time to complete the maneuver	≤ 21 s	$21 \sim 25$ s

Table 9.7 Turn-to-target performance requirements

Specification	Desired level	Adequate level
Longitudinal position maintaining range	≤ 2 m	$2 \sim 4$ m
Lateral position maintaining range	≤ 2 m	$2 \sim 4$ m
Altitude maintaining range	≤ 2 m	$2 \sim 4$ m
Heading error	≤ 3 deg	≤ 3 deg
Time to complete the maneuver	≤ 5 s	$5 \sim 10$ s

9.2.7 Turn-to-Target

Turn-to-target is similar to hovering turn but with a rapid yaw rate.

1. Objectives: We aim to verify the existence of any undesirable handling qualities or inter-axis coupling in the rapid hovering turn and to test the precision of heading maintenance after recovering from a rapid hovering turn.
2. Maneuver description: The turn-to-target maneuver starts from hovering at a good-eyesight attitude. The helicopter is required to complete a 180-degree rapid turn either clockwise or counter-clockwise and re-achieve a stable hover at the same point where it started.
3. Performance requirements: See Table 9.7.

9.2.8 Slalom

The slalom flight simultaneously examines the forward and sideslip control performances with moderate aggressiveness.

1. Objectives: We aim to examine both agility and tracking performance, the coordination of the lateral movement in moderately aggressive forward flight, and the existence of any undesirable inter-axis coupling in the selected maneuver.
2. Maneuver description: The slalom operation starts from a stable hover, followed by acceleration to a predefined cruising speed (6 m/s). The rotorcraft then per-

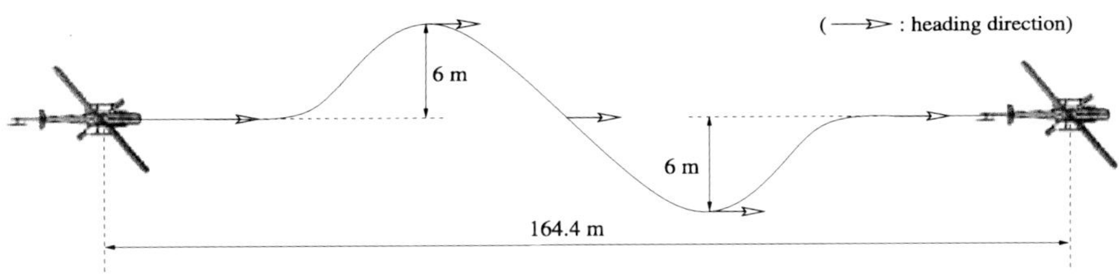

Fig. 9.1 Illustration of the slalom maneuver

Table 9.8 Slalom performance requirements

Specification	Desired level	Adequate level
Maintaining forward speed	≥ 6 m/s	$3 \sim 5$ m/s
Longitudinal position error	≤ 2 m	$2 \sim 4$ m
Lateral position error	≤ 2 m	$2 \sim 4$ m
Altitude error	< 3 m	$3 \sim 6$ m
Heading error	≤ 10 deg	$10 \sim 15$ deg

forms one round of the slalom movement as depicted in Fig. 9.1. The slalom path is 164.4 m in length and ± 6 m in the lateral shift. During this maneuver, the heading angle is kept unchanged. After completing the slalom path, the rotorcraft is decelerated to a stable hover.

3. Performance requirements: See Table 9.8.

9.2.9 Pirouette

Pirouette motion is an essential maneuver in reconnaissance, rescue, and ground target tracking.

1. Objective: We aim to examine the ability of the helicopters to accomplish precision control in the roll, pitch, yaw, and heave axes simultaneously.
2. Maneuver description: The pirouette maneuver starts from hovering, at an altitude of good eyesight. As shown in Fig. 9.2, the helicopter is required to complete a lateral translation of a circle (clockwise or counter-clockwise) with a radius of 10 m with a constant lateral speed in the body frame. The nose of the helicopter should be kept pointing to the center of the circle throughout the maneuver.
3. Performance requirements: See Table 9.9.

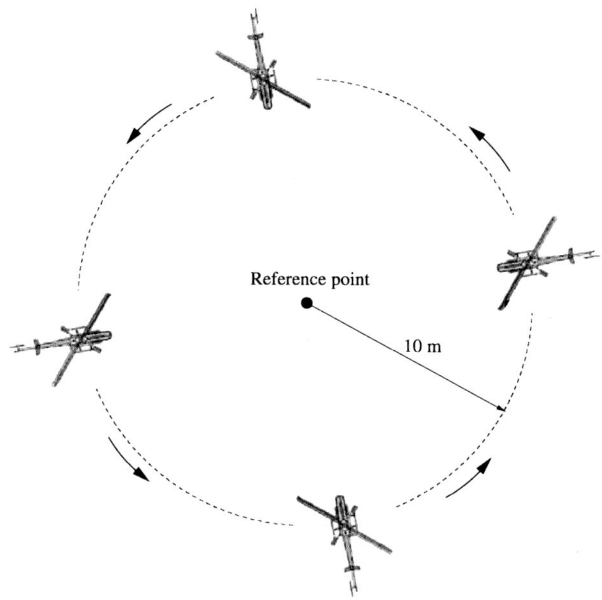

Fig. 9.2 Illustration of pirouette maneuver

9.2.10 MTE Concatenation

All the selected MTEs are concatenated to form the complete flight trajectory. We intentionally add several transient stages (either hover or heading angle adjustment) for easy extraction of data for post-flight analysis. The total operation time of the entire flight experiment lasts 265 s, with the specific time allocations being listed in Table 9.10.

9.3 Hardware-in-the-Loop Simulation Setup

Before conducting actual flight tests, we have to carry out a series of hardware-in-the-loop (HITL) simulations, in which the overall unmanned system is maximally activated. Using such a framework, we can effectively evaluate the reliability and

Table 9.9 Pirouette performance requirements

Specification	Desired level	Adequate level
Longitudinal position error	≤ 2 m	$2 \sim 4$ m
Lateral position error	≤ 2 m	$2 \sim 4$ m
Altitude error	≤ 2 m	$2 \sim 4$ m
Heading error	≤ 10 deg	$10 \sim 15$ deg
Time to complete the maneuver	≤ 45 s	$45 \sim 60$ s
Duration to maintain the re-stabilized hover	≥ 5 s	≥ 5 s

Table 9.10 Time allocation of the flight experiment

Flight operation	Time allocation
Depart/abort (forward flight)	25 s
Hover	35 s
Depart/abort (backward flight)	25 s
Transient hover	8 s
Hovering turn	13.5 s
Transient hover	8 s
Vertical maneuver	10 s
Transient hover	8 s
Lateral reposition	21 s
Transient hover	8 s
Transient hovering turn (90 deg)	11 s
Transient hover	5 s
Turn-to-target	5 s
Transient hover	8 s
Slalom	34.5 s
Transient hover	8 s
Pirouette	25 s
Transient hover	7 s

performance of the overall system. As a result, any harmful deficiencies or improper designs can be discovered and the probability of flight accidents can then be minimized. The general framework of the HITL simulation is shown in Fig. 9.3, in which

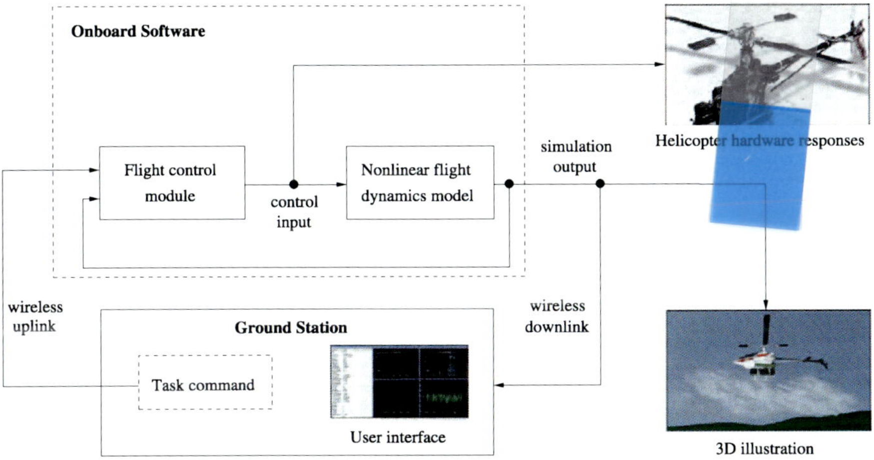

Fig. 9.3 General framework of hardware-in-the-loop simulation experiment

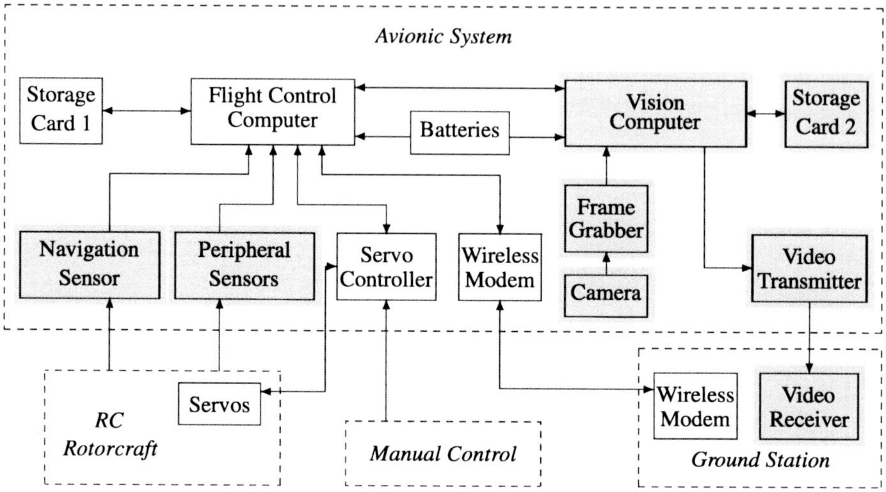

Fig. 9.4 Hardware configuration of hardware-in-the-loop simulation

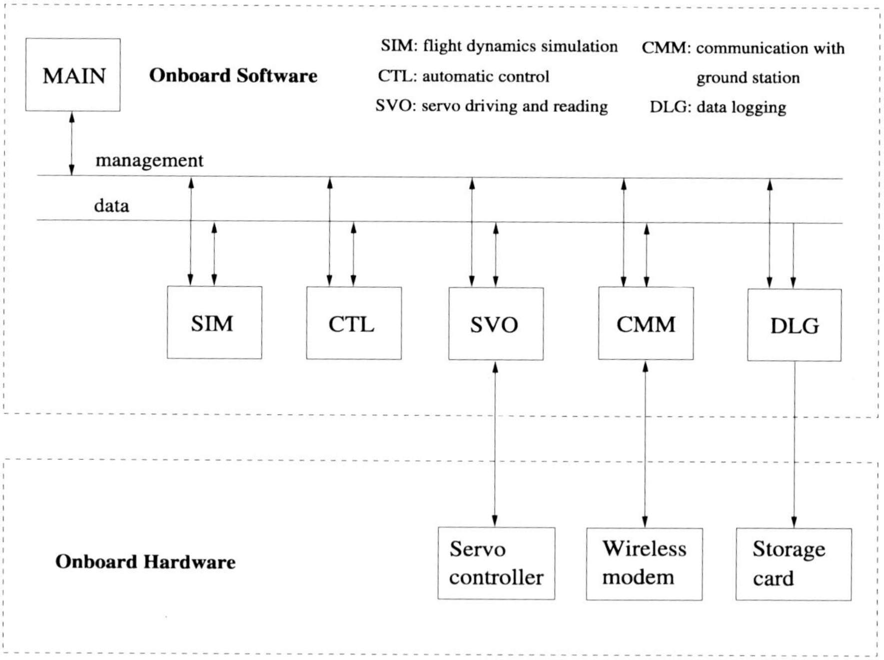

Fig. 9.5 Software configuration of hardware-in-the-loop simulation

all the key components of the unmanned system, including the RC helicopter, the avionic system, and the ground control station, are activated to maximumly emulate the rotorcraft UAV maneuvering in real flight. To carry out the HITL simulation,

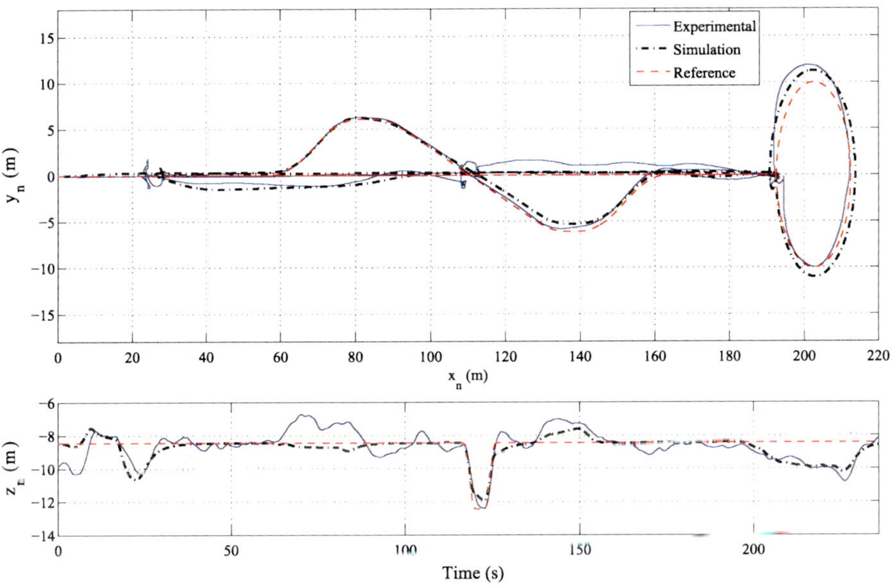

Fig. 9.6 Responses of the wide-envelope flight—position

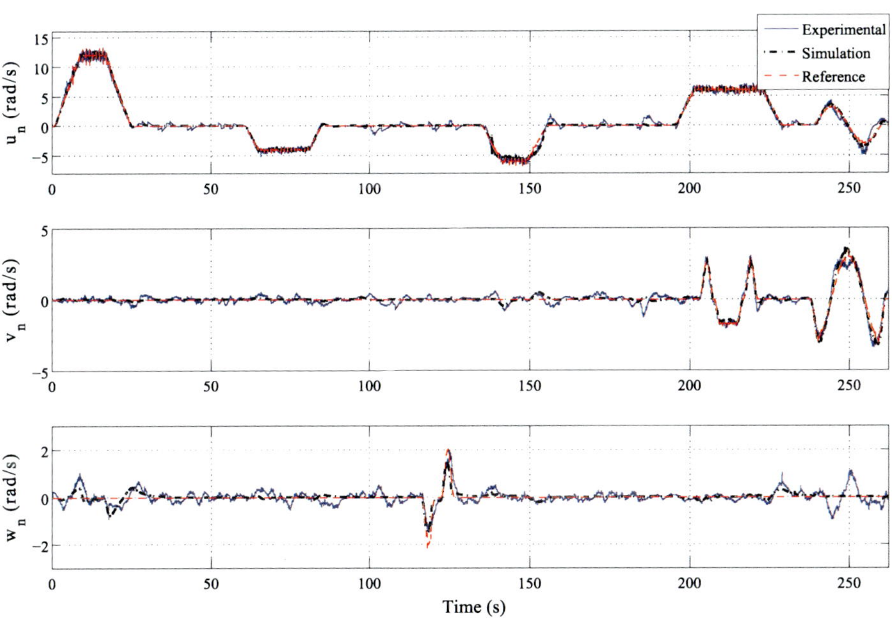

Fig. 9.7 Responses of the wide-envelope flight—velocities

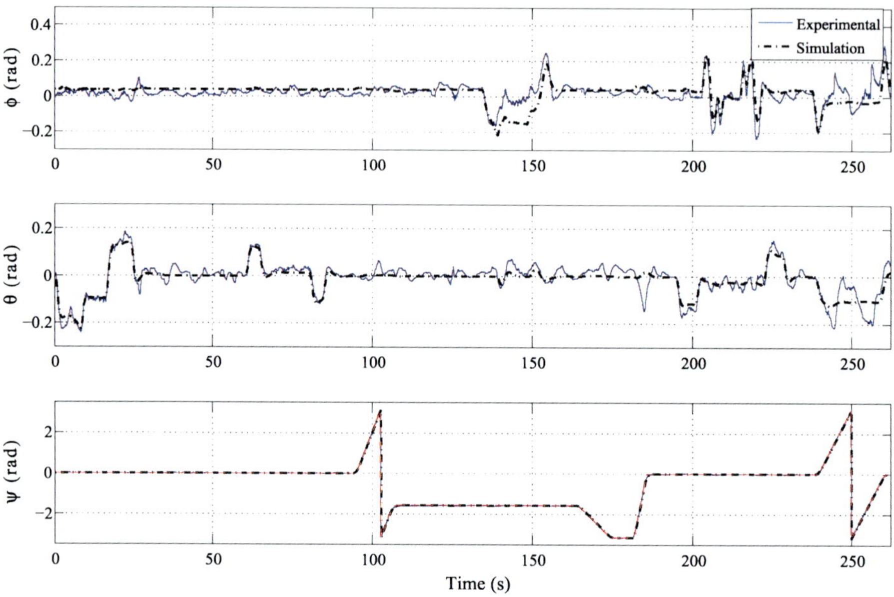

Fig. 9.8 Responses of the wide-envelope flight—Euler angles

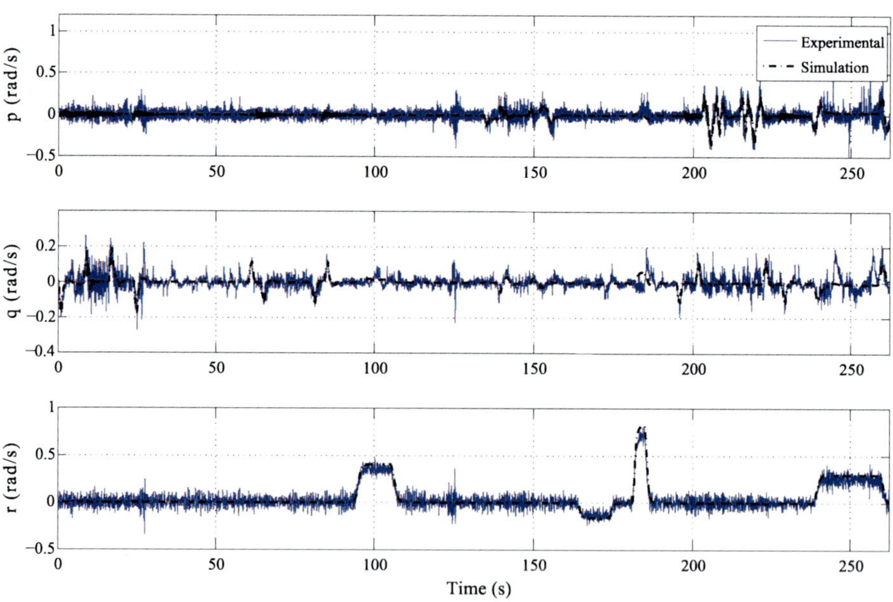

Fig. 9.9 Responses of the wide-envelope flight—angular rates

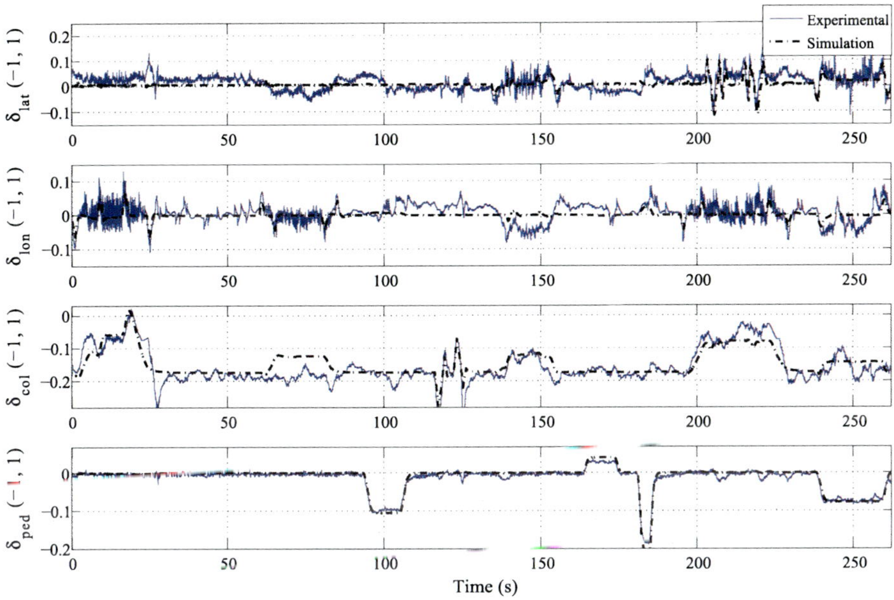

Fig. 9.10 Responses of the wide-envelope flight—control inputs

we modify the hardware configuration of our unmanned system as in Fig. 9.4. The shaded blocks are kept inactive during the HITL simulation.

1. The navigation and peripheral sensors are not activated. The associated data are generated by a flight dynamics software module, named **SIM**, that is programmed based on the nonlinear dynamic model of Chap. 6. The **SIM** module is executed in the onboard flight control computer instead of another simulation computer. With such an arrangement, we can avoid involving extra hardware components and the corresponding data I/O and software programming.
2. The hardware modules related to the vision processing part are kept inactive.
3. Servo actuator activation is essential to the HITL simulation. The deflecting direction and amplitude of the actuator surface are the primary reflection of the performance of automatic flight control law.
4. The manual control function is retained in the HITL simulation. Safety is always one of the most essential issues that shall be faced in conducting actual flight tests. As such, the manual switching-back function (to manual control) is required to be activated and repeatedly examined in the simulation process.
5. The ground control station is naturally included for data display, status monitoring, and command/trajectory uploading purposes.

In accordance with the hardware reconfiguration, we have slightly modified our software system as depicted in Fig. 9.5. Besides the replacement of the **NAV** and **DAQ** modules with **SIM**, the overall structure and execution scheme of the flight control software remain unchanged. We note that the vision processing software is not executed at all at this stage.

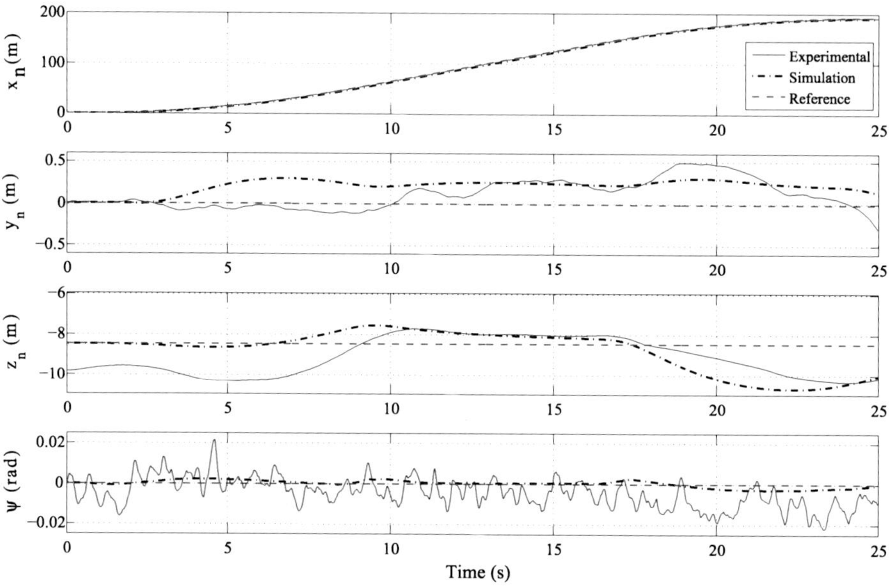

a. Simulation and experimental responses of depart/abort forward flight

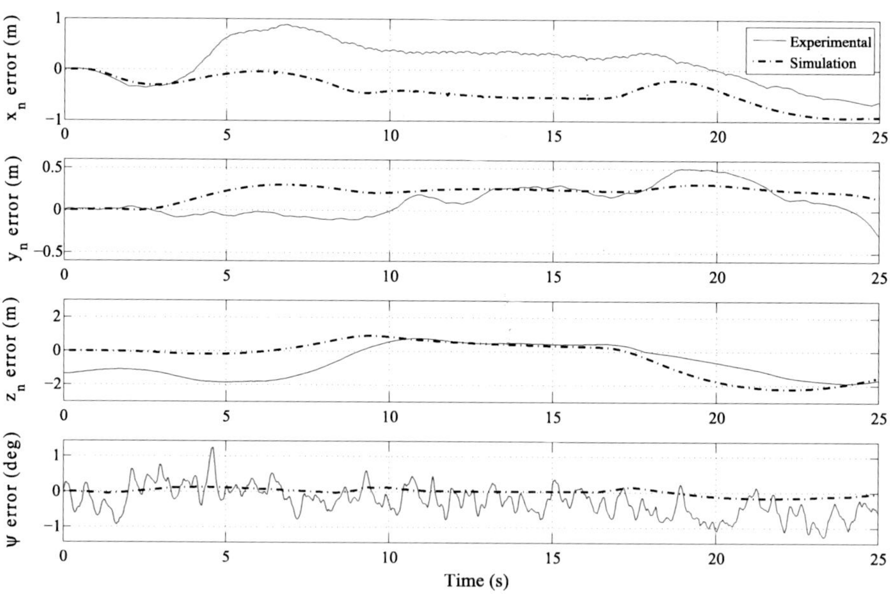

b. Tracking errors of depart/abort forward flight

Fig. 9.11 Depart/abort forward flight results

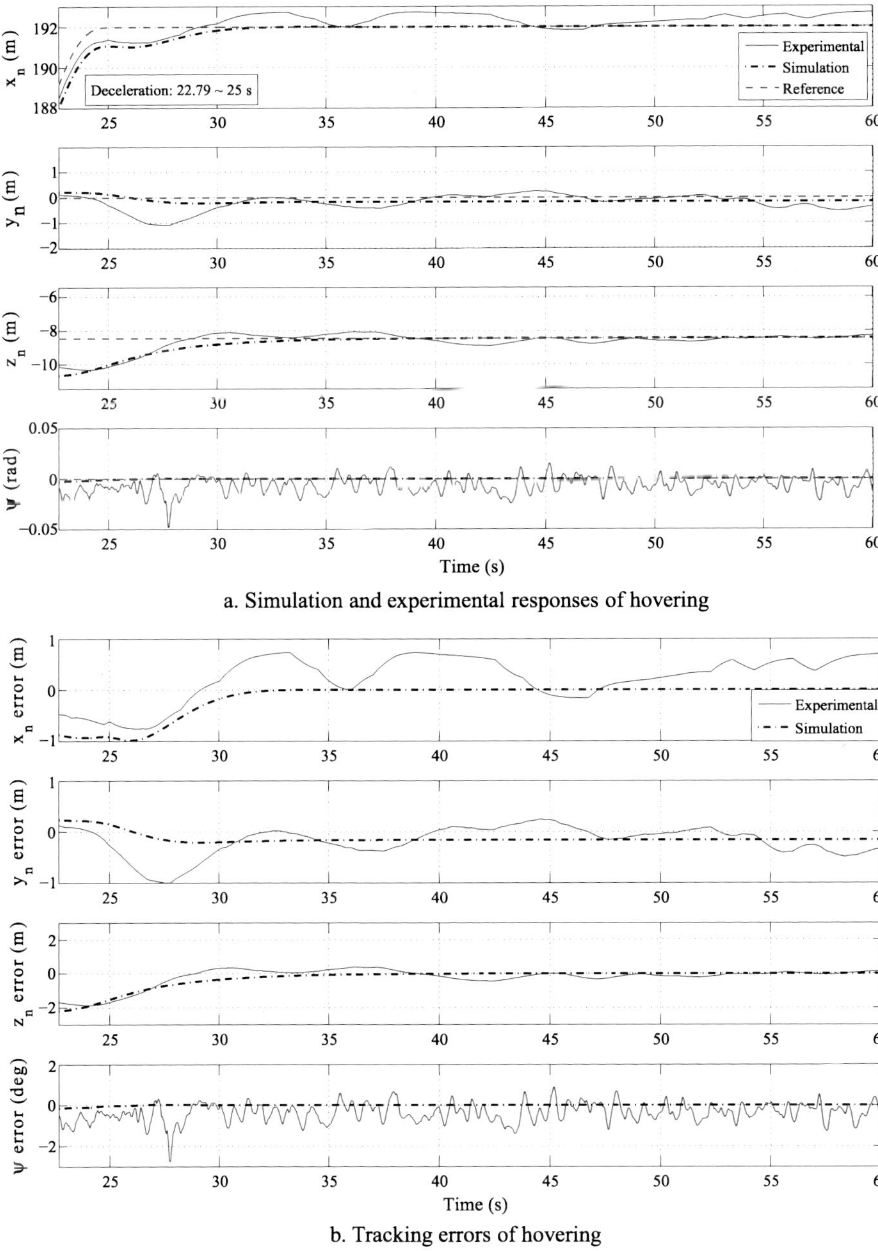

Fig. 9.12 Hovering flight results

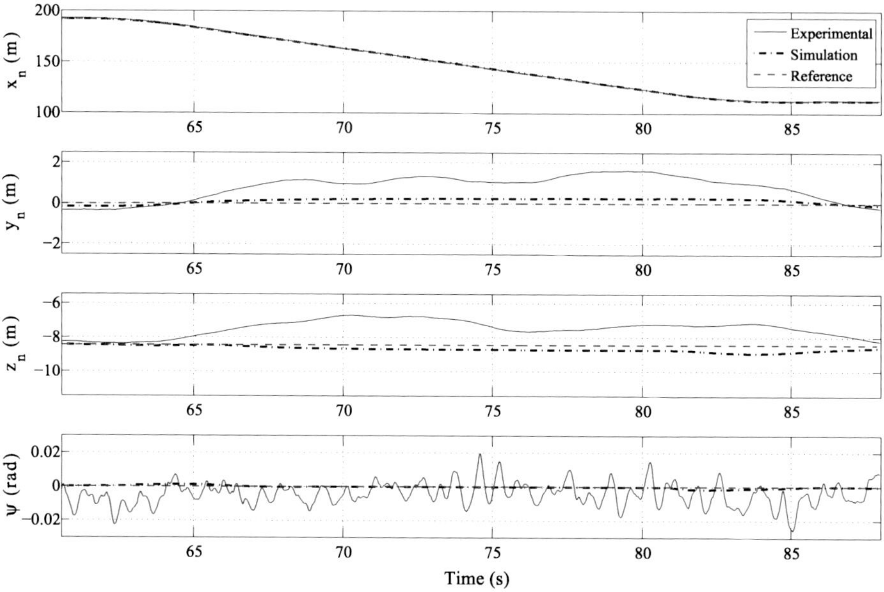

a. Simulation and experimental responses of depart/abort backward flight

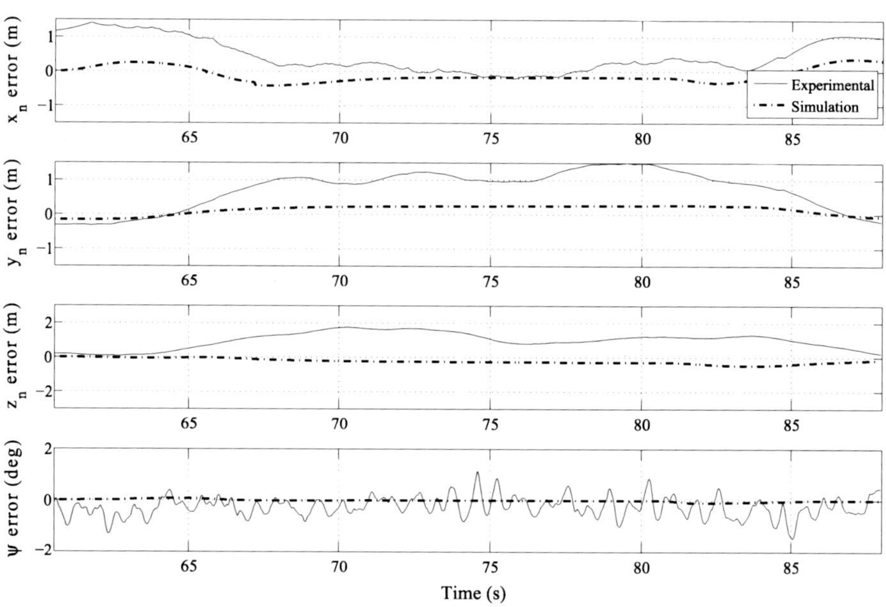

b. Tracking errors of depart/abort backward flight

Fig. 9.13 Depart/abort backward flight results

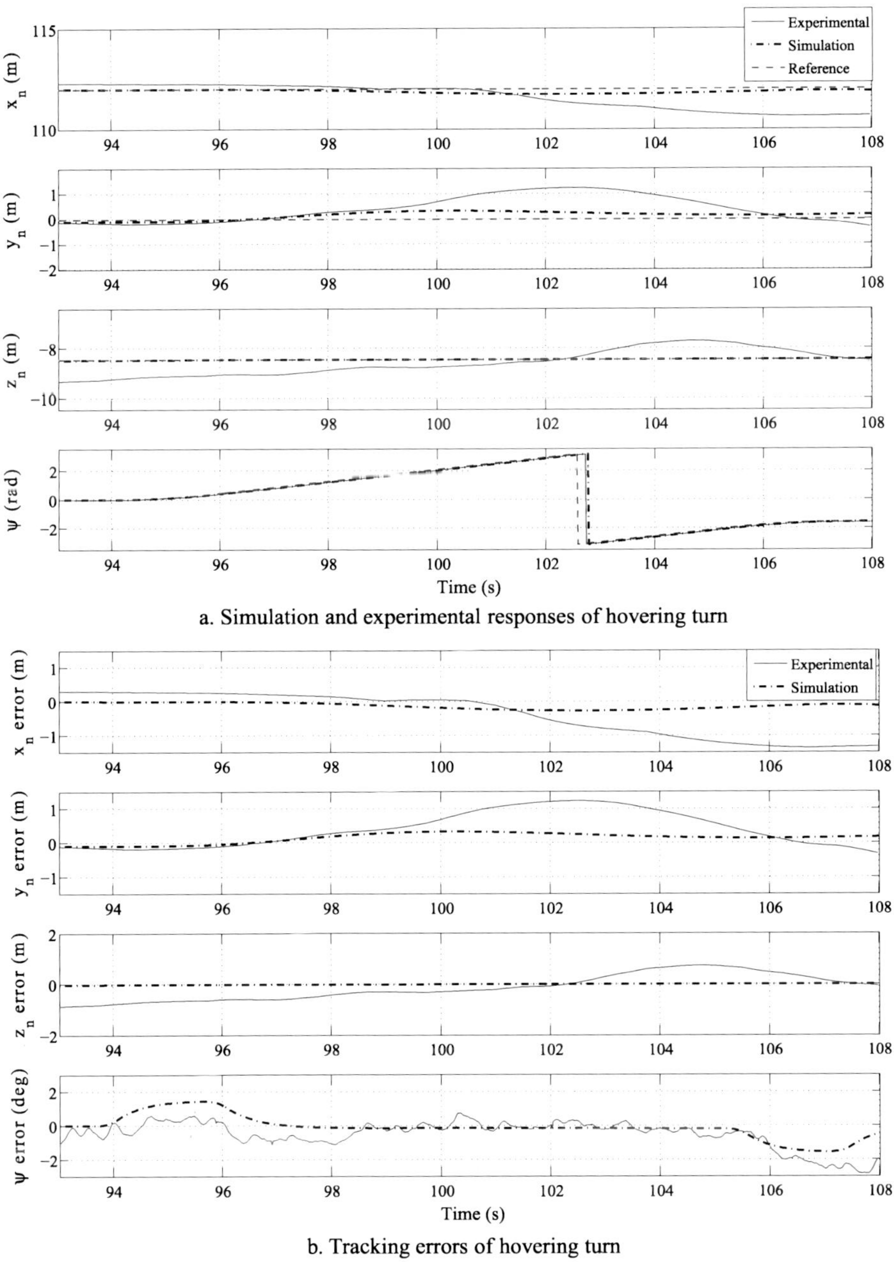

Fig. 9.14 Hovering turn flight results

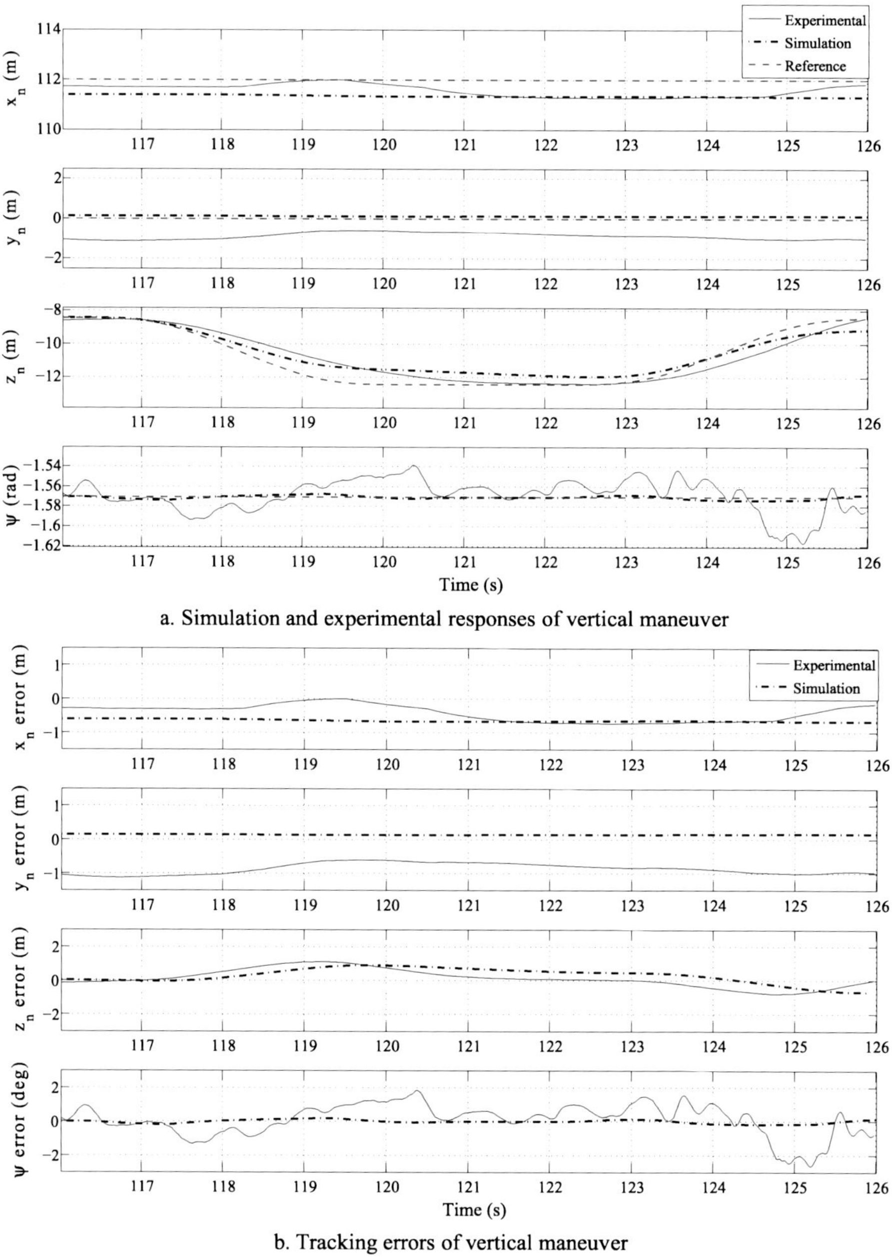

Fig. 9.15 Vertical maneuver flight results

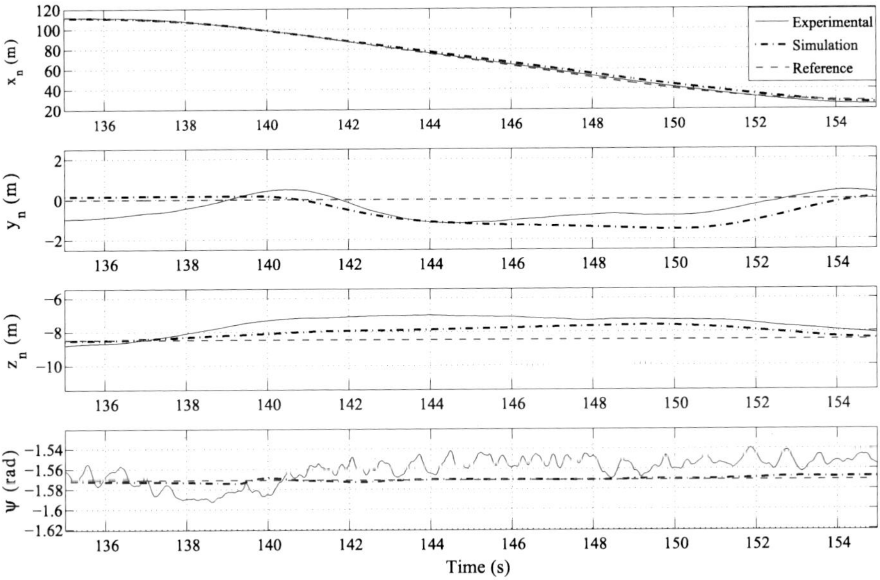

a. Simulation and experimental responses of lateral reposition

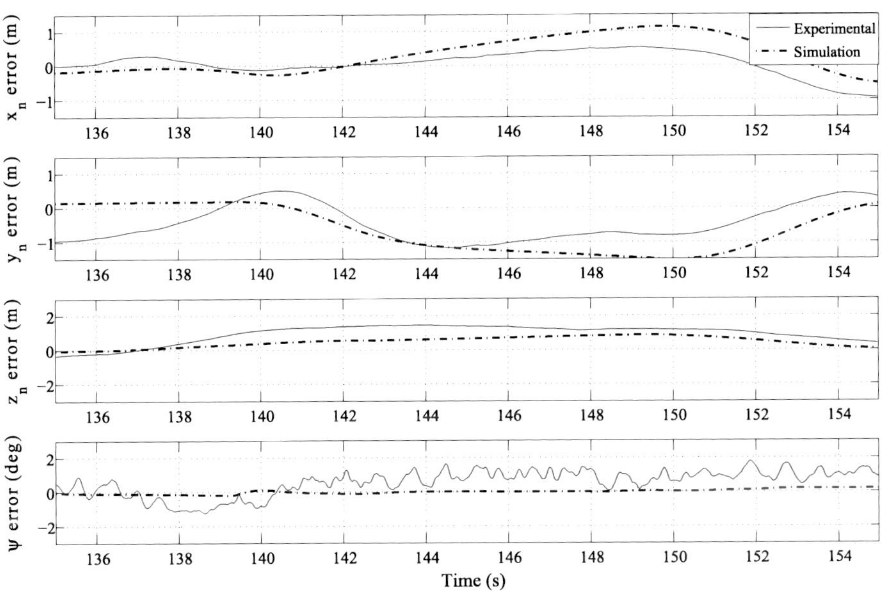

b. Tracking errors of lateral reposition

Fig. 9.16 Lateral reposition flight results

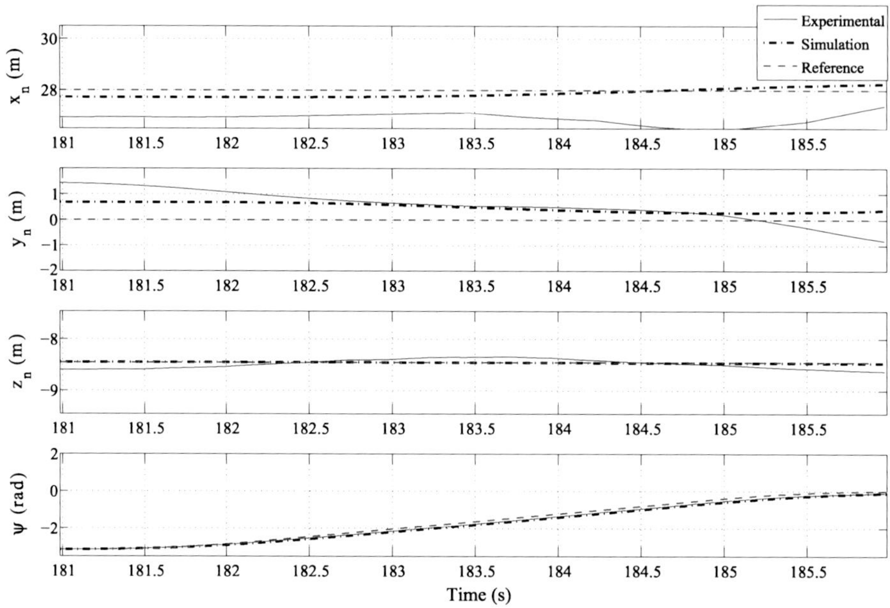

a. Simulation and experimental responses of turn-to-target

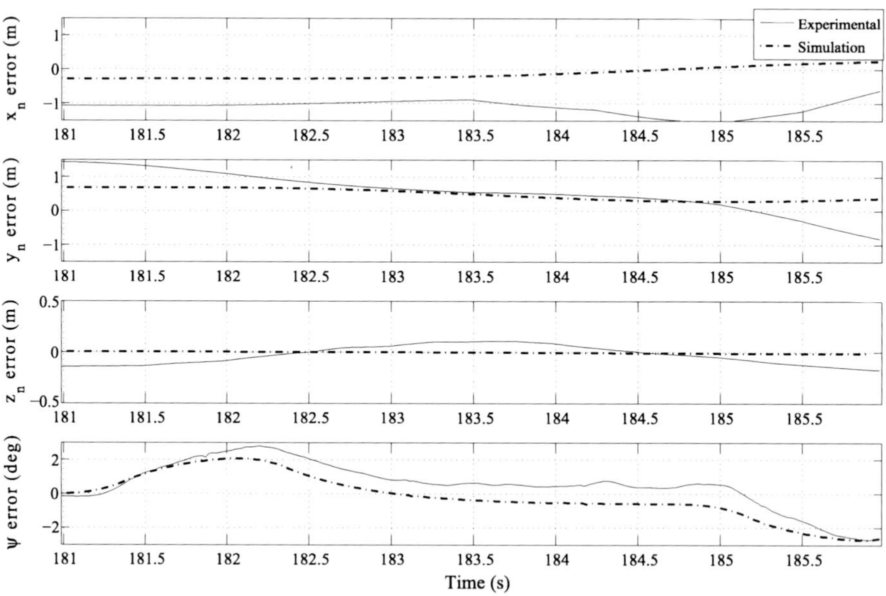

b. Tracking errors of turn-to-target

Fig. 9.17 Turn-to-target flight results

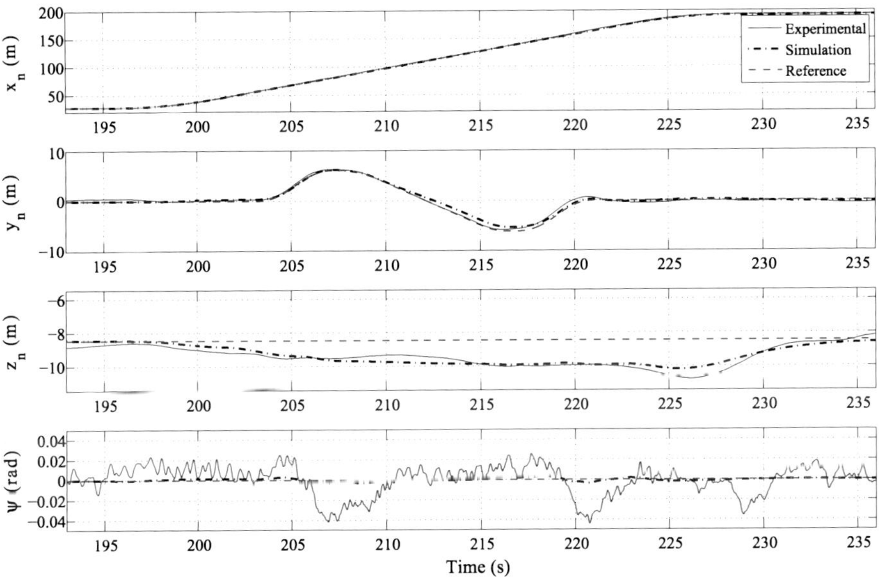

a. Simulation and experimental responses of slalom

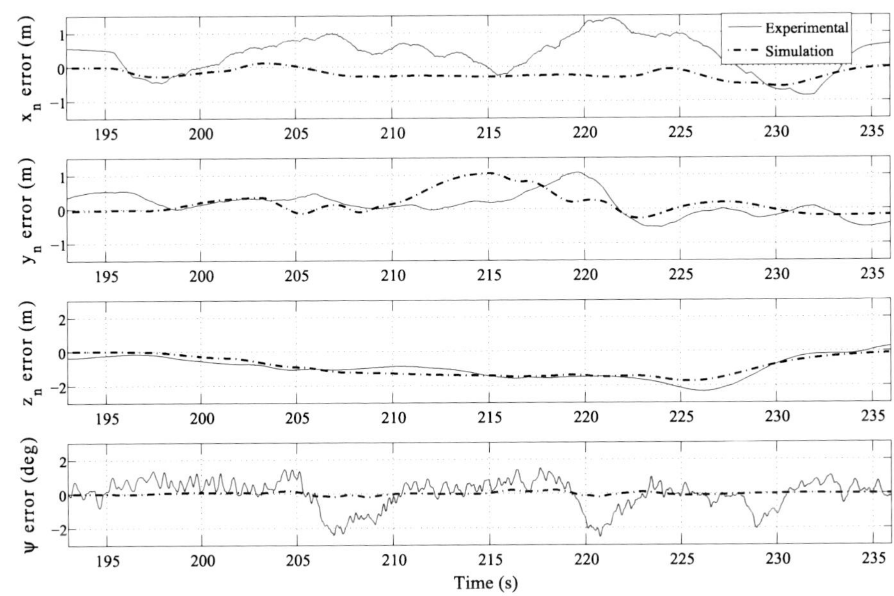

b. Tracking errors of slalom

Fig. 9.18 Slalom flight results

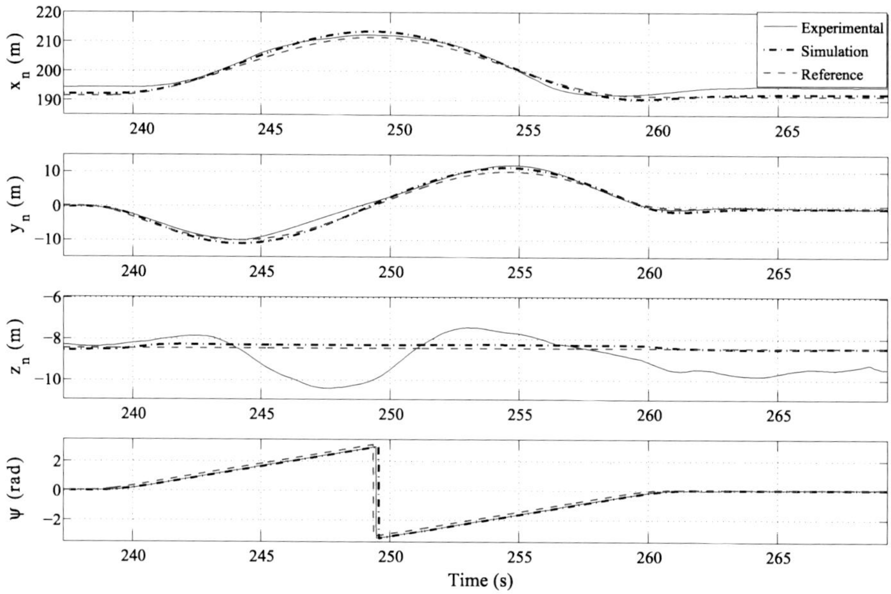

a. Simulation and experimental responses of pirouette

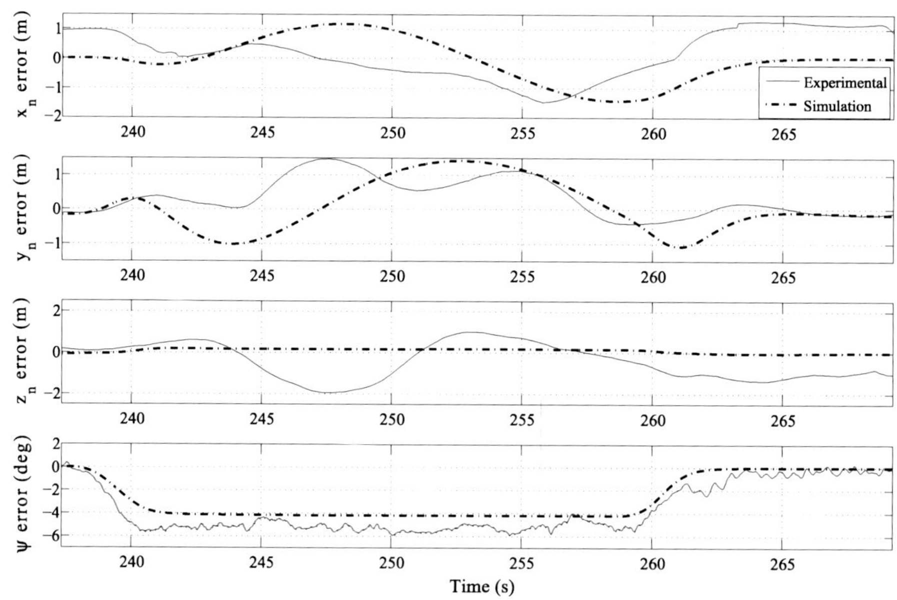

b. Tracking errors of pirouette

Fig. 9.19 Pirouette flight results

Table 9.11 Depart/abort (forward flight) performance evaluation

Specification	Desired level	Simulation	Actual test
Longitudinal position error	≤ 3 m	≤ 0.93 m	≤ 0.88 m
Lateral position error	≤ 3 m	≤ 0.31 m	≤ 0.51 m
Altitude error	≤ 3 m	≤ 2.16 m	≤ 1.85 m
Heading error	≤ 10 deg	≤ 0.14 deg	≤ 1.24 deg
Time to complete the maneuver	≤ 25 s	25 s	25 s

Table 9.12 Hover performance evaluation

Specification	Desired level	Simulation	Actual test
Time to decelerate to a stabilized hover	≤ 3 s	2.21 s	2.21 s
Duration in maintaining a stabilized hover	≥ 30 s	35 s	35 s
Longitudinal position maintaining range	≤ 1.5 m	≤ 0.98 m	≤ 0.74 m
Lateral position maintaining range	≤ 1.5 m	≤ 0.11 m	≤ 0.97 m
Altitude maintaining range	≤ 3 m	≤ 2.17 m	≤ 1.71 m
Heading maintaining range	≤ 5 deg	≤ 0.12 deg	≤ 2.71 deg

9.4 Simulation and Flight Test Results

Shown in Figs. 9.6 to 9.10 are the simulation and experimental results of the wide envelope path defined in Table 9.10, for the local NED position, local NED velocity, Euler angles, and angular rates, and control inputs, respectively. We would like to highlight the following key observations:

1. The actual flight test results perfectly match both the desired references (for position and velocity) and those obtained in the HITL simulation. The maximum tracking errors for the local NED position, local NED velocity, Euler angles, and angular rates are 1.5 m, 1 m/s, 0.1 rad, and 0.02 rad/s, respectively. Such small deviations persuasively demonstrate the good tracking performance of the automatic control system developed in Chaps. 7 and 8.
2. We have intentionally conducted the flight test in the situation with strong wind gusts (about 4 to 5 m/s in the horizontal plane, roughly measured by a handheld anemometer). With the existence of such strong disturbances, accurate tracking performance can be successfully achieved. Particularly, no jumps or oscillations can be observed in the responses of the velocity, Euler angles, and angular rate in Figs. 9.7, 9.8 and 9.9, which show the good robustness of our flight control system.
3. The tight matching between the practical and simulation responses indicates the high fidelity of our nonlinear flight dynamics model.

To examine the control performance in more detail, we break the overall results into individual intervals related to each pre-selected MTE, shown in Figs. 9.11

Table 9.13 Depart/abort (backward flight) performance evaluation

Specification	Desired level	Simulation	Actual test
Longitudinal position error	≤ 3 m	≤ 0.42 m	≤ 1.38 m
Lateral position error	≤ 3 m	≤ 0.26 m	≤ 1.48 m
Altitude error	≤ 3 m	≤ 0.21 m	≤ 1.71 m
Heading error	≤ 10 deg	≤ 0.08 deg	≤ 1.17 deg
Time to complete the maneuver	≤ 25 s	25 s	25 s

Table 9.14 Hovering turn performance evaluation

Specification	Desired level	Simulation	Actual test
Longitudinal position error	≤ 1.5 m	≤ 0.28 m	≤ 1.33 m
Lateral position error	≤ 1.5 m	≤ 0.32 m	≤ 1.24 m
Altitude maintaining range	≤ 3 m	≤ 0.01 m	≤ 0.85 m
Heading error	≤ 3 deg	≤ 1.44 deg	≤ 2.81 deg
Time to complete the maneuver	≤ 15 s	13.5 s	13.5 s

Table 9.15 Vertical maneuver performance evaluation

Specification	Desired level	Simulation	Actual test
Longitudinal position error	≤ 1.5 m	≤ 0.66 m	≤ 0.72 m
Lateral position error	≤ 1.5 m	≤ 0.15 m	≤ 1.12 m
Altitude error	≤ 2 m	≤ 0.89 m	≤ 1.11 m
Heading change	≤ 3 deg	≤ 0.19 deg	≤ 2.62 deg
Time to complete the maneuver	≤ 10 s	10 s	10 s

Table 9.16 Lateral reposition performance evaluation

Specification	Desired level	Simulation	Actual test
Longitudinal position error	≤ 3 m	≤ 1.12 m	≤ 0.49 m
Lateral position error	≤ 3 m	≤ 1.49 m	≤ 0.95 m
Altitude error	≤ 3 m	≤ 0.84 m	≤ 1.45 m
Heading error	≤ 10 deg	≤ 0.09 deg	≤ 1.65 deg
Time to complete the maneuver	≤ 21 s	21 s	21 s

to 9.19. The local NED position and heading angles, the key performance indicators in the MTE evaluation, are provided together with the resulting tracking errors. We summarize in Tables 9.11 to 9.19 the detailed performance evaluation of both the simulation and experimental results. It is clear that we have achieved the de-

Table 9.17 Turn-to-target performance evaluation

Specification	Desired level	Simulation	Actual test
Longitudinal position maintaining range	≤ 2 m	≤ 0.28 m	≤ 1.48 m
Lateral position maintaining range	≤ 2 m	≤ 0.68 m	≤ 1.42 m
Altitude maintaining range	≤ 2 m	≤ 0.01 m	≤ 0.17 m
Heading error	≤ 3 deg	≤ 2.04 deg	≤ 2.78 deg
Time to complete the maneuver	≤ 5 s	5 s	5 s

Table 9.18 Slalom performance evaluation

Specification	Desired level	Simulation	Actual test
Maintaining forward speed	> 6 m/s	6 m/s	6 m/s
Longitudinal position error	≤ 2 m	≤ 0.56 m	≤ 1.41 m
Lateral position error	≤ 2 m	≤ 1.05 m	≤ 1.06 m
Altitude error	≤ 3 m	≤ 1.73 m	≤ 2.33 m
Heading error	≤ 10 deg	≤ 0.18 deg	≤ 2.54 deg

Table 9.19 Pirouette performance evaluation

Specification	Desired level	Simulation	Actual test
Longitudinal position error	≤ 2 m	≤ 1.16 m	≤ 1.47 m
Lateral position error	≤ 2 m	≤ 1.41 m	≤ 1.45 m
Altitude error	≤ 2 m	≤ 0.18 m	≤ 1.95 m
Heading error	≤ 10 deg	≤ 4.18 deg	≤ 5.89 deg
Time to complete the maneuver	≤ 45 s	25 s	25 s
Duration to maintain the re-stabilized hover	≥ 5 s	7 s	7 s

sired performance in every category. Interested readers can access the web link at http://uav.ece.nus.edu.sg/ for video clips recorded during the actual flight tests.

Chapter 10
Flight Formation of Multiple UAVs

10.1 Introduction

Formation and cooperation phenomena have been long observed in the natural world. For example, many birds and insects as well as fishes are observed to fly or swim in a variety of formation behaviors. Studies have proven that their formation strategies can increase the efficiency of group performance and decrease the chance of being captured by predators (see, e.g., [158, 191]). When it comes to unmanned systems, the cooperation of multiple UAVs is an interesting and popular topic nowadays as multiple UAVs are sometimes necessary for achieving some complicated missions in real-life applications.

In this chapter, we present some preliminary results of the formation control of multiple unmanned systems. More specifically, we study and implement a simple formation flight strategy, based on the well-known leader-follower flight pattern, which aims to maintain a fixed geometrical formation of the unmanned systems, while navigating the UAVs to follow some desired trajectories. Our unmanned rotorcraft systems, HeLion and SheLion, are used throughout for the implementation of flight tests. An effective collision avoidance scheme is also utilized to guarantee the safety of the overall flight formation.

There are various options for realizing formation cooperation. Each method has its advantages and disadvantages. Generally speaking, there are two different types of strategies adopted in UAV formation, i.e., model-based and nonmodel-based strategies. The former is relatively traditional and has been proven in numerous manned or unmanned vehicle applications. As a result, it plays a dominant role in both theoretical study and practical implementations.

The model-based formation strategy can be further categorized into three groups including so-called (i) trajectory tracking, (ii) leader-follower station-keeping, and (iii) virtual structure approaches.

Trajectory tracking is the simplest formation strategy. It is the most suitable choice for realizing trajectory following with simple maneuvers, in which the flight trajectory is required to be predefined. Trajectory tracking focuses on the maintenance of the actual flight path as close as possible to the predefined trajectory. Because of its simplicity in implementation, such a method has been adopted in many

G. Cai et al., *Unmanned Rotorcraft Systems*, Advances in Industrial Control,
DOI 10.1007/978-0-85729-635-1_10, © Springer-Verlag London Limited 2011

applications (see, e.g., [94]). The following advantages of the trajectory tracking formation scheme include (i) ease in predicting actual flight performance and behavior, (ii) loose requirement on the relative position between UAVs, and (iii) less requirement on information and data exchange and communication among UAVs. Its disadvantages are (i) inflexibility of the cooperative operation and (ii) lack of reliable references for all agents involved in formation.

Leader-follower strategy is the most popular formation methodology adopted in practical implementations. It was first presented by Larson [106] and further explored by many others (see, e.g., [29, 72, 74, 147, 219]). A well-known project that aims to evaluate the leader-follower formation benefit is the NASA autonomous formation flight program (see [73, 83]), which has successfully demonstrated that the airflow from the wing tips of a fixed-wing UAV aircraft (an F/A-18) is capable of providing energy to its follower UAV aircraft(s). The leader-follower formation scheme maintains a fixed geometric position structure among all the agents in formation. Such a static position structure can be pre-determined by either aerodynamic theory or experimental data. There are mainly two types of leader-follower station keeping, i.e., the *front mode*, in which the relative position between two neighboring UAVs is held in the vortex of the wingman directly ahead of the trailing aircraft, and the *leader mode*, in which each aircraft maintains a relative position to a designated leader aircraft. For the fixed-wing UAV aircraft, the former is well investigated since it is beneficial in saving power and energy for the trailing aircraft. For the small-scale UAV rotorcraft formation flight, the main focus is on maintaining the relative position. The leader mode is more suitable and thus adopted in our work presented in this chapter.

In the virtual structure approach, the entire formation is considered as a single rigid structure. To achieve the desired global coordination for the entire formation flight, the control strategy of this approach is first to translate the structure motion into the motions of individual UAV agents and then to control each agent using an appropriate flight control law. For such a formation scheme, an important concept is the so-called formation geometry center, which is actually an imaginary point determined by a variety of factors such as the shape, moving speed, and heading of the entire team. It was originally brought forward based on the observation of the natural behavior of birds or fishes in maintaining certain geometrical shapes. The formation geometry center is essential in maintaining the overall geometry and balancing the deviations from a pre-determined formation geometry or flight trajectory. In the virtual structure approach, if a UAV unit loses its position in formation, the overall flock can modify the flight trajectory to assist the lost one. It has the capability of maintaining the formation geometry and tracking a prescribed path for each agent. It is an effective combination of pure trajectory tracking and a variant of the station-keeping. However, such an approach is relatively hard to implement in actual flight tests.

10.2 Leader-Follower Formation

To understand better the mechanism of the leader-follower formation scheme, we present in this section the coordinate systems and the kinematics model involved in flight formation. We should note that the matching of velocities and the accelerations between the leader and the follower are highly important when they are commanded to perform missions requiring complicated maneuvers. The flight control system designed using the robust and perfect tracking (RPT) technique in Chap. 8 has rendered the control mission of flight formation a trivial task. Under the RPT control framework given in Chap. 8, the follower has already had the capability of following the leader not only in position but also velocity and acceleration, if necessary. There is no need whatsoever to design a separate formation flight controller. The whole task involved in the flight formation control of multiple UAVs is to properly generate necessary command references, i.e., the position, velocity, and acceleration reference signals, for the followers. This is exactly what we will investigate in this section. We focus on situation in which there is only a single follower in the formation as the design to be implemented on actual formation flight tests using our unmanned systems, HeLion (leader) and SheLion (follower). Some of the results given below are also reported in Wang et al. [195].

10.2.1 Coordinate Systems in Formation Flight

There are four coordinate systems involved in the leader-follower flight formation, namely, the local NED coordinate system, the vehicle-carried NED coordinate system, the body coordinate system, and the formation coordinate system. The first three coordinate systems have been described in Chap. 2 in detail whereas the last one is illustrated below. We should note that the local NED coordinate system is taken as the inertial reference frame for both the leader and follower. We assume as usual that there is no directional difference between the local NED coordinate system and the vehicle-carried NED frames of the leader and the follower. To distinguish the leader and follower, in this chapter we denote the vectors (or states) corresponding to the leader and follower with subscripts LD and FL, respectively.

We need to define in more detail the formation coordinate system (see Fig. 10.1), which is particularly useful for establishing the leader-follower kinematics model. It is carried by the leader UAV with the origin and axes defined as the following:

1. The origin is located at the CG of the leader.
2. The X-axis (denoted by X_f) coincides with the projected vector of the X-axis of its body coordinate system on the X-Y plane of the leader-carried NED frame (parallel to the local NED horizontal plane).
3. The Z-axis (denoted by Z_f) points downward along the ellipsoid normal of the earth.
4. The Y-axis (denoted by Y_f) is orthogonal to the X- and Z-axes with the right-hand rule.

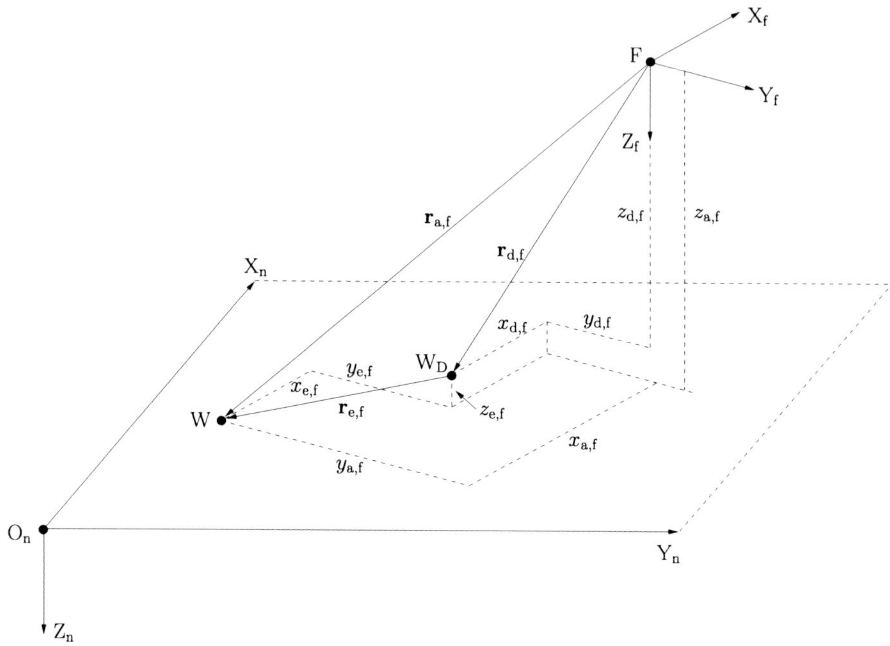

Fig. 10.1 Geometry of leader-follower UAV formation strategy

As usual, the coordinate vectors expressed in the formation coordinate system are denoted with a subscript f.

It is straightforward to verify that the rotation transformation from the formation coordinate system to the local NED frame can be described by a single Euler rotation along their parallel Z-axis with the leader's heading angle ψ_{LD} and is given by

$$\mathbf{R}_{\mathrm{n/f}} = \begin{bmatrix} \cos\psi_{\mathrm{LD}} & -\sin\psi_{\mathrm{LD}} & 0 \\ \sin\psi_{\mathrm{LD}} & \cos\psi_{\mathrm{LD}} & 0 \\ 0 & 0 & 1 \end{bmatrix}. \tag{10.1}$$

Furthermore, we can easily define the relative angular velocity. Taking the rotation of the leader-carried NED frame with respect to the formation coordinate system projected onto the leader-carried NED frame, we have

$$\boldsymbol{\omega}^{\mathrm{f}}_{\mathrm{f/n}} = \begin{pmatrix} 0 \\ 0 \\ \dot{\psi}_{\mathrm{LD}} \end{pmatrix}, \tag{10.2}$$

where $\dot{\psi}_{\mathrm{LD}}$ can be expressed as

$$\dot{\psi}_{\mathrm{LD}} = q_{\mathrm{LD}} \sin\phi_{\mathrm{LD}} / \cos\theta_{\mathrm{LD}} + r_{\mathrm{LD}} \cos\phi_{\mathrm{LD}} / \cos\theta_{\mathrm{LD}}, \tag{10.3}$$

and where q_{LD} and r_{LD} are, respectively, the pitch and yaw angular rates of the leader, and ϕ_{LD} and θ_{LD} are the Euler angles of the leader. The detailed definition of Euler angles can be found in Chap. 2.

10.2.2 Kinematics Model

The leader-follower station-keeping scheme is shown in Fig. 10.1. We note that for simplicity, the vehicle-carried NED coordinate systems of two UAVs are not shown in the figure. We also note that (i) Point F is the CG position of the leader UAV, (ii) Point W is the CG position of the follower UAV, and (iii) Point $\mathrm{W_D}$ is the desired position of the follower.

Let $\mathbf{r}_{\mathrm{e,f}}$ be the distance error vector from the desired position to the actual location of the follower in the formation coordinate system. Similarly, let $\mathbf{r}_{\mathrm{a,f}}$ and $\mathbf{r}_{\mathrm{d,f}}$ be, respectively, the actual and desired distance vectors of the follower in the formation coordinate frame. We have

$$\mathbf{r}_{\mathrm{e,f}} = \mathbf{r}_{\mathrm{a,f}} - \mathbf{r}_{\mathrm{d,f}}, \tag{10.4}$$

where $\mathbf{r}_{\mathrm{e,f}}$, $\mathbf{r}_{\mathrm{a,f}}$, and $\mathbf{r}_{\mathrm{d,f}}$ are the vector projections onto the formation frame and can be further expressed by

$$\mathbf{r}_{\mathrm{e,f}} = \begin{pmatrix} x_{\mathrm{e,f}} \\ y_{\mathrm{e,f}} \\ z_{\mathrm{e,f}} \end{pmatrix}, \qquad \mathbf{r}_{\mathrm{a,f}} = \begin{pmatrix} x_{\mathrm{a,f}} \\ y_{\mathrm{a,f}} \\ z_{\mathrm{a,f}} \end{pmatrix}, \qquad \mathbf{r}_{\mathrm{d,f}} = \begin{pmatrix} x_{\mathrm{d,f}} \\ y_{\mathrm{d,f}} \\ z_{\mathrm{d,f}} \end{pmatrix}. \tag{10.5}$$

For an ideal leader-follower formation, $\mathbf{r}_{\mathrm{e,f}} = 0$.

We can also express the distance vector from the leader to the actual location of the follower in the local NED frame as

$$\mathbf{r}_{\mathrm{a,n}} = \mathbf{P}_{\mathrm{n,FL}} - \mathbf{P}_{\mathrm{n,LD}}, \tag{10.6}$$

where $\mathbf{P}_{\mathrm{n,FL}}$ and $\mathbf{P}_{\mathrm{n,LD}}$ are, respectively, the position vectors of the follower and the leader in the local NED frame. Both can be measured directly by the navigation sensors equipped in our unmanned systems.

We are ready now to derive the kinematics model of the leader-follower formation scheme, which is then used to generate the position, velocity, and acceleration references for the follower in the local NED frame. In order to achieve a perfect formation, in view of (10.4) and (10.6), we need to ensure that

$$\mathbf{P}_{\mathrm{nr,FL}} = \mathbf{P}_{\mathrm{n,LD}} + \mathbf{r}_{\mathrm{d,n}} = \mathbf{P}_{\mathrm{n,LD}} + \mathbf{R}_{\mathrm{n/f}} \mathbf{r}_{\mathrm{d,f}}, \tag{10.7}$$

where $\mathbf{P}_{\mathrm{nr,FL}}$ is the position reference in the local NED coordinate system for the follower. The velocity reference in the local NED frame for the follower, $\mathbf{V}_{\mathrm{nr,FL}}$, can be obtained from the first-order derivatives of (10.7) and is given as

$$\begin{aligned} \mathbf{V}_{\mathrm{nr,FL}} &= \mathbf{V}_{\mathrm{n,LD}} + \mathbf{R}_{\mathrm{n/f}} \dot{\mathbf{r}}_{\mathrm{d,f}} + \mathbf{R}_{\mathrm{n/f}} (\boldsymbol{\omega}^{\mathrm{f}}_{\mathrm{f/n}} \times \mathbf{r}_{\mathrm{d,f}}) \\ &= \mathbf{V}_{\mathrm{n,LD}} + \mathbf{R}_{\mathrm{n/f}} (\boldsymbol{\omega}^{\mathrm{f}}_{\mathrm{f/n}} \times \mathbf{r}_{\mathrm{d,f}}), \end{aligned} \tag{10.8}$$

where $\mathbf{V}_{\mathrm{n,LD}}$ is the actual velocity of the leader in the local NED frame. As mentioned earlier, we are only focusing on the leader-follower formation pattern with constant $\mathbf{r}_{\mathrm{d,f}}$. Thus, $\dot{\mathbf{r}}_{\mathrm{d,f}} = 0$.

Similarly, we can obtain the acceleration reference for the follower in the local NED frame as the following:

$$\mathbf{a}_{\mathrm{nr,FL}} = \mathbf{a}_{\mathrm{n,LD}} + \mathbf{R}_{\mathrm{n/f}} (\dot{\boldsymbol{\omega}}^{\mathrm{f}}_{\mathrm{f/n}} \times \mathbf{r}_{\mathrm{d,f}}) + \mathbf{R}_{\mathrm{n/f}} [\boldsymbol{\omega}^{\mathrm{f}}_{\mathrm{f/n}} \times (\boldsymbol{\omega}^{\mathrm{f}}_{\mathrm{f/n}} \times \mathbf{r}_{\mathrm{d,f}})], \tag{10.9}$$

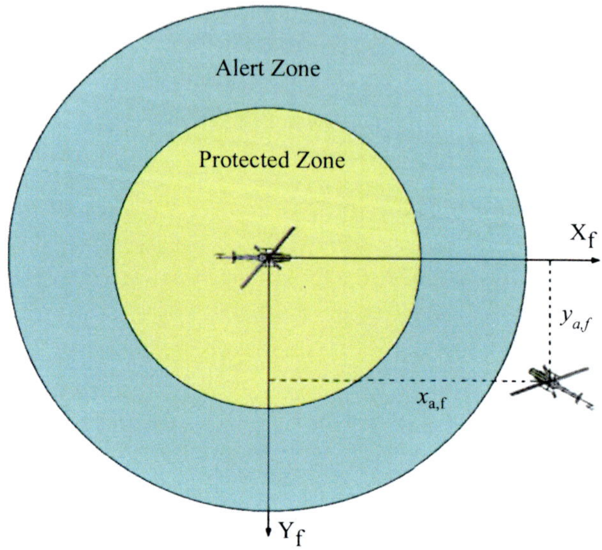

Fig. 10.2 Protected and alert zones surrounding a UAV rotorcraft

where $\mathbf{a}_{n,LD}$ is the actual local NED acceleration of the leader, and $\dot{\boldsymbol{\omega}}^{f}_{f/n}$ is given by

$$\dot{\boldsymbol{\omega}}^{f}_{f/n} = \begin{pmatrix} 0 \\ 0 \\ \ddot{\psi}_{LD} \end{pmatrix} \tag{10.10}$$

with

$$\begin{aligned} \ddot{\psi}_{LD} = {} & \dot{q}_{LD} \sin\phi_{LD} / \cos\theta_{LD} + q_{LD}\dot{\theta}_{LD} \sin\phi_{LD} \sin\theta_{LD} / \cos\theta^{2}_{LD} \\ & + q_{LD}\dot{\phi}_{LD} \cos\phi_{LD} / \cos\theta_{LD} + \dot{r}_{LD} \cos\phi_{LD} / \cos\theta_{LD} \\ & - r_{LD}\dot{\phi}_{LD} \sin\phi_{LD} / \cos\theta_{LD} + r_{LD}\dot{\theta}_{LD} \cos\phi_{LD} \sin\theta_{LD} / \cos\theta^{2}_{LD}. \end{aligned} \tag{10.11}$$

We note that $\dot{q}_{LD}$, $\dot{r}_{LD}$, $\dot{\phi}_{LD}$, and $\dot{\theta}_{LD}$ cannot be directly measured. We need to estimate these parameters online in actual implementation. For formation maneuvers with small changes in $\dot{\boldsymbol{\omega}}^{f}_{f/n}$, the second term in (10.9) can also be safely ignored.

10.3 Collision Avoidance

As a safety measure, we implement a collision avoidance scheme, which is commonly adopted in the literature and in which we define two imaginary safety spheres, namely, the protected zone and the alert zone for each involved UAV member (see Fig. 10.2). When an agent enters the alert zone of the other, a predefined protocol for conflict resolution is activated and executed by either or both UAVs. On the other hand, no agent is allowed to enter the protected zone of any other member in the formation.

The determination of the zone area has great importance. It has to take into consideration many factors of the unmanned systems, such as velocities, sensor accuracy, and data communication rates among all members in formation. Generally, a systematic determination of an optimal and safe size of the protected space is still an open problem. For our unmanned systems, HeLion and SheLion, we decide to choose 2 m and 4 m radiuses for the protected and alert zones.

Multiple proposals for conflict prediction and resolution can be found in the open literature (see, e.g., [174, 178]). We adopt a decentralized approach that permits each UAV to optimize its own trajectory. When any pair (or group) of UAVs is overlapped with their alert zones, coordination among multiple UAVs is activated to avoid any possible collision. In our proposed collision avoidance scheme (see Figs. 10.1 and 10.2), we focus on the following geometric relationship expressed in the local NED coordinate system:

$$\begin{cases} \mathbf{r}_{\mathrm{a,n}} = \mathbf{P}_{\mathrm{n,FL}} - \mathbf{P}_{\mathrm{n,LD}}, \\ \psi_{\mathrm{dif}} = \psi_{\mathrm{FL}} - \psi_{\mathrm{LD}}, \end{cases} \tag{10.12}$$

where ψ_{dif} is the difference between the heading angles of the leader and follower. We intend to resolve any potential vehicle conflict through the following steps:

1. First of all, based on the inflight data obtained, we need to judge whether it is necessary to execute the collision avoidance scheme. Two threshold functions have been defined and both of them are required to be examined in every execution loop of the flight control software systems for both the leader and follower. More specifically, the first one evaluates the leader-follower relative distance and is given by

$$\ell(\mathbf{r}_{\mathrm{a,n}}) = |\mathbf{r}_{\mathrm{a,n}}|^2 - \alpha_{\mathrm{TH}}^2, \tag{10.13}$$

where α_{TH} is the radius of the alert zone (i.e., 4 m for HeLion and SheLion), $|\mathbf{r}_{\mathrm{a,n}}|$ is the amplitude of the actual distance vector between the leader and the follower, i.e.,

$$|\mathbf{r}_{\mathrm{a,n}}| = \sqrt{(x_{\mathrm{n,FL}} - x_{\mathrm{n,LD}})^2 + (y_{\mathrm{n,FL}} - y_{\mathrm{n,LD}})^2 + (z_{\mathrm{n,FL}} - z_{\mathrm{n,LD}})^2}, \tag{10.14}$$

and $\ell(\mathbf{r}_{\mathrm{a,n}}) \le 0$ indicates that one UAV enters the alert zone of the other. In this situation, we need to further evaluate the threshold function by finding its rate of change, i.e.,

$$\begin{aligned} \dot{\ell}(\mathbf{r}_{\mathrm{a,n}}) = {} & 2(x_{\mathrm{n,FL}} - x_{\mathrm{n,LD}})(u_{\mathrm{n,FL}} - u_{\mathrm{n,LD}}) + 2(y_{\mathrm{n,FL}} - y_{\mathrm{n,LD}})(v_{\mathrm{n,FL}} - v_{\mathrm{n,LD}}) \\ & + 2(z_{\mathrm{n,FL}} - z_{\mathrm{n,LD}})(w_{\mathrm{n,FL}} - w_{\mathrm{n,LD}}), \end{aligned} \tag{10.15}$$

where all the parameters involved are the components of the position and velocity vectors of the leader and the follower in the local NED frame. If $\dot{\ell}(\mathbf{r}_{\mathrm{a,n}}) > 0$, no action is required. If $\dot{\ell}(\mathbf{r}_{\mathrm{a,n}}) \le 0$, we proceed to the next two steps to execute collision avoidance maneuvers. Physically, they represent that the intruder UAV has entered a protected zone of the other UAV and they are moving toward each other. We should point out that once the collision avoidance maneuver is executed, the outer-loop control of the UAV involved is disabled. Only the inner-loop part in the flight control system remains activated.

2. When a possible collision between the UAVs is detected, the leader is commanded to perform only a translational maneuver for collision avoidance. For this purpose, we define two switching functions, s_1 and s_2, as the following:
$$\begin{cases} s_1 = \cos\psi_{\text{LD}}(x_{\text{n,FL}} - x_{\text{n,LD}}) + \sin\psi_{\text{LD}}(y_{\text{n,FL}} - y_{\text{n,LD}}), \\ s_2 = \cos\psi_{\text{LD}}(y_{\text{n,FL}} - y_{\text{n,LD}}) - \sin\psi_{\text{LD}}(x_{\text{n,FL}} - x_{\text{n,LD}}). \end{cases} \tag{10.16}$$
To effectively avoid any collision, we apply the following solution to generate attitude references for the inner-loop control of the leader:
$$\theta_{\text{r,LD}} = \begin{cases} \theta_{\text{rmax,LD}} & \text{if } s_1 < 0, \\ \theta_{\text{rmin,LD}} & \text{if } s_1 \geq 0, \end{cases} \tag{10.17}$$
and
$$\phi_{\text{r,LD}} = \begin{cases} \phi_{\text{rmax,LD}} & \text{if } s_2 < 0, \\ \phi_{\text{rmin,LD}} & \text{if } s_2 \geq 0, \end{cases} \tag{10.18}$$
where the parameters for HeLion are set as follows:
$$\theta_{\text{rmax,LD}} = -2.5 \text{ deg}, \qquad \theta_{\text{rmin,LD}} = 2.5 \text{ deg}, \tag{10.19}$$
and
$$\phi_{\text{rmax,LD}} = 5.5 \text{ deg}, \qquad \phi_{\text{rmin,LD}} = -0.5 \text{ deg}. \tag{10.20}$$
These values are adopted for the follower UAV (SheLion) as well since HeLion and SheLion are almost physically identical. The asymmetrical setting of $\phi_{\text{r,LD}}$ is due to the leaning effect of the helicopter. Geometrically, taking the first situation, i.e., $s_1 < 0$, as an example, it indicates that the leader is ahead of the follower along the formation frame X-axis. Also, $\ell(\mathbf{r}_{\text{a,n}}) < 0$ and $\dot{\ell}(\mathbf{r}_{\text{a,n}}) < 0$ imply that an alert zone (of either the leader or the follower) has been intruded up and the follower is approaching the leader. In such a case, we issue $\phi_{\text{rmax,LD}}$ to drive the leader to fly away from the follower with its maximum pitch-down attitude, i.e., -2.5 degrees.
3. Similarly, for the heave direction, we define a switching function
$$s_3 = z_{\text{n,FL}} - z_{\text{n,LD}} \tag{10.21}$$
and adopt the following collision avoidance solution:
$$\delta_{\text{col,LD}} = \delta_{\text{trim,LD}} - \Delta\delta_{\text{col,max}} \tag{10.22}$$
if $s_3 < 0$, where $\delta_{\text{col,LD}}$ is the actual collective input to the leader UAV, $\delta_{\text{trim,LD}}$ is its trimmed value, and $\Delta\delta_{\text{col,max}}$ is a predefined maximum offset to realize a reasonable heave-up motion. For HeLion and SheLion, it is set to be 0.03. We note that in our collision avoidance scheme, the UAV is commanded to perform only a heave-up motion since any downward flight is potentially dangerous and can result in a crash. Also, the situation $s_3 \geq 0$ is not considered for the leader. It is handled by the follower instead.
4. A similar collision avoidance is also adopted by the follower. The switching functions are identical to those defined for the leader, whereas the command references are given by

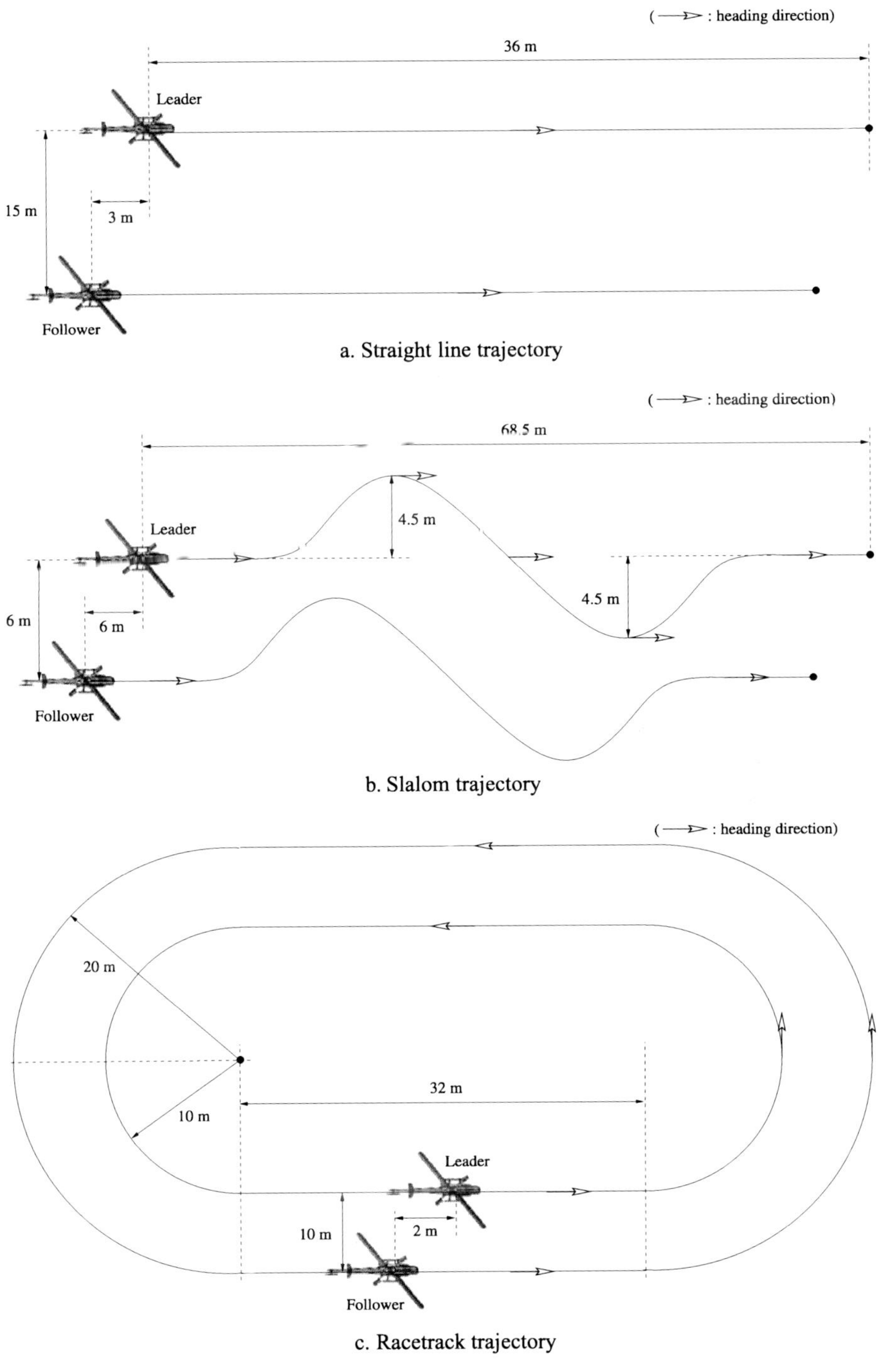

Fig. 10.3 Illustration of formation flight trajectories

Table 10.1 Desired distance vector settings of formation flight tests

Flight trajectory	$x_{d,f}$	$y_{d,f}$	$z_{d,f}$
Straight line	−3 m	15 m	−4 m
Slalom	−6 m	6 m	−4 m
Racetrack	−2 m	10 m	−4 m

$$\theta_{r,FL} = \begin{cases} \theta_{rmin,FL} & \text{if } s_1 < 0, \\ \theta_{rmax,FL} & \text{if } s_1 \geq 0, \end{cases} \tag{10.23}$$

$$\phi_{r,FL} = \begin{cases} \phi_{rmin,FL} & \text{if } s_2 < 0, \\ \phi_{rmax,FL} & \text{if } s_2 \geq 0, \end{cases} \tag{10.24}$$

where

$$\theta_{rmax,FL} = -2.5 \text{ deg}, \qquad \theta_{rmin,FL} = 2.5 \text{ deg}, \tag{10.25}$$

and

$$\phi_{rmax,FL} = 5.5 \text{ deg}, \qquad \phi_{rmin,FL} = -0.5 \text{ deg}, \tag{10.26}$$

and if $s_3 \geq 0$,

$$\delta_{col,FL} = \delta_{trim,FL} - \Delta\delta_{col,max}, \tag{10.27}$$

where $\delta_{col,FL}$ is the actual collective input to the follower UAV, and $\delta_{trim,FL}$ is its trimmed value.

We note that it is for simplicity that we do not adjust the heading angle and yaw rate for collision avoidance, which is, however, commonly seen in fixed-wing UAV applications (see, e.g., [178]). Finally, after executing all of the above collision avoidance procedures, we need to further examine (10.13) with α_{TH} being replaced by the radius of the protected zone. A negative $\ell(\mathbf{r}_{a,n})$ means that the protected zone is intruded upon and a collision will occur. An emergency alert signal is sent immediately to the ground control station. The human pilots are called to take control of the UAVs.

10.4 Flight Test Results

In this section, we present the experiment results for the proposed leader-follower flight formation. As mentioned earlier, HeLion is assigned to be the leader and SheLion is the follower. Three formation trajectories have been performed for the evaluation purpose, which include the following:

1. STRAIGHT-LINE TRAJECTORY: Both HeLion (leader) and SheLion (follower) are commanded to start with a stable automatic hover and then complete a 36 m straight-line path with a forward speed of 2 m/s.
2. SLALOM TRAJECTORY: It is similar to that defined in Sect. 9.2.8 of Chap. 9. We reduce the top forward speed and the total length of the flying path to 2.5 m/s and 68.5 m, respectively.

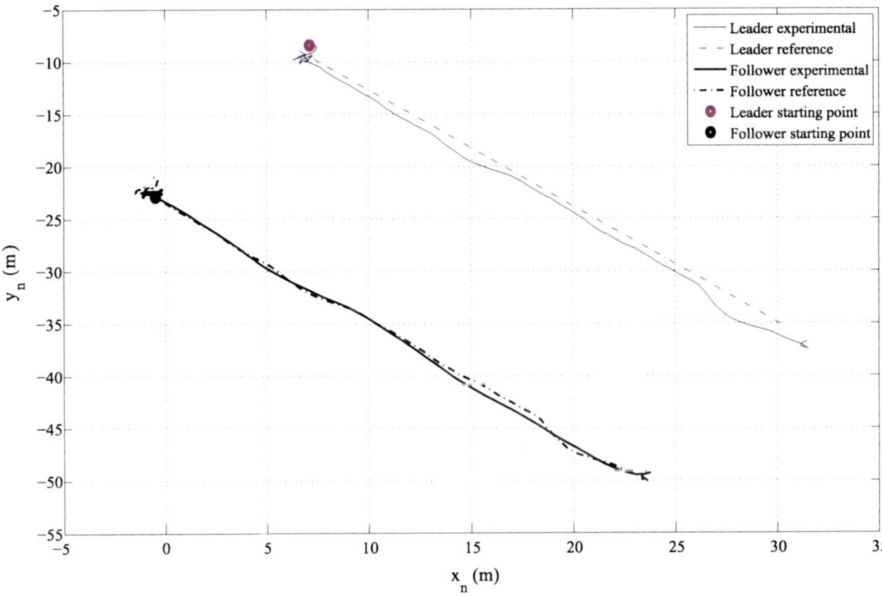

a. Straight-line formation results—local NED horizontal position

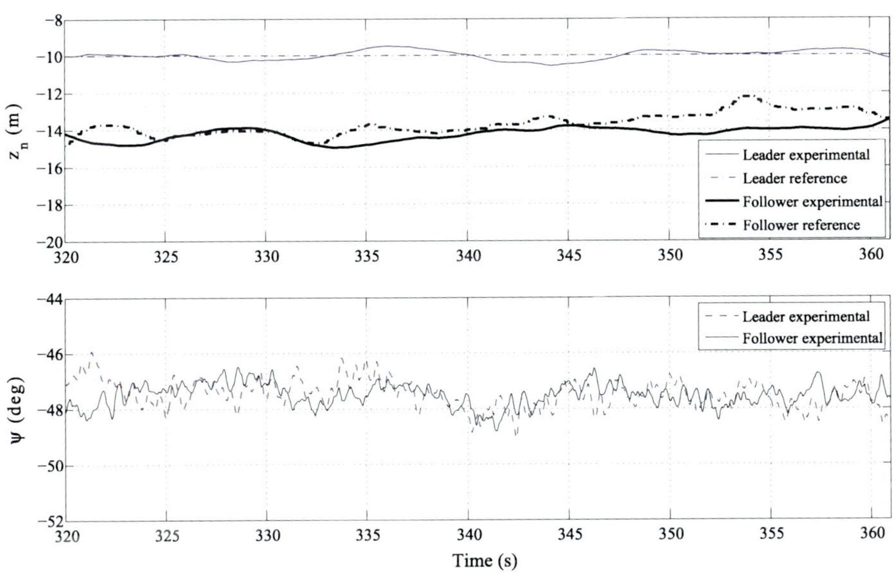

b. Straight-line formation results—local NED vertical position and heading angle

Fig. 10.4 Straight-line formation results—position

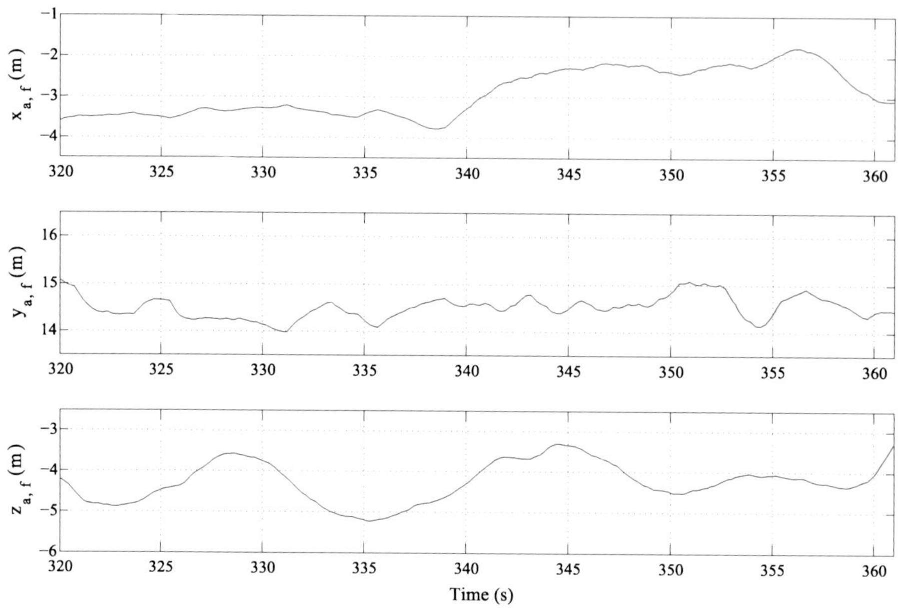

Fig. 10.5 Straight-line formation results—desired separation keeping

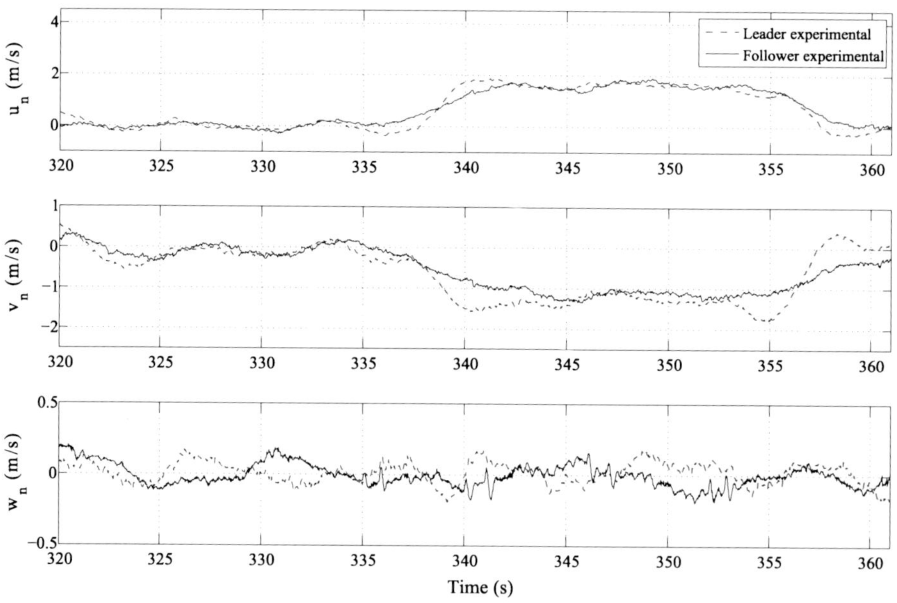

Fig. 10.6 Straight-line formation results—velocity

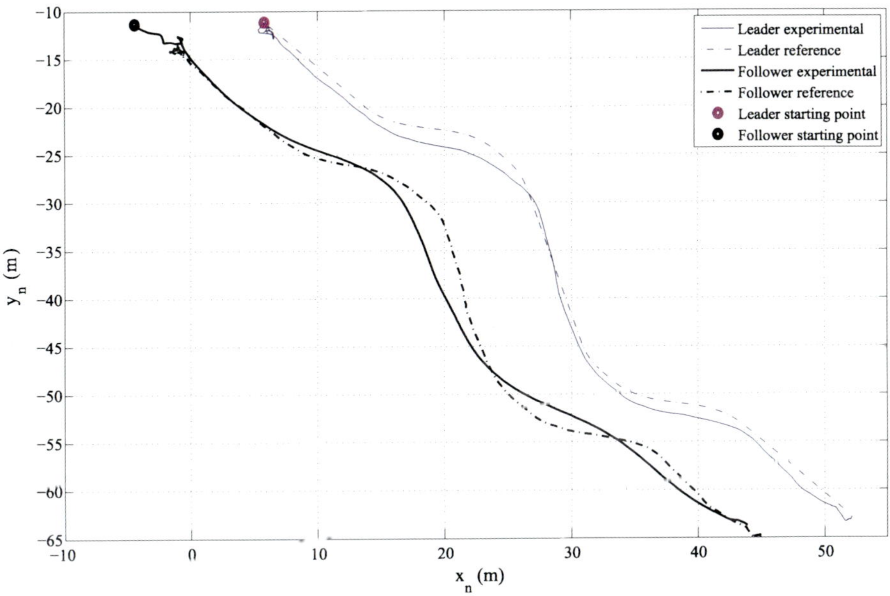

a. Slalom formation results—local NED horizontal position

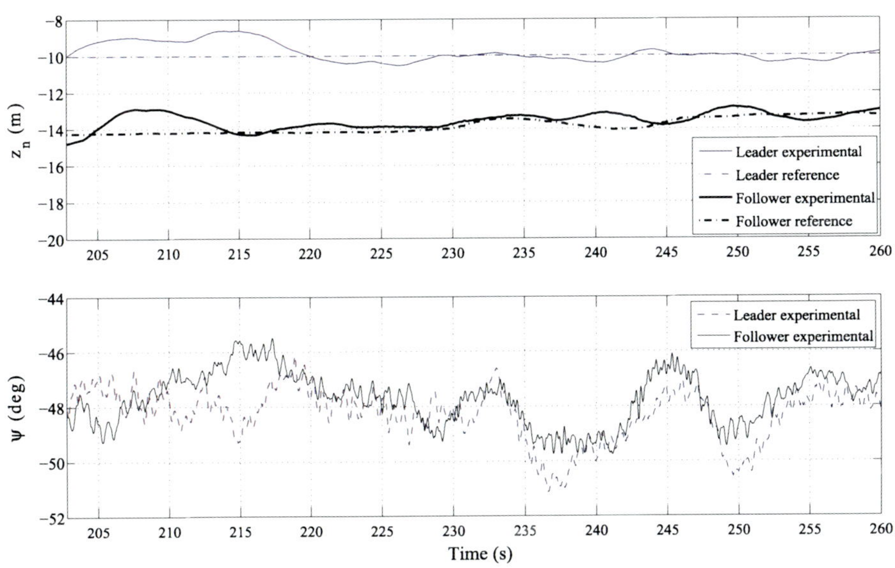

b. Slalom formation results—local NED vertical position and heading angle

Fig. 10.7 Slalom formation results—position

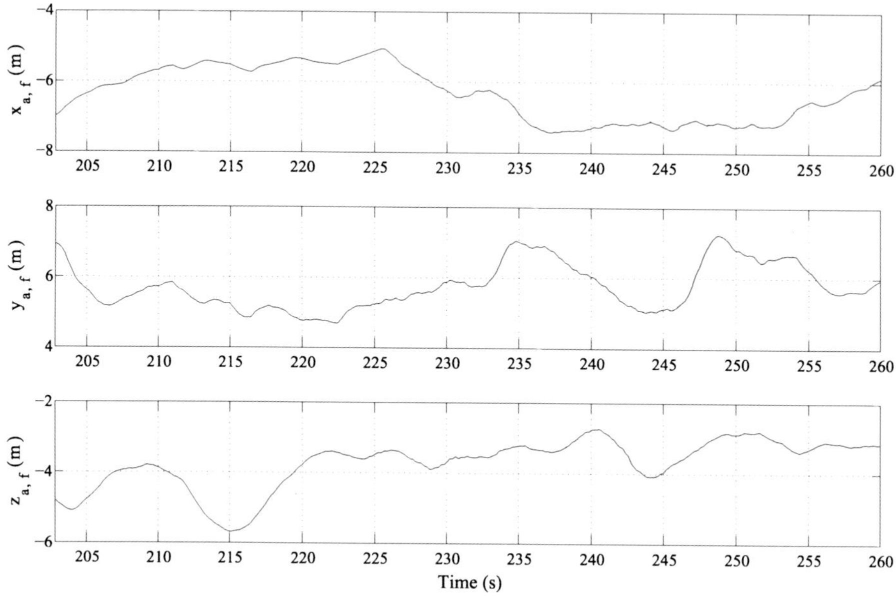

Fig. 10.8 Slalom formation results—desired separation keeping

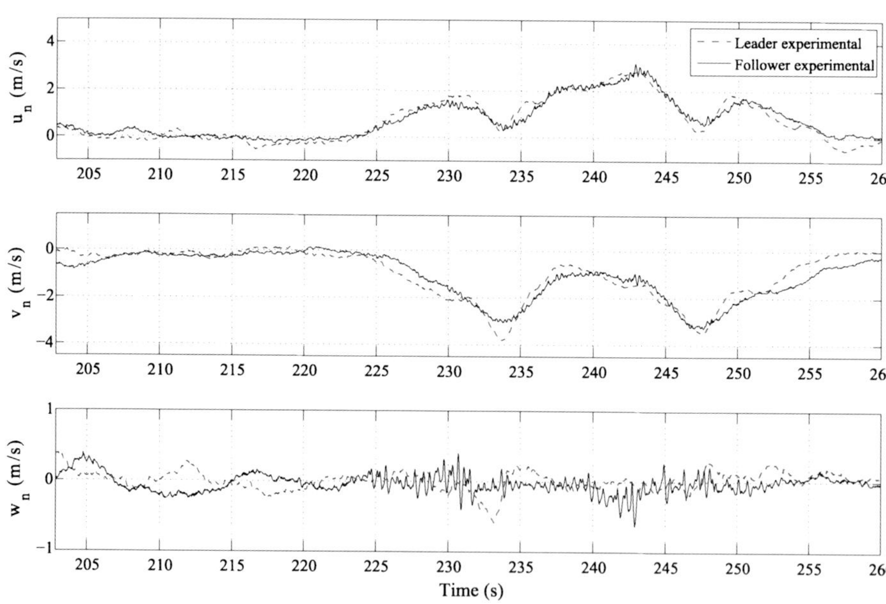

Fig. 10.9 Slalom formation results—velocity

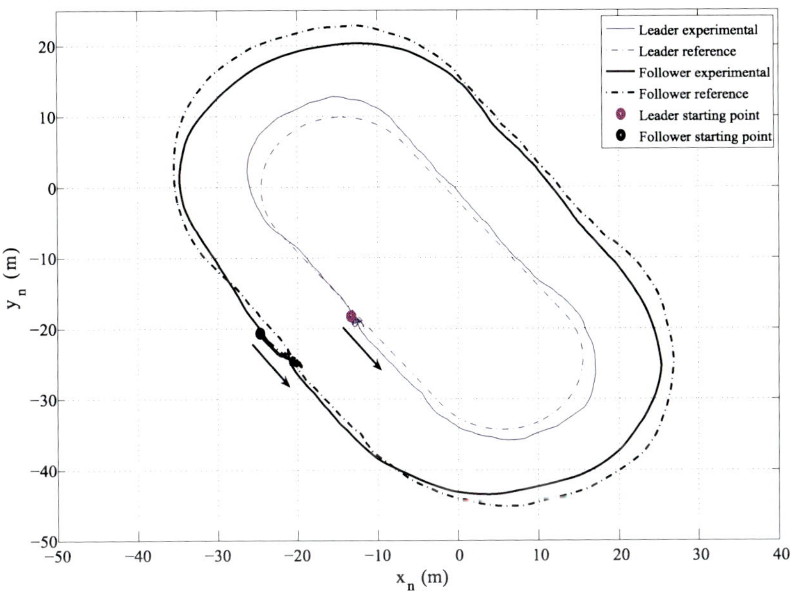

a. Racetrack formation results—local NED horizontal position

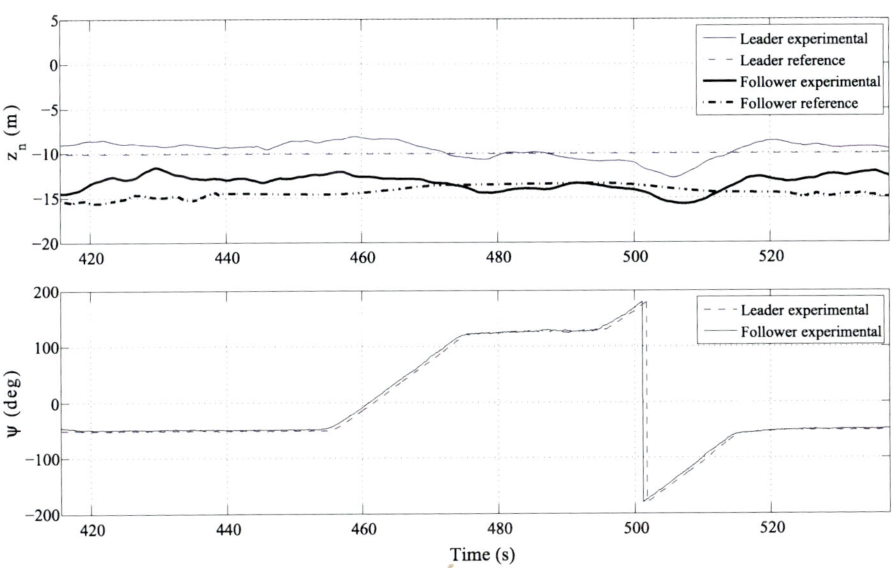

b. Racetrack line formation results—local NED vertical position and heading angle

Fig. 10.10 Racetrack formation results—position

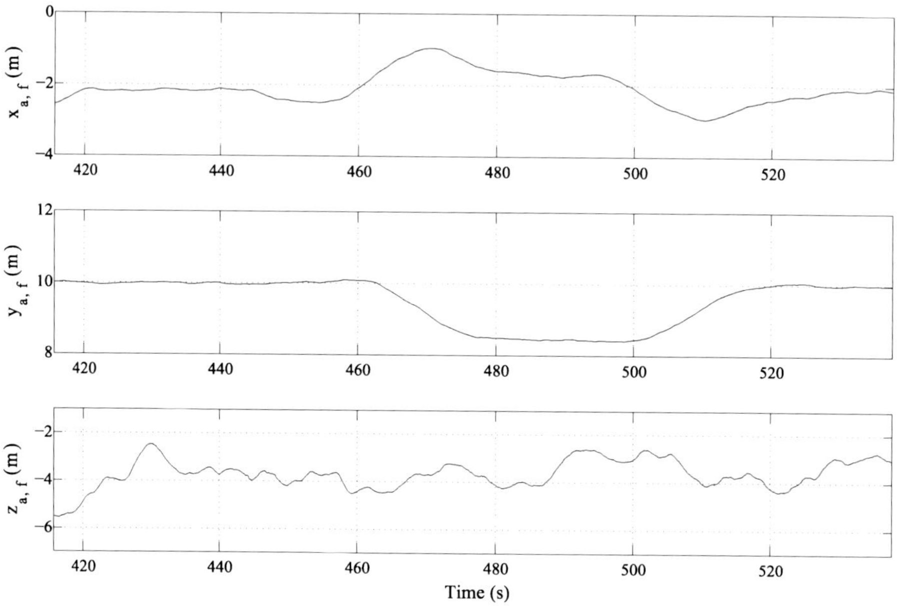

Fig. 10.11 Racetrack formation results—desired separation keeping

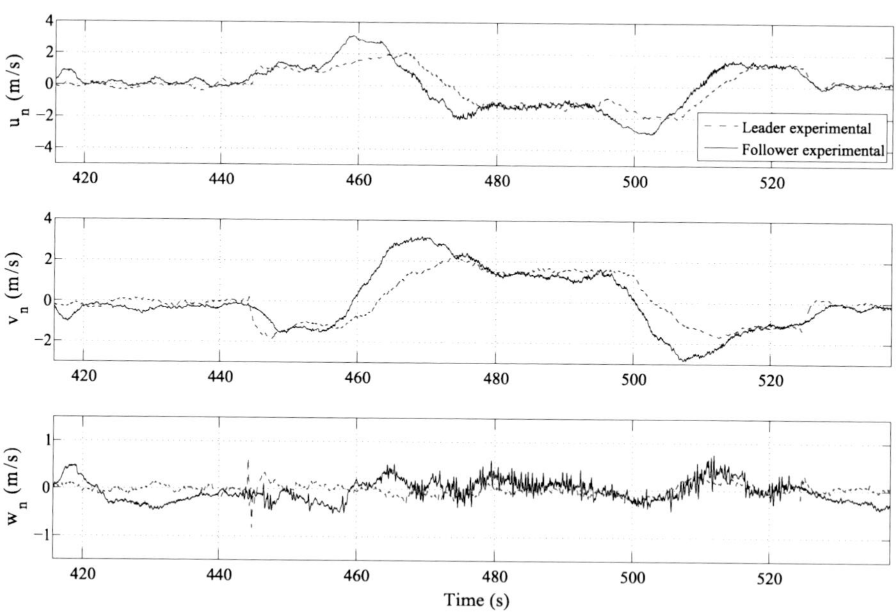

Fig. 10.12 Racetrack formation results—velocity

3. RACETRACK TRAJECTORY: It consists of both a straight-line path and heading turn motion. Again, the cruise speed of the leader is set at 2 m/s.

For clarity, we illustrate the detailed settings of the above formation flight trajectories in Fig. 10.3.

Listed in Table 10.1 are the desired distance vectors adopted in the actual flight tests. The experiment results shown in Figs. 10.4 to 10.12 capture the resulting positions, heading angles, separation distances, and velocities of the leader (HeLion) and the follower (SheLion). We note that with the RPT outer-loop controllers as given in Chap. 8, the implementation of formation flight tests are done with almost no additional cost. Interested readers can access the website of our UAV Group [135] for video clips recorded during the actual tests.

Finally, we note that the results presented in this chapter are rather preliminary at this stage. Our research team is currently investigating issues related to robust air-to-air communications between unmanned rotorcraft. We are also working on the development of more sophisticated techniques for flight formation and cooperative control of multiple unmanned aerial and ground systems.

Chapter 11
Vision-Based Target Following

11.1 Introduction

Vision systems have become an indispensable part of a UAV system to successfully complete tasks in many applications. A variety of vision systems using off-the-shelf components have been reported in the literature (see, e.g., [20, 60, 77]). By integrating information from the vision sensors and other adopted sensors, functions of the unmanned systems can be fully extended to perform missions such as surveillance (see, e.g., [151]), vision-aided flight control (see, e.g., [4, 71, 155, 159–161]), tracking (see, e.g., [110, 120]), terrain mapping (see, e.g., [119]), and navigation (see e.g., [82, 90, 97]). These missions are derived from both military and civilian requirements, such as aiding soldiers in urban operations to effectively spot, identify, designate, and destroy targets, and providing emergency relief workers a better view of damage in search and rescue missions after natural disasters. Compared to other traditional navigation systems and sensors, vision systems have the following unique features:

1. They are capable of providing rich information of objects of interest and the surrounding environments, including shape and appearance.
2. They require only natural light and do not depend on any other signal source, such as beacon stations or satellite signals.
3. They are generally of low cost and lightweight compared to other related sensing systems such as radars. As such, they are well suited for small-size unmanned vehicles, which have limited space and payload.
4. They do not emit any energy, so that the whole system is almost undetectable and safer in special conditions, such as battlefields.

In this chapter, we follow the results reported in [108–110] to present the design and implementation of a comprehensive vision system for an unmanned rotorcraft. More specifically, we focus on issues related to vision-based ground target following. Interested readers are referred to Lin [108] for more information on the development and applications of vision systems for unmanned helicopters.

G. Cai et al., *Unmanned Rotorcraft Systems*, Advances in Industrial Control,
DOI 10.1007/978-0-85729-635-1_11,

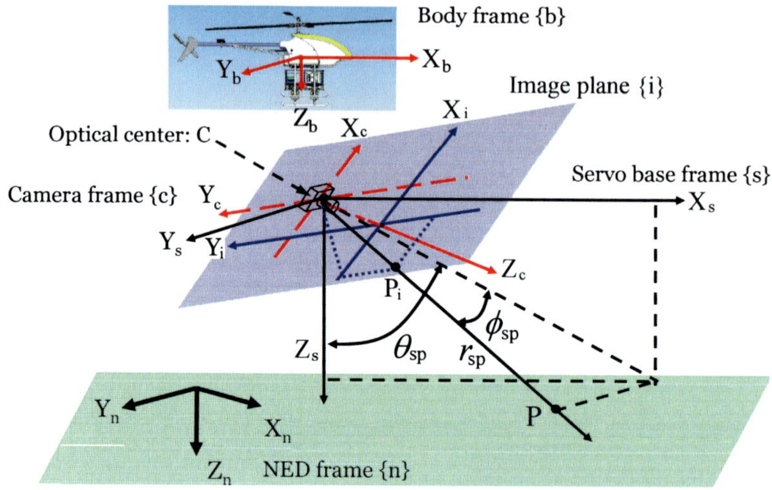

Fig. 11.1 Coordinate frames used in unmanned vision systems

11.2 Coordinate Frames Used in Vision Systems

Depicted in Fig. 11.1 are coordinate systems adopted in the UAV vision systems. More specifically, we have the following:

1. The local north-east-down (NED) coordinate system (labeled with a subscript n) is the same as the one defined in Chap. 2. We recall it here again for easy reference. It is an orthogonal frame on the surface of the earth, whose origin is the launching point of the aircraft on the surface of the earth. The X_n-Y_n plane of the NED frame is tangent to the surface of the earth at the origin of the NED frame. The coordinate X_n-, Y_n-, and Z_n-axes of the NED frame are specified to point toward the north, east, and down (toward into the earth, vertically to the surface of the Earth), respectively.
2. The body coordinate system (labeled with a subscript b; see also Chap. 2) is aligned with the shape of the fuselage of the aircraft. The X_b-axis points through the nose of the aircraft, and the Y_b-axis points to the right of the X_b-axis (facing the direction of the view of the pilot), perpendicular to the X_b-axis. Finally, the Z_b-axis points down through the bottom of the craft, perpendicular to the X_b-Y_b plane.
3. The world coordinate system or the world reference frame (labeled with a subscript w), which is not shown in Fig. 11.1, is the projection of the body frame of the aircraft on the ground.
4. The servo-based coordinate system (labeled with a subscript s) is attached to the base of the pan/tilt servo mechanism, which is aligned with the body coordinate system of the UAV. The rotation between the servo-based coordinate system and the body coordinate system can be ignored, and the translation is fixed.
5. The spherical coordinate system (labeled with a subscript sp) is also attached to the base of the pan/tilt servo mechanism. It is used to define the orientation of

the camera and the target with respect to the UAV. Given a generic point $\mathbf{P}_s$ in the servo-based coordinate system, say

$$\mathbf{P}_s = \begin{pmatrix} x_s \\ y_s \\ z_s \end{pmatrix}, \tag{11.1}$$

its position can be defined in the spherical coordinate system by three numbers: radius r_{sp}, azimuth angle θ_{sp}, and elevation angle ϕ_{sp}, which is given by

$$\mathbf{P}_{sp} = \begin{pmatrix} r_{sp} \\ \theta_{sp} \\ \phi_{sp} \end{pmatrix}, \tag{11.2}$$

where

$$r_{sp} = \sqrt{x_s^2 + y_s^2 + z_s^2}, \tag{11.3}$$

and

$$\theta_{sp} = \tan^{-1}\left(\frac{x_s}{z_s}\right), \qquad \phi_{sp} = \sin^{-1}\left(\frac{y_s}{r_{sp}}\right). \tag{11.4}$$

6. The camera coordinate system or camera frame (labeled with a subscript c) describes the orientation of the camera, which is attached to the end of the pan/tilt servo mechanism, with the pan/tilt rotation with respect to the servo-based frame being given as

$$\mathbf{x}_{c,sp} = \begin{pmatrix} \phi_{c,sp} \\ \theta_{c,sp} \end{pmatrix}. \tag{11.5}$$

   The origin of the camera coordinate system is the optical center of the camera. The Z_c-axis is aligned with the optical axis of the camera and points from the optical center C toward the image plane. When the camera frame coincides with the servo-based frame, $\mathbf{x}_{c,sp} = 0$. Otherwise, $\mathbf{x}_{c,sp}$ corresponds to the rotation of the camera frame first about the Y_s-axis by an angle of $\theta_{c,sp}$, and then about the X_s-axis by an angle of $\phi_{c,sp}$. This rotation sequence is defined in terms of the structure of the pan/tilt servo mechanism. The object coordinate system or object frame (labeled with a subscript o) is attached to a fixed ground landmark.
7. The image frame (or the principle image coordinate system) (appended with a subscript i) is a 2D image plane perpendicular to the optical axis of the camera. The origin is the intersection of the optical axis and the image plane, namely, the principle point. The coordinate axes, X_i and Y_i, are aligned with the camera coordinate axes, X_c and Y_c, respectively.

11.3 Camera Calibration

This section presents the calibration of the adopted camera. As a measurement sensor, a camera is used to map the points in the 3D space to the projected points in

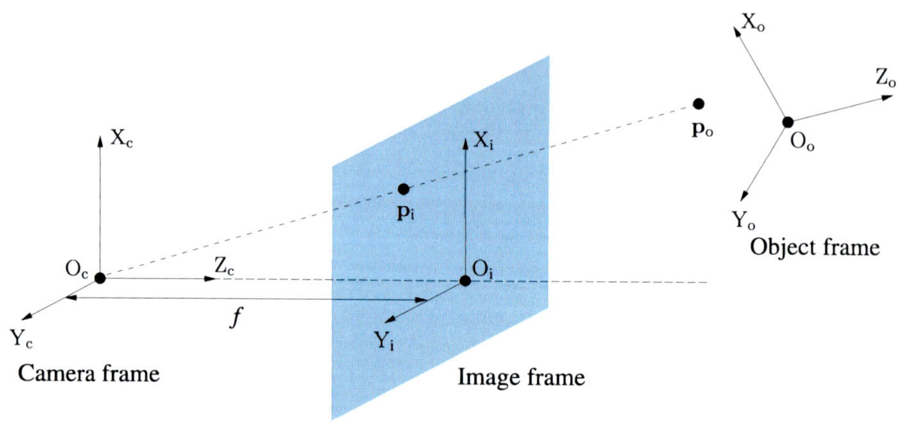

Fig. 11.2 Frontal pinhole camera model

the 2D image frame. Such a mapping can be described using a suitable mathematic model with intrinsic and extrinsic parameters. The computation of the intrinsic and extrinsic parameters is called camera calibration. Calibration results strongly affect performance of a vision system. In this section, we first introduce a camera model and then explain how to estimate the its parameters. A compensation approach is also utilized to cope with distortions in the model.

11.3.1 Camera Model

The frontal pinhole projection [113] is a commonly used camera model, illustrated in Fig. 11.2. It is simple and convenient by placing the image plane in front of the optical center. Given a generic point, say **P**, utilizing the model of the pinhole projection and geometric structure of the pan/tilt servo mechanism, we can obtain the transformations among the camera coordinate system, the image coordinate system, and the object coordinate system (see Fig. 11.1), which are given by

$$\lambda \begin{pmatrix} \mathbf{p}_{\mathrm{i}} \\ 1 \end{pmatrix} = \begin{bmatrix} f_{\mathrm{x}} & s_{\theta} & o_{\mathrm{x}} \\ 0 & f_{\mathrm{y}} & o_{\mathrm{y}} \\ 0 & 0 & 1 \end{bmatrix} \begin{bmatrix} 1 & 0 & 0 & 0 \\ 0 & 1 & 0 & 0 \\ 0 & 0 & 1 & 0 \end{bmatrix} \begin{bmatrix} \mathbf{R}_{\mathrm{c/o}} & \mathbf{t}_{\mathrm{c/o}} \\ 0 & 1 \end{bmatrix} \begin{pmatrix} \mathbf{p}_{\mathrm{o}} \\ 1 \end{pmatrix}, \quad (11.6)$$

or in the matrix format

$$\lambda \begin{pmatrix} \mathbf{p}_{\mathrm{i}} \\ 1 \end{pmatrix} = \mathbf{K}_{\mathrm{s}} \Pi_0 \mathbf{G} \begin{pmatrix} \mathbf{p}_{\mathrm{o}} \\ 1 \end{pmatrix} = \mathbf{K}_{\mathrm{s}} \Pi_0 \begin{pmatrix} \mathbf{p}_{\mathrm{c}} \\ 1 \end{pmatrix}, \quad (11.7)$$

where

$$\mathbf{p}_{\mathrm{o}} = \begin{pmatrix} x_{\mathrm{o}} \\ y_{\mathrm{o}} \\ z_{\mathrm{o}} \end{pmatrix}, \qquad \mathbf{p}_{\mathrm{c}} = \begin{pmatrix} x_{\mathrm{c}} \\ y_{\mathrm{c}} \\ z_{\mathrm{c}} \end{pmatrix}, \quad (11.8)$$

are the coordinates of $\mathbf{P}$ with respect to the object frame and the camera frame, respectively;

$$\mathbf{p}_\mathrm{i} = \begin{pmatrix} x_\mathrm{i} \\ y_\mathrm{i} \end{pmatrix} \tag{11.9}$$

is the projection of $\mathbf{P}$ in the image plane; $\mathbf{R}_{\mathrm{o/c}}$ and $\mathbf{t}_{\mathrm{o/c}}$ are, respectively, the rotation matrix and the translation vector, which define the rigid-body transformation from the object frame to the camera frame; λ is a scaling factor, which is the depth of $\mathbf{P}$ in the camera coordinate system; s_θ is the so-called skew factor, which is very close to zero; $[o_\mathrm{x}, o_\mathrm{y}]^\mathrm{T}$ are the coordinate of the principal point in pixels; and finally $f_\mathrm{x} = f s_\mathrm{x}$ and $f_\mathrm{y} = f s_\mathrm{y}$ are, respectively, the vertical and horizontal focal lengths measured in pixels (f is the focal length measured in a metric unit, and s_x and s_y are the numbers of pixels per unit length of the sensor along the x and y directions).

The pinhole projection model in (11.6) is an approximation of the real image formation process. It is, however, not accurate enough in demanding situations, as the image distortion introduced by optical lenses always severely affects its accuracy. A more precise pinhole projection model, extended by using radial and slight tangential distortion [80], is given as

$$\left.\begin{aligned}
\begin{pmatrix} \mathbf{p}_\mathrm{p} \\ 1 \end{pmatrix} &= \begin{pmatrix} x_\mathrm{p} \\ y_\mathrm{p} \\ 1 \end{pmatrix} = \frac{1}{\lambda}\Pi_0 \begin{pmatrix} \mathbf{p}_\mathrm{c} \\ 1 \end{pmatrix} = \frac{1}{\lambda}\Pi_0 \mathbf{G} \begin{pmatrix} \mathbf{p}_\mathrm{o} \\ 1 \end{pmatrix}, \\
\mathbf{p}_\mathrm{d} &= (1 + k_1 d^2 + k_2 d^4 + k_5 d^6)\mathbf{p}_\mathrm{p} + \mathbf{K}_\mathrm{d}, \\
d &= \sqrt{x_\mathrm{p}^2 + y_\mathrm{p}^2}, \\
\mathbf{K}_\mathrm{d} &= \begin{pmatrix} 2k_3 x_\mathrm{p} y_\mathrm{p} + k_4(d^2 + 2x_\mathrm{p}^2) \\ k_3(d^2 + 2y_\mathrm{p}^2) + 2k_4 x_\mathrm{p} y_\mathrm{p} \end{pmatrix}, \\
\begin{pmatrix} \mathbf{p}_\mathrm{i} \\ 1 \end{pmatrix} &= \mathbf{K}_\mathrm{s} \begin{pmatrix} \mathbf{p}_\mathrm{d} \\ 1 \end{pmatrix},
\end{aligned}\right\} \tag{11.10}$$

where $\mathbf{p}_\mathrm{p}$ is the normalized projected image point, and $\mathbf{p}_\mathrm{d}$ is the normalized distorted image point; k_1, k_2, k_5 are radial distortion coefficients; and k_3, k_4 are tangential distortion coefficients.

The intrinsic parameters f_x, f_y, o_x, o_y and the distortion coefficients k_1 to k_5 are referred to as the physical camera parameters. The objective of the calibration is to search for the optimal values of these parameters in terms of the image observations of known feature points.

11.3.2 Intrinsic Parameter Estimation

Based on the camera model given in (11.10), the camera calibration toolbox proposed in [23] can be used to estimate the intrinsic parameters for the camera adopted in SheLion. The procedure of the calibration includes the following steps:

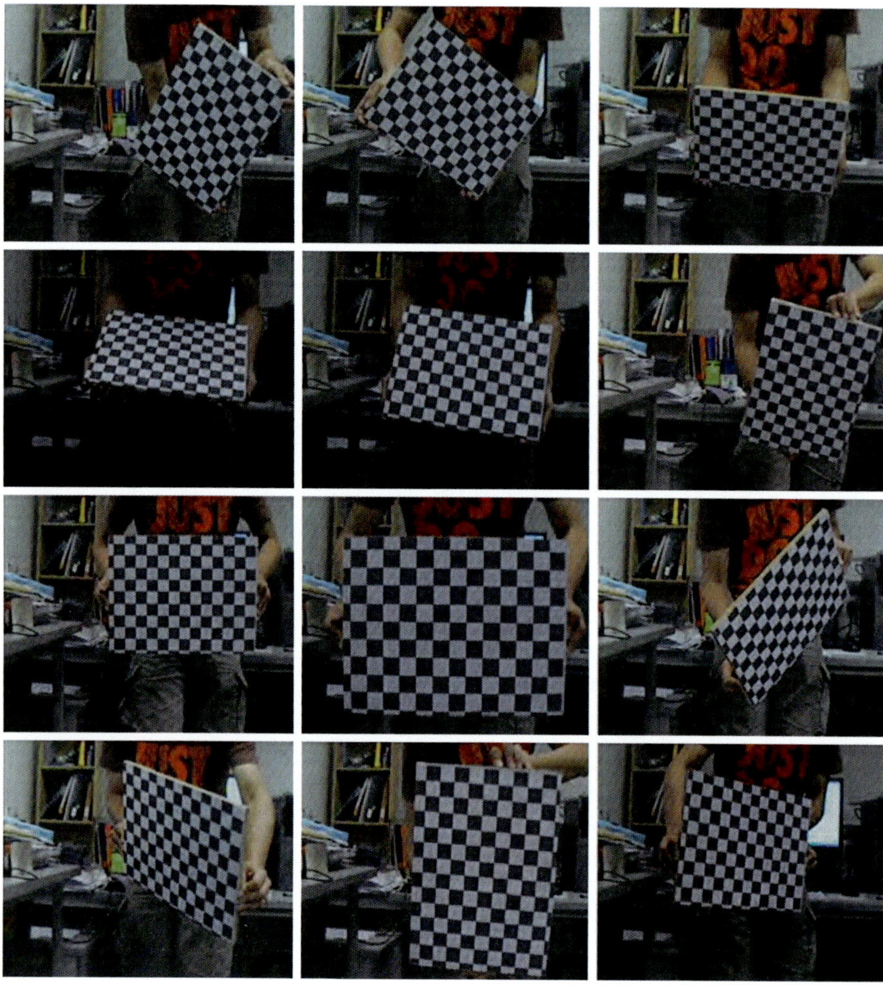

Fig. 11.3 Images for camera calibration

1. Capture images of the chessboard pattern using the onboard camera with the varying rotation and the translation of the chessboard with respect to the camera. The examples of the calibration images are shown in Fig. 11.3.
2. Extract grid corners of the chessboard pattern using the calibration toolbox in Matlab. The extracted corners are illustrated in Fig. 11.4.
3. Calculate the intrinsic parameters of the camera using the toolbox, which employs a two-step optimization approach to minimize the total re-projection error of all calibration parameters [224]. First, the direct linear transformation (DLT) algorithm is used to obtain a linear solution. Then, using this solution as the initial values, an optimal solution can be achieved by an iterative searching scheme. The calibration parameters comprise 9 DOF intrinsic parameters and $n \times 6$ DOF

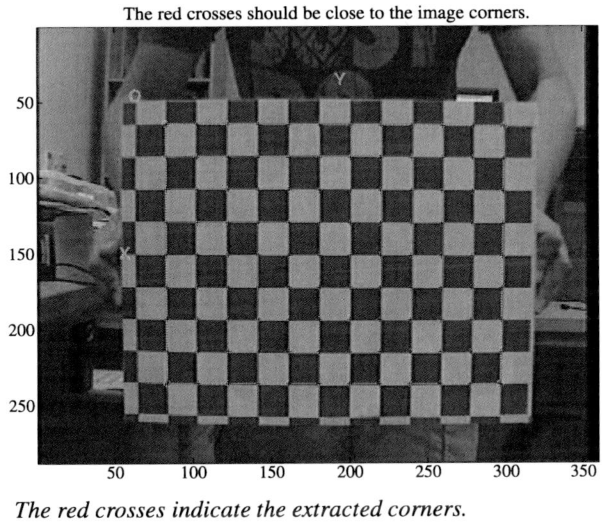

Fig. 11.4 Grid corner extraction for camera calibration

extrinsic parameters, where n is the number of the images. More detailed descriptions on the camera calibration can be found in [23].

The final calibrated intrinsic parameters of the onboard camera used on SheLion are given as: the focal length

$$\begin{pmatrix} f_x \\ f_y \end{pmatrix} = \begin{pmatrix} 605.67766 \\ 661.84569 \end{pmatrix} \pm \begin{pmatrix} 2.65276 \\ 2.91701 \end{pmatrix}, \tag{11.11}$$

the principal point

$$\begin{pmatrix} o_x \\ o_y \end{pmatrix} = \begin{pmatrix} 176.26743 \\ 141.21784 \end{pmatrix} \pm \begin{pmatrix} 3.33796 \\ 3.18607 \end{pmatrix}, \tag{11.12}$$

the skew $s_\theta = 0$, and finally, the distortion coefficients

$$\begin{pmatrix} k_1 \\ k_2 \\ k_3 \\ k_4 \\ k_5 \end{pmatrix} = \begin{pmatrix} -0.42687 \\ 0.50745 \\ 0.00349 \\ -0.00146 \\ 0.00000 \end{pmatrix} \pm \begin{pmatrix} 0.02702 \\ 0.34637 \\ 0.00097 \\ 0.00123 \\ 0.00000 \end{pmatrix}. \tag{11.13}$$

11.3.3 Distortion Compensation

The distorted images can be corrected using the estimated intrinsic parameters of the camera obtained in the previous section. Generally, the equation in (11.10) cannot be solved analytically from a distorted image to a pinhole projection; an iteration process is required. The experiment of the distortion compensation is conducted using the intrinsic parameters of the camera in (11.11) to (11.13). The comparison of the images before and after the correction is shown in Fig. 11.5.

a. The original image

b. The image after distortion compensation

Fig. 11.5 Distortion compensation

11.3.4 Simplified Camera Model

In image processing, the distortion of the lens is compensated, and the origin of the image plane is set as the principle point. Thus, o_x and o_y can be set to be zeros. Moreover, since $s_\theta \approx 0$, we can obtain a simplified pinhole projection model based on (11.7), which is given as

$$\lambda \begin{pmatrix} \mathbf{p}_i \\ 1 \end{pmatrix} = \mathbf{K}_f \Pi_0 \begin{pmatrix} \mathbf{p}_c \\ 1 \end{pmatrix}, \tag{11.14}$$

where

$$\lambda = z_c, \qquad \mathbf{K}_f = \begin{bmatrix} f_x & 0 & 0 \\ 0 & f_y & 0 \\ 0 & 0 & 1 \end{bmatrix}. \tag{11.15}$$

This simplified pinhole camera model is adopted throughout the rest of this chapter. Following from (11.14), we can express $\mathbf{p}_c$ in terms of the location of $\mathbf{P}$ in the image plane as

$$\mathbf{p}_c = \lambda \begin{pmatrix} \bar{x}_i \\ \bar{y}_i \\ 1 \end{pmatrix}, \tag{11.16}$$

where

$$\bar{x}_i = \frac{x_i}{f_x}, \qquad \bar{y}_i = \frac{y_i}{f_y}. \tag{11.17}$$

11.4 Vision-Based Ground Target Following

In order to explore capabilities of the vision-based system of SheLion, a ground target following application is conducted. In the experiment, a sophisticated vision-based target detection and tracking scheme is proposed, which employs robust feature descriptors and the image tracking techniques. Based on the vision sensing and

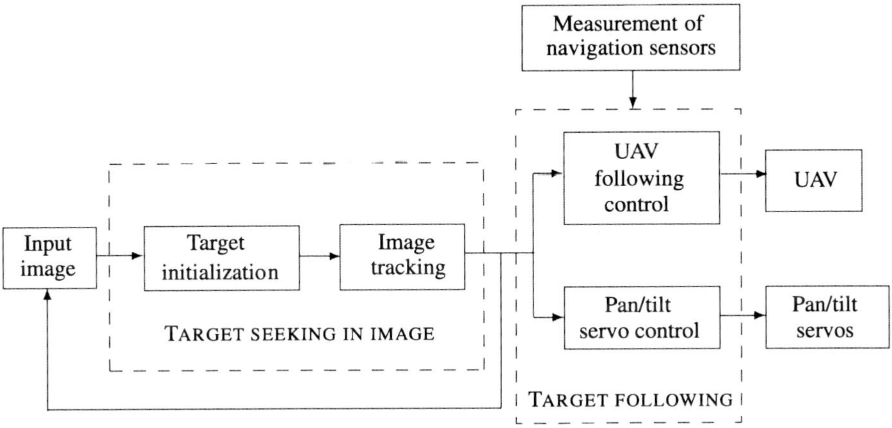

Fig. 11.6 Flow chart of the ground target following

navigation sensors, the relative distance to the target is estimated. Such vision feed back is integrated with the flight control system to guide the UAV to follow the ground target inflight.

In what follows, we propose a systematic approach to realize robust and efficient ground target following. As illustrated in Fig. 11.6, in the target initialization (or the target detection part), a predefined target is identified automatically by using the color segmentation and feature-based pattern recognition methods. A hierarchical tracking scheme is then used to track the target in the image. Lastly, the target following approach is implemented.

11.4.1 Target Detection

The purpose of target detection is to identify the target interested from the image automatically based on a database of the pre-selected targets. In the experiment, a toy car is chosen as the ground target. A classical pattern recognition procedure is used to identify the target automatically, which includes three main steps, i.e., segmentation, feature extraction, and pattern recognition.

11.4.1.1 Segmentation

The segmentation step aims to separate the objects of interest from the background. To simplify further processing, some assumptions are made. First, the target and environments exhibit *Lambertian reflectance*, in other words, their brightness is unchanged regardless of viewing directions. Second, the target has a distinct color distribution compared to the surrounding environments.

1. STEP 1: THRESHOLD IN COLOR SPACE. Based on the above assumptions, a fast threshold approach is applied to the input color image. To make the surface color of the target constant and stable under varying lighting conditions, the color image is represented in HSV space, which stands for hue (*hue*), saturation (*sat*), and value (*val*), introduced originally by Smith [169].

   In fact, there are a number of color models used in various applications involving color image processing, such as CIE, RGB, YUV, HSV, and CMYK (see, e.g., [201]). The HSV color space is selected in this work due to its two useful facts, i.e., (i) the intensity component is decoupled from chrominance information represented as hue and saturation; and (ii) the hue and saturation are intimately related to the way in which humans perceive chrominance [70, 145, 165].

   Given that each pixel has a set of RGB values $\{r, g, b\}$, defined $max_{\mathrm{rgb}} = \max\{r, g, b\}$ and $min_{\mathrm{rgb}} = \min\{r, g, b\}$, the HSV color model, conveniently represented by the hexcome model [59], is defined as follows:

$$hue = \begin{cases} \text{undefined}, & \text{if } max_{\mathrm{rgb}} = min_{\mathrm{rgb}}, \\ 60^\circ \times \frac{g-b}{max_{\mathrm{rgb}}-min_{\mathrm{rgb}}} + 0^\circ, & \text{if } max_{\mathrm{rgb}} = r \text{ and } g \geq b, \\ 60^\circ \times \frac{g-b}{max_{\mathrm{rgb}}-min_{\mathrm{rgb}}} + 360^\circ, & \text{if } max_{\mathrm{rgb}} = r \text{ and } g < b, \\ 60^\circ \times \frac{b-r}{max_{\mathrm{rgb}}-min_{\mathrm{rgb}}} + 120^\circ, & \text{if } max_{\mathrm{rgb}} = g, \\ 60^\circ \times \frac{r-g}{max_{\mathrm{rgb}}-min_{\mathrm{rgb}}} + 240^\circ, & \text{if } max_{\mathrm{rgb}} = b, \end{cases} \tag{11.18}$$

$$sat = \begin{cases} 0, & \text{if } max_{\mathrm{rgb}} = 0, \\ 1 - \frac{min_{\mathrm{rgb}}}{max_{\mathrm{rgb}}}, & \text{otherwise}, \end{cases} \tag{11.19}$$

   and

$$val = max_{\mathrm{rgb}}. \tag{11.20}$$

   Pre-calculated threshold ranges are applied to the *hue*, *sat*, and *val* channels:

$$hue_{\mathrm{r}} = [hue_1, hue_2], \qquad sat_{\mathrm{r}} = [sat_1, sat_2], \qquad val_{\mathrm{r}} = [val_1, val_2]. \tag{11.21}$$

   Only the pixel values falling in these color ranges are identified as the foreground points, and pixels of the image that fall out of the specified color range are removed. The procedure of the image pre-process is illustrated in Fig. 11.7. The raw image captured will be converted into the HSV color space shown in Fig. 11.7.b. The converted image is then segmented by using the method mentioned earlier based on hue_{r}, sat_{r}, and val_{r}—the three channels shown in Fig. 11.7.c.

2. STEP 2: MORPHOLOGICAL OPERATIONS. As shown in Fig. 11.7, normally, the segmented image is not smooth and has many noise points. Morphological operations are then employed to filter out noise, fuse narrow breaks and gulfs, eliminate small holes, and fill gaps in the contours. Next, a contour detection approach is used to obtain the complete boundary of the objects in the image, which will be used in the feature extraction.

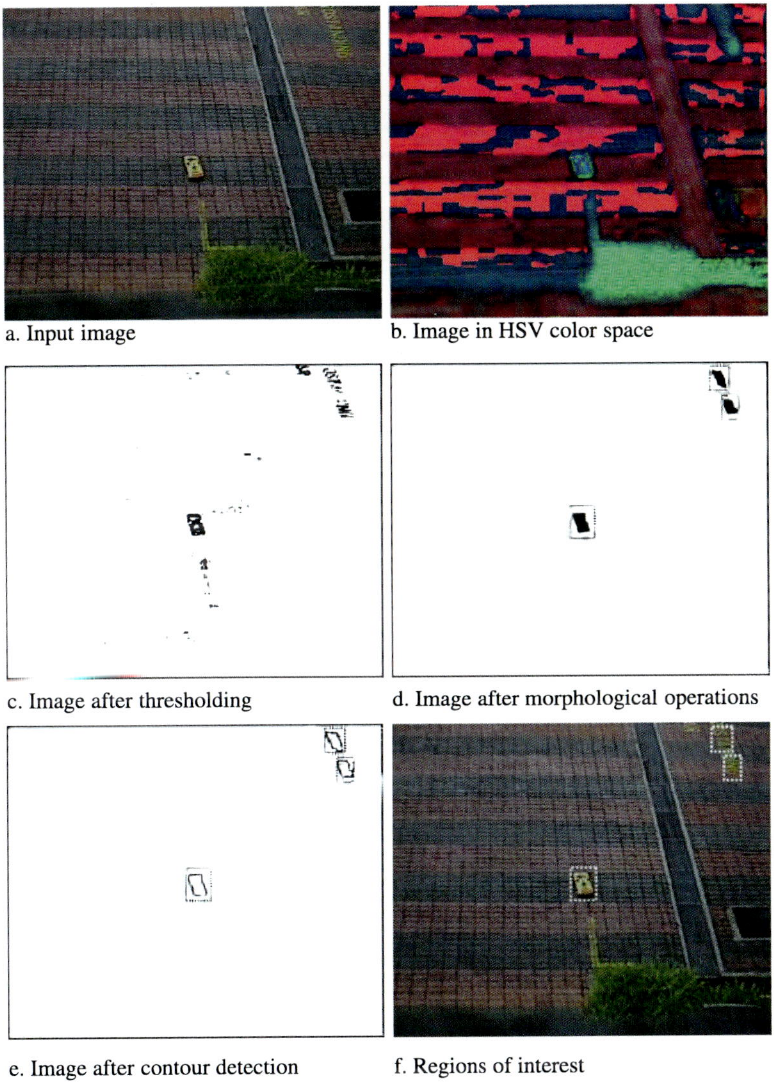

Fig. 11.7 Illustration of segmentation

11.4.1.2 Feature Extraction

Generally, multiple objects will be found in the segmented images, including the true target and false objects. The geometric and color features are used as the descriptors to identify the true target.

1. GEOMETRIC FEATURE EXTRACTION: To describe the geometric features of the objects, the four lowest moment invariants proposed in [110] are employed, since

they are independent of position, size, and orientation in the visual field. The four lowest moment invariants, defined in the segmented image $I(x, y)$, are given by

$$\phi_1 = \eta_{20}^{\mathrm{m}} + \eta_{02}^{\mathrm{m}}, \tag{11.22}$$

$$\phi_2 = (\eta_{20}^{\mathrm{m}} - \eta_{02}^{\mathrm{m}})^2 + 4(\eta_{11}^{\mathrm{m}})^2, \tag{11.23}$$

$$\phi_3 = (\eta_{30}^{\mathrm{m}} - 3\eta_{12}^{\mathrm{m}})^2 + (\eta_{03}^{\mathrm{m}} - 3\eta_{21}^{\mathrm{m}})^2, \tag{11.24}$$

$$\phi_4 = (\eta_{30}^{\mathrm{m}} + \eta_{12}^{\mathrm{m}})^2 + (\eta_{03}^{\mathrm{m}} + \eta_{21}^{\mathrm{m}})^2, \tag{11.25}$$

where η_{pq}^{m}, for $p + q = 2, 3, \ldots$, is the improved normalized central moment defined as

$$\eta_{pq}^{\mathrm{m}} = \frac{\mu_{pq}^{\mathrm{c}}}{A^{(p+q+1)/2}} \tag{11.26}$$

and where A is the interior area of the shape, and μ_{pq}^{c} is the central moment defined as

$$\mu_{pq}^{\mathrm{c}} = \int_C (x - \bar{x})^p (y - \bar{y})^q \, ds, \quad p, q = 0, 1, \ldots. \tag{11.27}$$

Note that in (11.27), C is the boundary curve of the shape, $\int_C$ is a line integral along C, $ds = \sqrt{(dx)^2 + (dy)^2}$, and $[\bar{x}, \bar{y}]$ is the coordinate of the centroid of the shape in the image plane.

2. COLOR FEATURE EXTRACTION: To make the target detection and tracking more robust, we also employ a color histogram to represent the color distribution of the image area of the target, which is not only independent of the target orientation, position, and size but also robust to partial occlusion of the target and easy to implement. Due to the stability in outdoor environments, only *hue* and *val* are employed to construct the color histogram for object recognition, which is defined as

$$H = \{hist(i, j)\}, \quad i = 1, \ldots, N_{\mathrm{hue}}, \; j = 1, \ldots, N_{\mathrm{val}}, \tag{11.28}$$

where

$$hist(i, j) = \sum_{(x,y)\in\Omega} \delta\left(i, \left[\frac{hue(x, y)}{N_{\mathrm{hue}}}\right]\right) \delta\left(j, \left[\frac{val(x, y)}{N_{\mathrm{val}}}\right]\right), \tag{11.29}$$

and where N_{hue} and N_{val} are the partition numbers of *hue* and *val* channels, respectively, Ω is the region of the target, $[\cdot]$ is the nearest integer operator, and $\delta(a, b)$ is the Kronecker delta function given by

$$\delta(a, b) = \begin{cases} 1, & \text{if } a = b, \\ 0, & \text{otherwise}. \end{cases} \tag{11.30}$$

We illustrate the concept of the color histogram of a sample image in Fig. 11.8.

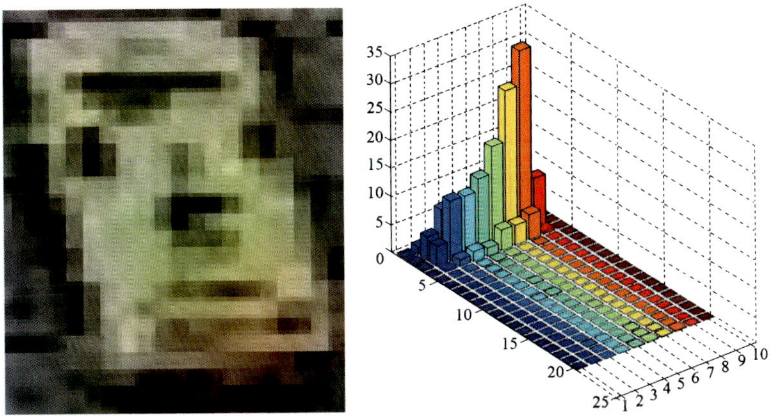

a. The sample image

b. Color histogram

Fig. 11.8 Color histogram extraction

11.4.1.3 Object Representation

The purpose of object representation is to arrange features of an object in a compact and identifiable form [146]. A straightforward way is to convert these features in a high dimensional vector. For example, the feature vector of the i-th object is given by

$$\alpha_i = [\beta_{c,i}, \phi_{1,i}, \phi_{2,i}, \phi_{3,i}, \phi_{4,i}, H_i] = \{\alpha_{k,i}\}, \quad k = 1, \ldots, d, \qquad (11.31)$$

where d is the dimension of the feature vector. Theoretically, all the states or the features of an object template in a database should be updated based on certain rules. To reduce the computational cost, the appearance of objects is assumed to be constant.

11.4.1.4 Pattern Recognition

The purpose of pattern recognition is to identify the target from the extracted foreground objects in terms of the extracted features in (11.31). The straightforward classifier is to use the nearest-neighbor rule. It calculates a metric or "distance" between an object and a template in a feature space and assigns the object to the class with the highest scope. But to take advantage of a priori knowledge of the feature distribution, the classification problem is formulated under the model-based framework and solved by using a probabilistic classifier. A discriminant function, derived from Bayes' theorem, is employed to identify the target. This function is computed based on the measured feature values of each object and the known distribution of features obtained from training data. The pattern recognition will be described in the following parts:

1. STEP 1. PRE-FILTER: Before classifying the objects, a pre-filter is carried out to remove the objects whose feature values are outside certain regions determined by a priori knowledge. This step aims to improve the robustness of the pattern recognition and speed up the calculation.
2. STEP 2. DISCRIMINANT FUNCTION: It is assumed that only one target appears in the image. The discriminant function, derived from Bayes' theorem, is employed to determinate the target based on the measured feature values of each object and known distribution of features of the target obtained from training data. We assume that these features are independent and fulfill normal distributions. Thus, we can define the simplified discriminant function and classifier with weightings as

$$f(\alpha_i) = \sum_{k=1}^{5} w_k \left(\frac{\alpha_{k,i} - \mu_k}{\sigma_k} \right)^2 + w_6 \left(\frac{d_c(H_i, G) - \mu_6}{\sigma_6} \right)^2 \tag{11.32}$$

and

$$h(\alpha_i) = \begin{cases} \text{target}, & \text{if } f(\alpha_i) \le \Gamma, \\ \text{undetectable object}, & \text{if } f(\alpha_i) > \Gamma, \end{cases} \tag{11.33}$$

where

$$d_c(H_i, G) = \frac{\sum_{p=1}^{N_{\text{hue}}} \sum_{q=1}^{N_{\text{val}}} \min(H_i(p,q), G(p,q))}{\min(|H_i|, |G|)}, \tag{11.34}$$

and $\alpha_{k,i}$ is the k-th element of the feature vector of the object i; w_1 to w_6 are the weighting scalars of the corresponding features; Γ is a threshold value to be chosen empirically. Based on this classifier, an object is assigned to the target class, if $f(\alpha_i) \le \Gamma$. Otherwise, it is assigned to the undetectable object class. If there are multiple objects in the target class, the one with the smallest value of the discriminant function is selected. In the situation that the target class is empty, no target is found in the current frame.

11.4.2 Image Tracking

As shown in Fig. 11.6, after initialization, the image tracking techniques are employed. The purpose of image tracking is to find the corresponding region or point to the given target. Unlike detection, the entire image search is not required. Thus, the processing speed of image tracking is faster than the detection. The image tracking problem can be solved by using two main approaches: (i) filtering and data association, and (ii) target representation and localization [34].

1. FILTERING AND DATA ASSOCIATION: The filtering and data association approach can be considered as a top-down process. The purpose of filtering is to estimate the state of the target, such as static appearance and location. Typically, the state estimation is achieved by using filtering techniques [216, 225]. It is

known (see, e.g., [107]) that most of the tracking algorithms are model based because a good model-based tracking algorithm will greatly outperform any model-free tracking algorithm if the underlying model turns out to be a good one. If the measurement noise satisfies the Gaussian distribution, the optimal solution can be achieve by the Kalman filtering techniques [10]. In the more general case, particle filters are more suitable and robust [84]. However, the computational cost increases and sample degeneracy is also a problem. When multiple targets are tracked in the image sequence, the validation and association of the measurements become a critical issue. The association techniques, such as the probabilistic data association filter and joint probabilistic data association filter, are widely used (see, e.g., [189]).

2. TARGET REPRESENTATION AND LOCALIZATION: Besides using motion prediction to find the corresponding region or point, *target representation and localization* is considered another efficient way and is referred to as a bottom-up approach. Among the searching methods, the mean shift approach using the density gradient is commonly used [8], which is trying to search the peak value of the object probability density. However, the efficiency will be limited when the space movement of the target becomes significant.

To take advantage of the aforementioned approaches, multiple trackers were widely adopted in applications of image tracking. In [189], the tracking scheme by integrating motion, color, and geometric features was proposed to realize robust image tracking. In conclusion, combining the motion filtering and advanced searching algorithms definitely make the tracking processing more robust, but the computational load is heavier.

In our approach, instead of using multiple trackers simultaneously, a hierarchical tracking scheme is proposed to balance the computational cost and performance, which is illustrated in Fig. 11.9. In the model-based image tracking, the Kalman filtering technique is employed to provide accurate estimation and prediction of the position and velocity of a single target, referred to as dynamic information. If the model-based tracker fails to find the target, a mean shift-based image tracking method will be activated to retrieve the target in the image.

11.4.2.1 Model-Based Image Tracking

Model-based image tracking is a top-down method. It will predict the possible location of the target in the subsequent frames and then do the data association based on an updated likelihood function. Several methods are employed to make the tracking more robust and efficient, which are given by the following:

1. Narrow the search window in terms of the prediction of the Kalman filter.
2. Integrate the spatial information with appearance and set the different weightings for the discriminant function. For example, it is meaningful to give higher weightings to spatial information, when the geometric features may change significantly under noise conditions.

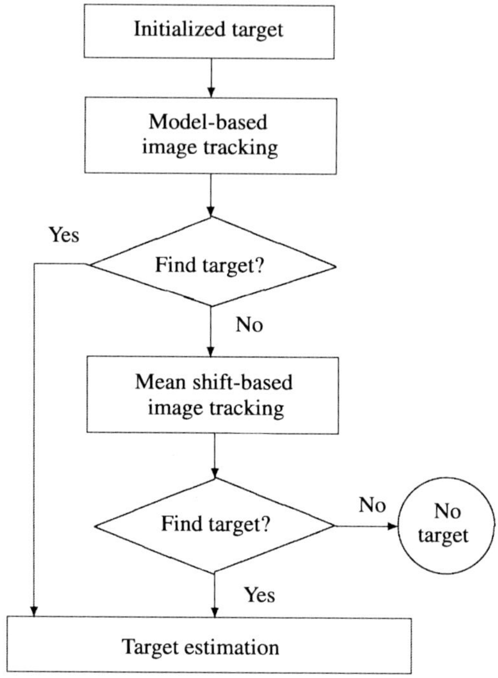

Fig. 11.9 Flow chart of image tracking

To predict the target location in the image, a motion model is required. It is well known that the motion of a point mass in the two-dimensional plane can be defined by its two-dimensional position and velocity vector. Let $\mathbf{x} = [\bar{x}, \dot{\bar{x}}, \bar{y}, \dot{\bar{y}}]^{\mathrm{T}}$ be the state vector of the centroid of the tracked target in the Cartesian coordinate system. Non-maneuvering motion of the target is defined by it having zero acceleration: $[\ddot{\bar{x}}, \ddot{\bar{y}}]^{\mathrm{T}} = [0, 0]^{\mathrm{T}}$. Strictly speaking, the motion of the intended ground targets may be maneuvering with unknown inputs. Nevertheless, we assume the standard 4th-order non-maneuvering motion model by setting the acceleration as $[\ddot{\bar{x}}, \ddot{\bar{y}}]^{\mathrm{T}} = \mathbf{w}(t)$, where $\mathbf{w}(t)$ is a white noise process [107]. The resulting continuous-time model is

$$\dot{\mathbf{x}} = \mathbf{A}\mathbf{x} + \mathbf{B}\mathbf{w}, \tag{11.35}$$

$$\mathbf{z} = \mathbf{C}\mathbf{x} + \mathbf{v}, \tag{11.36}$$

where $\mathbf{w}$ and $\mathbf{v}$ denote the input and measurement zero-mean Gaussian noises, and

$$\mathbf{A} = \begin{bmatrix} 0 & 1 & 0 & 0 \\ 0 & 0 & 0 & 0 \\ 0 & 0 & 0 & 1 \\ 0 & 0 & 0 & 0 \end{bmatrix}, \quad \mathbf{B} = \begin{bmatrix} 0 & 0 \\ 1 & 0 \\ 0 & 0 \\ 0 & 1 \end{bmatrix},$$
$$\mathbf{C} = \begin{bmatrix} 1 & 0 & 0 & 0 \\ 0 & 0 & 1 & 0 \end{bmatrix}. \tag{11.37}$$

The motion of the centroid of the target, $\mathbf{x} = [\bar{x}, \dot{\bar{x}}, \bar{y}, \dot{\bar{y}}]^{\mathrm{T}}$, in the two-dimensional image coordinate is tracked using a standard 4th-order Kalman filter, which predicts the possible location of the target in the successive frames. The discrete-time model of the target motion can be expressed as

$$\mathbf{x}(k|k-1) = \mathbf{\Phi}\mathbf{x}(k-1) + \mathbf{\Lambda}\mathbf{w}(k-1), \tag{11.38}$$

$$\mathbf{z}(k) = \mathbf{H}x(k) + \mathbf{v}(k), \tag{11.39}$$

where

$$\mathbf{H} = \begin{bmatrix} 1 & 0 & 0 & 0 \\ 0 & 0 & 1 & 0 \end{bmatrix}, \tag{11.40}$$

and

$$\mathbf{\Phi} = \begin{bmatrix} 1 & \mathrm{T_s} & 0 & 0 \\ 0 & 1 & 0 & 0 \\ 0 & 0 & 1 & \mathrm{T_s} \\ 0 & 0 & 0 & 1 \end{bmatrix}, \qquad \mathbf{\Lambda} = \begin{bmatrix} \frac{\mathrm{T_s^2}}{2} & 0 \\ \mathrm{T_s} & 0 \\ 0 & \frac{\mathrm{T_s^2}}{2} \\ 0 & \mathrm{T_s} \end{bmatrix}, \tag{11.41}$$

and $\mathrm{T_s}$ is the sampling period of the vision-based tracking system. A Kalman filter can then be designed based on the above motion model to estimate the states of the target in the image plane. The filter consists of the following stages:

1. Predicted state

$$\hat{\mathbf{x}}(k|k-1) = \Phi\hat{\mathbf{x}}(k-1). \tag{11.42}$$

2. Updated state estimate

$$\hat{\mathbf{x}}(k) = \hat{\mathbf{x}}(k|k-1) + \mathbf{K}(k)(\mathbf{z}(k) - \mathbf{H}\hat{\mathbf{x}}(k|k-1)), \tag{11.43}$$

where $\mathbf{K}(k)$ is the optimal Kalman gain.

The distance between the location of each object $\mathbf{z}_i$ and the predicted location of the target $\hat{\mathbf{z}}$ is employed as the dynamic feature defined by

$$\tilde{\mathbf{z}}_i = \mathbf{z}_i(k) - \hat{\mathbf{z}}(k) = \mathbf{z}_i(k) - \mathbf{H}\hat{\mathbf{x}}(k|k-1). \tag{11.44}$$

Both of the static and dynamic features of an object are employed in image tracking. Thus, the discriminant functions, which includes appearance and spatial information, are shown as follows:

$$\begin{aligned} f(\alpha_i, \tilde{\mathbf{z}}_i) = {} & \sum_{k=1}^{5} w_k \left(\frac{\alpha_{k,i} - \mu_k}{\sigma_k}\right)^2 + w_6 \left(\frac{d_\mathrm{c}(H_i, G) - \mu_6}{\sigma_6}\right)^2 \\ & + w_7 \left(\frac{\|\tilde{\mathbf{z}}_i\| - \mu_7}{\sigma_7}\right)^2 \end{aligned} \tag{11.45}$$

and

$$h(\alpha_i, \tilde{\mathbf{z}}_i) = \begin{cases} \text{target}, & \text{if } f(\alpha_i, \tilde{\mathbf{z}}_i) \le \Gamma, \\ \text{undetectable object}, & \text{if } f(\alpha_i, \tilde{\mathbf{z}}_i) > \Gamma. \end{cases} \tag{11.46}$$

Most of the time, the model-based tracker can lock the target in the image sequence, but sometimes it may fail due to noise or disturbance, such as partial occlusion. Thus, a scheme is required to check whether the target is still in the image, and then other trackers are activated.

11.4.2.2 Switching Mechanism

The purpose of the switching mechanism is to check whether the target is still in the image at the moment that the target is lost by the model-based tracker. If yes, the mean shift tracker will be activated. The loss of the target can be attributed to the poor match of features due to noise, distortion, or occlusion in the image. An alternative reason may be the maneuvering motion of the target, and the target is out of the image. Therefore, in order to know the reason and take the special way to find the target again, it is necessary to formulate the decision making as the following hypothesis testing problem:

H_0: The target is still in the image.

H_1: The target is not in the image due to maneuvers.

The estimation error is considered as a random variable, which is defined by

$$\varepsilon = \|\mathbf{H}\hat{\mathbf{x}}_{k-1} - \mathbf{z}_{k-1}\|^2_{\Sigma^{-1}} = (\mathbf{H}\hat{\mathbf{x}}_{k-1} - \mathbf{z}_{k-1})'\Sigma^{-1}(\mathbf{H}\hat{\mathbf{x}}_{k-1} - \mathbf{z}_{k-1}), \quad (11.47)$$

where $\mathbf{H}\hat{\mathbf{x}}_{k-1} - \mathbf{z}_{k-1}$ is assumed to be $N(0, \Sigma)$-distributed. ε is chi-square distributed with 2 degrees of freedom (x and y directions) under H_0.

$$\begin{cases} \varepsilon < \lambda = \chi^2_2(\alpha), & \text{if } H_0 \text{ is true,} \\ \varepsilon \geq \lambda = \chi^2_2(\alpha), & \text{if } H_1 \text{ is true,} \end{cases}$$

where $1 - \alpha$ is the level of confidence, which should be sufficiently high (for our system, $1 - \alpha = 99\%$). If H_0 is true, the chi-square testing-based switching declares that the target is still in the image and enables the mean shift-based tracker.

11.4.2.3 Mean Shift-Based Image Tracking

If the target is still in the image, the algorithm of Continuously Adaptive Mean Shift (CAMSHIFT) [8] is employed, which is shown in Fig. 11.9. This algorithm uses the mean shift searching method to efficiently obtain the optimal location of the target in the search window. The principle idea is to search the dominated peak in the feature space based on the previous information and certain assumptions. The detected target is verified by comparing it with an adaptive target template. The CAMSHIFT algorithm consists of three main steps: back projection, mean shift searching, and search window adaptation, which is illustrated in Fig. 11.10.

1. STEP 1. BACK PROJECTION: In order to search the target in the image, the probability distribution image needs to be constructed based on the color distribution of the target. The color distribution of the target is defined in the *hue* channel and given by

$$hist_{\mathrm{tg}}(i) = \sum_{(x,y)\in\Omega} \delta\left(i, \left[\frac{hue_{\mathrm{tg}}(x, y)}{N_{\mathrm{hue}}}\right]\right), \quad i = 1, \ldots, N_{\mathrm{hue}}. \quad (11.48)$$

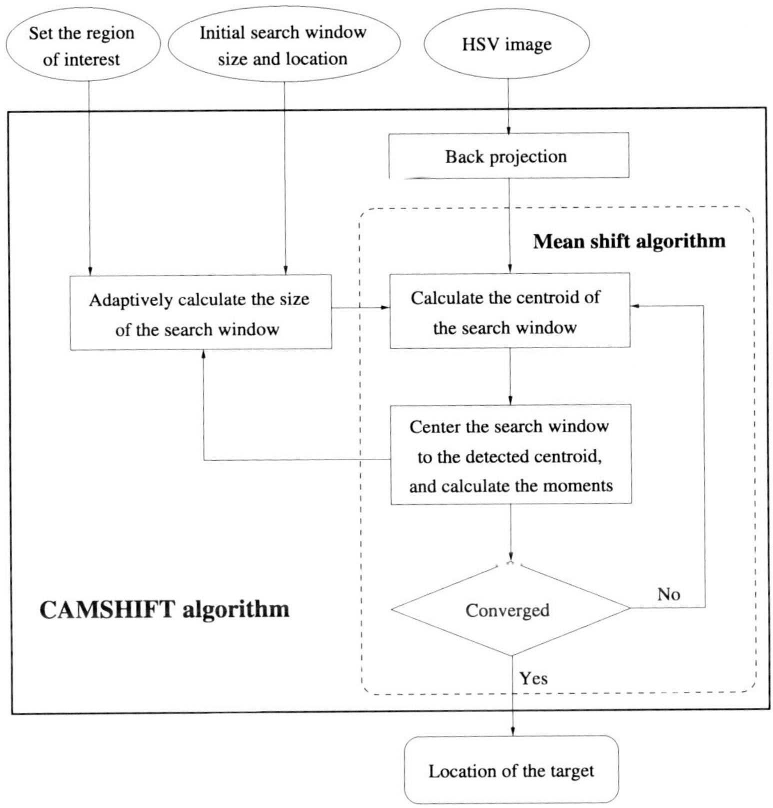

Fig. 11.10 Block diagram of the CAMSHIFT algorithm

Based on the color model of the target, the back projection algorithm is employed to convert the color image to the color probability distribution image. The probability of each pixel $\mathbf{I}_{\text{p}}(x, y)$ in the region of interest Ω_{r} is calculated based on the model of the target, which is used to map the histogram results and given by

$$\mathbf{I}_{\text{p}}(x, y) = hist_{\text{tg}}\left(\left[\frac{\mathbf{I}_{\text{hue}}(x, y)}{N_{\text{hue}}}\right]\right), \tag{11.49}$$

where $\mathbf{I}_{\text{hue}}$ is the pixel values of the image in the *hue* channel.

2. STEP 2. MEAN SHIFT ALGORITHM: Based on the obtained color density image, a robust non-parametric method, the mean shift algorithm is used to search the dominated peak in the feature space. The mean shift algorithm is an elegant way of identifying these locations without estimating the underlying probability density function [31].

   Recalling the discrete 2D image probability distributions in (11.49), the mean location (the centroid) of the search window is computed by

$$x_{\text{c}}(k) = \frac{\mathbf{M}_{10}}{\mathbf{M}_{00}}, \qquad y_{\text{c}}(k) = \frac{\mathbf{M}_{01}}{\mathbf{M}_{00}},$$

where k is the number of iterations,

$$\mathbf{M}_{00} = \sum_{(x,y)\in\Omega_\mathrm{w}} \mathbf{I}_\mathrm{p}(x, y), \qquad \mathbf{M}_{10} = \sum_{(x,y)\in\Omega_\mathrm{w}} \mathbf{I}_\mathrm{p}(x, y)x,$$

$$\mathbf{M}_{01} = \sum_{(x,y)\in\Omega_\mathrm{w}} \mathbf{I}_\mathrm{p}(x, y)y,$$

and where Ω_w is the region of the search window; $\mathbf{M}_{00}$ is the first moment; and $\mathbf{M}_{10}$ and $\mathbf{M}_{01}$ are the zeroth moments for x and y, respectively. The search window is centered at the mean location $\mathbf{c}(k) = (x_\mathrm{c}(k), y_\mathrm{c}(k))$. Step 2 is to be repeated until $\|\mathbf{c}(k) - \mathbf{c}(k-1)\| < \varepsilon$.

3. STEP 3. SEARCH WINDOW ADAPTATION: The region of interest is calculated dynamically using the motion filtering given in Sect. 11.4.2.1. To improve the performance of the CAMSHIFT algorithm, multiple search windows in the region of interest are employed. The initial locations and sizes of the search windows are adopted from the centers and boundaries of the foreground objects, respectively. These foreground objects are obtained using the color segmentation in the region of interest. In the CAMSHIFT algorithm, the size of the search window will be dynamically updated according to the moments of the region inside the search window [8]. Generally, more than one target candidate will be detected due to multiple search windows adopted. To identify the true target, the similarity between the target model and the detected target candidate is measured using the intersection comparison (11.34). This verification can effectively reduce the risk of detecting the false target.

In this work, the CAMSHIFT algorithm is only utilized to handle a special case in which the target is partially occluded or the light condition is changed, since the iterative calculation is involved in the mean shift. It is not easy to define the stop criterion of the iteration in the real-time applications. One example of target tracking using the CAMSHIFT algorithm in the partially occluded condition is given in Fig. 11.11.

11.4.2.4 Supervisor

A finite state machine is employed as the supervisor in the proposed vision scheme, as depicted in Fig. 11.12. This finite state machine dynamically chooses necessary features and gives different weighting to each feature in the discriminant function under different tracking conditions. The detailed functions of each state in the finite state machine are illustrated as follows:

STATE 0 (S0): Since there is no target found in the image, only static features are used in the discriminant function of (11.45) to identify the target in the entire image.

STATE 1 (S1): The same target has continuously been found by the algorithm in fewer than n frames, thus, the target cannot be locked with confidence. The discriminant function of (11.46) still uses static features in the pattern recognition but

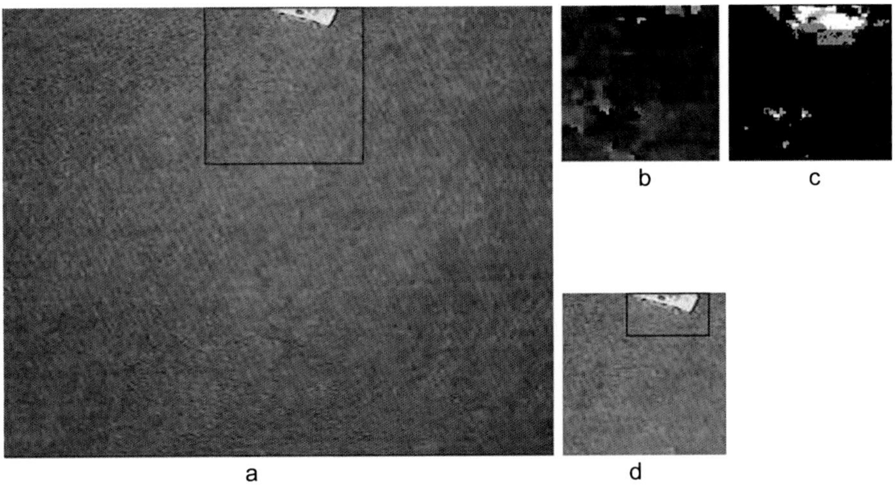

Fig. 11.11 Target tracking using CAMSHIFT algorithm. (**a**) The image with the search window; (**b**) the *hue* channel of the image inside the search window; (**c**) the back projection image; (**d**) the searched target using the CAMSHIFT algorithm

enables Kalman filtering to estimate the possible location of the target in the next frame.

STATE 2 (S2): The same target has continuously been found by the algorithm in more than n frames. We then have confidence to decide that it is the real target and activate model-based image tracking to lock the target in the successive frames.

STATE 3 (S3): The target is lost by the model-based tracking approach. If the partial occlusion detection, based on a chi-square test, indicates that the target is still in the image, it may be partially occluded. The mean shift-based tracker will be activated to iteratively search the target.

Shown in Fig. 11.13 is an example of the tracking of a toy car using the proposed vision detection algorithm in a ground test. The solid window is the measured location of the target, and the dashed window is the predicted location of the target in the image plane. The vision detection algorithm automatically initializes the detection and then tracks the target. When the target is partially occluded, the vision algorithm gives high weightings to the dynamic and color features in the discriminant function and also activates the mean shift-based tracking approach. Thus, the target still can be identified, even though it is partially occluded.

11.4.2.5 Experimental Verification of Target Detection and Visual Tracking

The proposed vision-based tracking algorithm is implemented in the onboard system of the unmanned helicopter, SheLion. The processing rate of the algorithm is 10 frames per second (FPS). During the real flight tests, the helicopter is manually

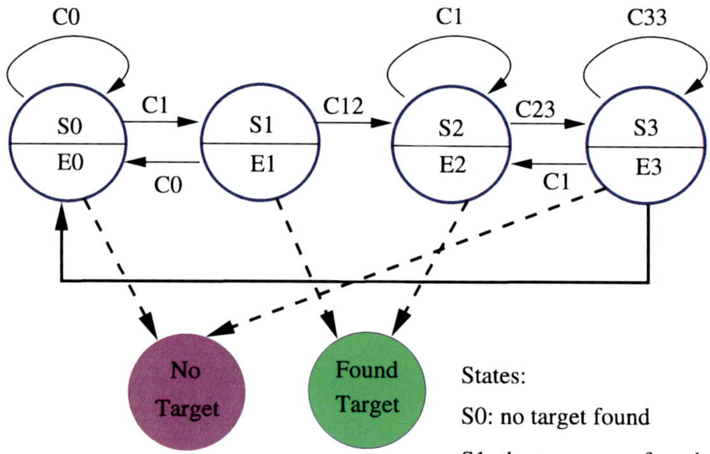

States:

S0: no target found

S1: the target was found

S2: found the target more than *n* frames

S3: lost the target fewer than *m* frames

Entry Actions:

E1: activate Kalman filter

E2: activate the tracking mode

In addition, E0, E1, E2, and E3 will define their own discriminate functions for the target recognition.

Conditions:

C0: did not find the target

C1: found the target

C12: continuously found the target more than *n* frames

C23: lost the target after locking it

C33: lost the target fewer than *m* frames

C40: lost the target more than *m* frames

Fig. 11.12 Decision making using the finite state machine

controlled to hover at a fixed position 10 meters above flat ground, and the onboard visual tracking system automatically identifies and tracks the ground moving target—a toy car, which is manually controlled to randomly move on flat ground.

We performed the visual tracking tests nine times and the tracking results are shown in Table 11.1. During these tests, the visual tracking system successfully tracked the ground target. One example of the tracking errors in vertical and horizontal directions is shown in Fig. 11.14, which indicates that the tracking error is bounded. The experimental results demonstrate the robustness and effectiveness of the visual tracking system, which can automatically identify and track the moving target in flight.

11.4.3 Target Following Control

We proceed to design a comprehensive target following system in this section. It consists of two main layers, the pan/tilt servo mechanism control and the UAV fol-

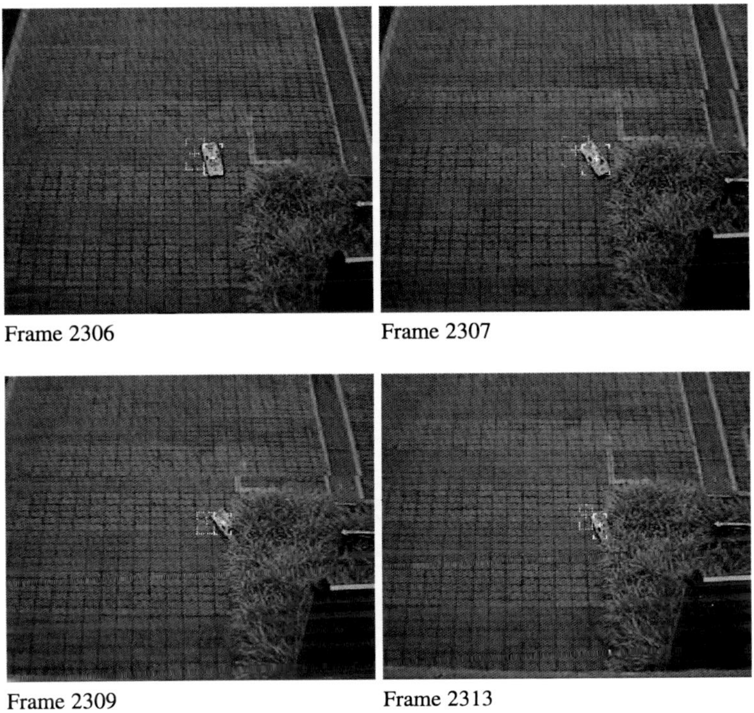

Fig. 11.13 Target detection with partial occlusion

Table 11.1 Experimental results of target detection and tracking

Times	Total frame	Detected frames	Accuracy
1	219	191	87.21%
2	284	209	73.59%
3	703	538	76.53%
4	375	295	78.67%
5	676	508	75.15%
6	431	311	72.16%
7	108	91	84.26%
8	1544	1162	75.26%
9	646	529	81.89%

lowing control. The overall structure of the target following control is depicted in Fig. 11.15. As mentioned earlier, a pan/tilt servo mechanism is employed in the first layer to control the orientation of the camera to keep the target in an optimal location in the image plane, which makes target tracking in the video sequence more robust and efficient. The parameters associated with the pan/tilt servo control in Fig. 11.15

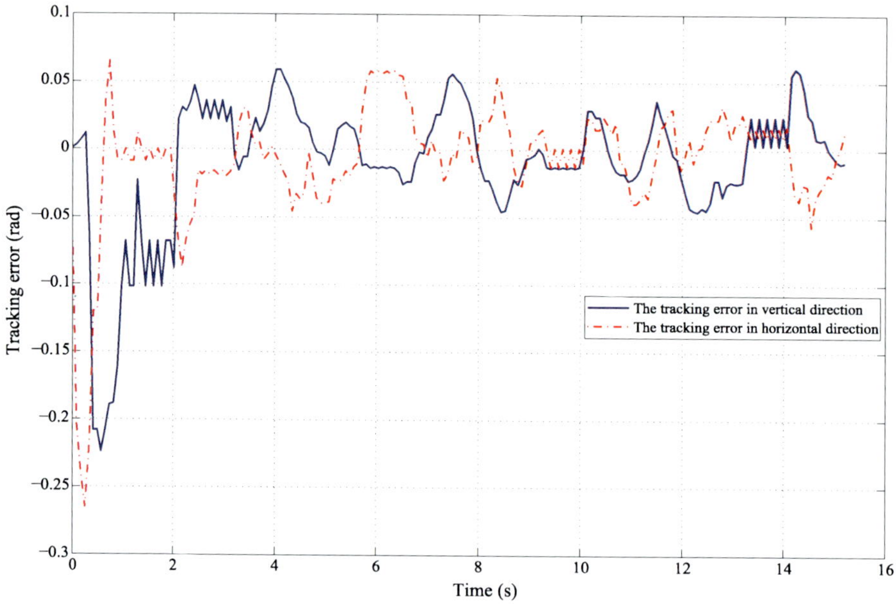

Fig. 11.14 Tracking error of θ_c and ϕ_c

will be explained in detail later. In the second layer, the UAV is controlled to maintain the constant relative distance between the moving target and the UAV in flight.

11.4.3.1 Control of the Pan/Tilt Servo Mechanism

The purpose of the control of the pan/tilt servo mechanism is to adjust the orientation of the onboard camera to keep the target in an optimal view (e.g., make the optical axis aligned with the line-of-sight vector of the target) in terms of the vision feedback, namely, eye-in-hand visual serving [25, 26]. A robust closed-loop tracking control scheme for the pan/tilt servo mechanism is designed using the vision information.

As depicted in Fig. 11.15, given a generic point **P**, $\mathbf{p}_\mathrm{i}$ and $\mathbf{p}_\mathrm{i}^*$ are the measured and desired locations of the projected point of **P** in the image plane, respectively. To make the vision-based detection more robust, $\mathbf{p}_\mathrm{i}^* = [x_\mathrm{i}^*, y_\mathrm{i}^*]^\mathrm{T}$ is set as the center of the image frame, which in general is not necessarily the principle point of the camera. $\mathbf{e} = [e_\phi, e_\theta]^\mathrm{T}$ is the tracking error, $\mathbf{u} = [u_\phi, u_\theta]^\mathrm{T}$ is the output of the tracking controller, and $\mathbf{v} = [v_\phi, v_\theta]^\mathrm{T}$ is the output of the pan/tilt servo mechanism. M is a nonlinear mapping between the coordinates of point **P** in the world frame and in the image plane under the current **v**. N is a nonlinear mapping between the location of point **P** in the image plane and its orientation with respect to the UAV under the current **v**. As mentioned in the definitions of the coordinate systems, the orientation of **P** with respect to the UAV can be defined using azimuth and elevation angles in

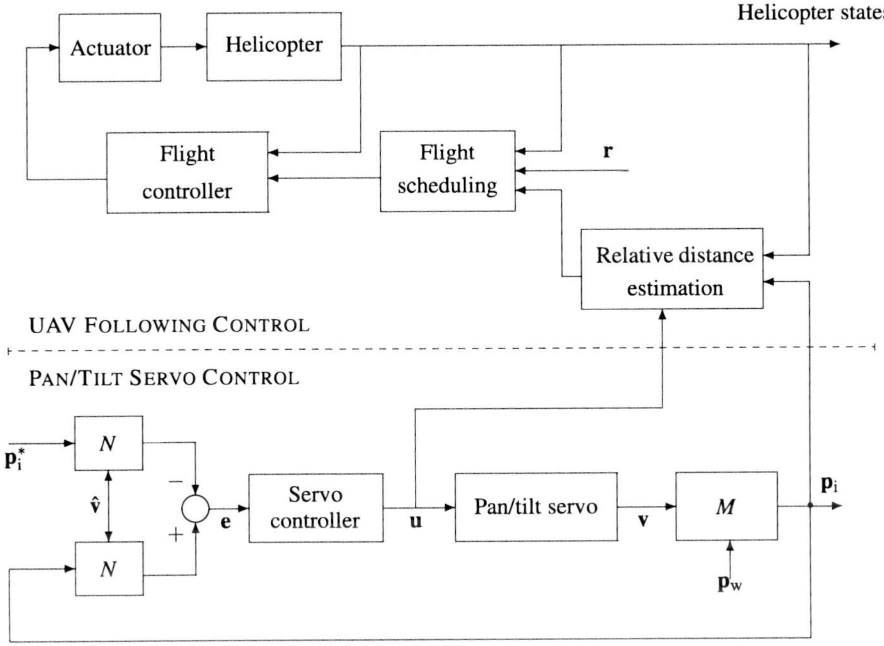

Fig. 11.15 Block diagram of the target following control scheme

the spherical coordinate system, which is described by two rotation angles, $\mathbf{p}_\mathrm{e} = [p_\phi, p_\theta]^\mathrm{T}$.

Given the generic point $\mathbf{P}$, recalling (11.14), the transformations among the camera coordinate system, the image coordinate system, and the world coordinate system (see Fig. 11.1) are given by

$$\begin{pmatrix} \mathbf{p}_\mathrm{i} \\ 1 \end{pmatrix} = \frac{1}{\lambda}\mathbf{K}_\mathrm{f}\Pi_0 \begin{pmatrix} \mathbf{p}_\mathrm{c} \\ 1 \end{pmatrix}, \tag{11.50}$$

and

$$\begin{pmatrix} \mathbf{p}_\mathrm{w} \\ 1 \end{pmatrix} = \begin{bmatrix} \mathbf{R}_\mathrm{w/c} & \mathbf{t}_\mathrm{w/c} \\ 0 & 1 \end{bmatrix} \begin{pmatrix} \mathbf{p}_\mathrm{c} \\ 1 \end{pmatrix}, \tag{11.51}$$

where $\mathbf{R}_\mathrm{w/c}$ and $\mathbf{t}_\mathrm{w/c}$ are, respectively, the rotation matrix and the translation vector, which define the rigid-body transformation from the camera frame to the world frame and will be described in detail in Sect. 11.4.3.2. For our applications, we can simplify (11.7) and define M as

$$\mathbf{p}_\mathrm{i} = \begin{pmatrix} x_\mathrm{i} \\ y_\mathrm{i} \end{pmatrix} = M(\mathbf{p}_\mathrm{w}, \mathbf{v}) = \frac{1}{\lambda}\begin{bmatrix} f_\mathrm{x} & 0 \\ 0 & f_\mathrm{y} \end{bmatrix}\begin{bmatrix} 1 & 0 & 0 \\ 0 & 1 & 0 \end{bmatrix}\left(\mathbf{R}_\mathrm{w/c}^{-1}\mathbf{p}_\mathrm{w} - \mathbf{R}_\mathrm{w/c}^{-1}\mathbf{t}_\mathrm{w/c}\right).$$

Next, to derive the nonlinear mapping N, we write the transformation between the camera coordinate system and the servo-based coordinate system as

$$\begin{pmatrix} \mathbf{p}_c \\ 1 \end{pmatrix} = \begin{bmatrix} \mathbf{R}_{c/s}(\mathbf{v}) & \mathbf{t}_{c/s} \\ 0 & 1 \end{bmatrix} \begin{pmatrix} \mathbf{p}_s \\ 1 \end{pmatrix}, \tag{11.52}$$

where

$$\begin{aligned} \mathbf{R}_{c/s}^{-1}(\mathbf{v}) &= \mathbf{R}_{s/c}(\mathbf{v}) \\ &= \begin{bmatrix} \cos v_\theta & 0 & \sin v_\theta \\ 0 & 1 & 0 \\ -\sin v_\theta & 0 & \cos v_\theta \end{bmatrix} \begin{bmatrix} 1 & 0 & 0 \\ 0 & \cos v_\phi & -\sin v_\phi \\ 0 & \sin v_\phi & \cos v_\phi \end{bmatrix}, \end{aligned} \tag{11.53}$$

$\mathbf{p}_s$ is the coordinate of point $\mathbf{P}$ with respect to the servo-based coordinate system, $(\mathbf{R}_{c/s}, \mathbf{t}_{c/s})$ describes the rigid-body transformations from the servo-based frame to the camera frame. Since only the rotation of the pan/tilt servo mechanism is considered in the above transformations and $\mathbf{t}_{c/s}$ is equal to zero, we can then simplify (11.52) as

$$\mathbf{p}_c = \mathbf{R}_{c/s}(\mathbf{v})\mathbf{p}_s. \tag{11.54}$$

It follows from (11.7) with all the necessary assumptions that

$$\lambda \begin{pmatrix} \frac{x_i}{f_x} \\ \frac{y_i}{f_y} \\ 1 \end{pmatrix} = \mathbf{p}_c, \tag{11.55}$$

which together with (11.54) and the fact that $\mathbf{R}_{c/s}(\mathbf{v}) = \mathbf{R}_{s/c}^{-1}(\mathbf{v})$ implies

$$\lambda \mathbf{R}_{s/c}(\mathbf{v}) \begin{pmatrix} \frac{x_i}{f_x} \\ \frac{y_i}{f_y} \\ 1 \end{pmatrix} = \mathbf{p}_s. \tag{11.56}$$

We then define

$$\bar{\mathbf{p}}_s = \begin{pmatrix} \bar{x}_s \\ \bar{y}_s \\ \bar{z}_s \end{pmatrix} = \mathbf{R}_{s/c}(\mathbf{v}) \begin{pmatrix} \frac{x_i}{f_x} \\ \frac{y_i}{f_y} \\ 1 \end{pmatrix}, \tag{11.57}$$

which together with (11.56) yields

$$\lambda \bar{\mathbf{p}}_s = \mathbf{p}_s. \tag{11.58}$$

Recalling the definitions of azimuth and elevation angles in (11.2) and (11.4), the nonlinear mapping N is then given as

$$\mathbf{p}_e = \begin{pmatrix} p_\phi \\ p_\theta \end{pmatrix} = N(\mathbf{p}_i, \mathbf{v}) = \begin{pmatrix} \sin^{-1}(\frac{y_s}{r_{sp}}) \\ \tan^{-1}(\frac{x_s}{z_s}) \end{pmatrix} = \begin{pmatrix} \sin^{-1}(\frac{\bar{y}_s}{\bar{r}_{sp}}) \\ \tan^{-1}(\frac{\bar{x}_s}{\bar{z}_s}) \end{pmatrix}, \tag{11.59}$$

where

$$\bar{r}_{sp} = \sqrt{\bar{x}_s^2 + \bar{y}_s^2 + \bar{z}_s^2}. \tag{11.60}$$

The pan/tilt servo mechanism can be approximately considered as two decoupled servo motors, which regulate the visual sensor for horizontal and vertical rotation, respectively. The dynamic model of the servo motor can be described by using a

Table 11.2 Parameters of the pan/tilt servos

Parameter	Tilt servo	Pan servo
DC gain: $\mathrm{K_d}$	1.1198	1.2945
Damping ratio: ζ	0.8143	0.8814
Natural frequency: ω_n	123.2525	130.676

standard second order system. To identify the parameters of the model, we inject step signals with different values to the pan/tilt servo mechanism. The parameters of the models of the vertical and horizontal servos are given in Table 11.2. Before proceeding to design the control law for the pan/tilt servo mechanism, we define the tracking error function as

$$\mathbf{e}(k) = \begin{pmatrix} e_\phi \\ e_\theta \end{pmatrix} = \mathbf{p}_e - \mathbf{p}_e^* = N(\mathbf{p}_i(k), \mathbf{v}(k)) - N(\mathbf{p}_i^*, \mathbf{v}(k)), \tag{11.61}$$

where $\mathbf{p}_e^*$ denotes the desired orientation of the camera. The control input will be sent to the pan/tilt servos after the vision-based target detection algorithm, which generally costs about one sampling period. To track the moving target efficiently, we calculate the pan/tilt servo control inputs using the predicted location of the target in the subsequent frame, which is derived from (11.38) and (11.39) and given by

$$\hat{\mathbf{p}}_i(k+1) = \begin{pmatrix} \hat{\mathbf{x}}_i \\ \hat{y}_i \end{pmatrix} = \hat{\mathbf{z}}(k+1|k) = \mathrm{H}\hat{\mathbf{x}}(k+1|k). \tag{11.62}$$

In implementation, it is not easy to measure the output of the pan/tilt servo $\mathbf{v}$ in (11.61). We assume that the bandwidth of the pan/tilt servo mechanism is much faster than that of the control system. We then can ignore the transient of the pan/tilt servos and consider them as scaling factors with one step delay. The estimate of $\mathbf{v}$ is defined as

$$\hat{\mathbf{v}}(k) = \mathrm{K_d}\mathbf{u}(k-1). \tag{11.63}$$

Replacing $\mathbf{v}$ and $\mathbf{p}_i$ with $\hat{\mathbf{v}}$ and $\hat{\mathbf{p}}_i$ in (11.61), we then can obtain the modified error function as

$$\mathbf{e}(k) = N(\hat{\mathbf{p}}_i(k+1), \hat{\mathbf{v}}(k)) - N(\mathbf{p}_i^*, \hat{\mathbf{v}}(k)). \tag{11.64}$$

The purpose of the design of the tracking control law is to minimize the tracking error function given in (11.64) by choosing a suitable control input $\mathbf{u}(k)$. Since the dynamics model of the pan/tilt servos is relatively simple, we employ a discrete-time proportional-integral controller (see, e.g., [62]), which is structurally simple but fairly robust. It is very suitable for our real-time application. The incremental implementation of the PI controller is given by

$$\Delta\mathbf{u}(k) = \mathrm{K_p}[\mathbf{e}(k) - \mathbf{e}(k-1)] + \frac{\mathrm{K_p T_s}}{\mathrm{T_i}}\mathbf{e}(k),$$

where the proportional gain and the integral time are chosen as $\mathrm{K_p} = 0.65$ and $\mathrm{T_i} = 0.8$, respectively. We note that two identical controllers are respectively used for the pan and tilt servos, since the dynamics of the two servos are very close.

11.4.3.2 Following Control of the Unmanned Aerial Vehicle

As illustrated in Fig. 11.15, to realize the UAV following control, a geometric approach is employed to estimate the relative distance, which uses the measured target location in the image plane and the pose of the UAV based on the flat ground assumption.

As illustrated in Fig. 11.1, to estimate the relative distance between the target and the UAV, we recall the transformation in (11.51) and simplify it as

$$\mathbf{p}_{\mathrm{w}} = \mathbf{R}_{\mathrm{w/c}}\mathbf{p}_{\mathrm{c}} + \mathbf{t}_{\mathrm{w/c}}, \tag{11.65}$$

which together with (11.55) and the fact that $z_{\mathrm{c}} = \lambda$ generates the overall geometric model from an ideal image to the world frame:

$$\mathbf{p}_{\mathrm{w}} = \begin{pmatrix} x_{\mathrm{w}} \\ y_{\mathrm{w}} \\ z_{\mathrm{w}} \end{pmatrix} = \mathbf{R}_{\mathrm{w/c}} \begin{pmatrix} \frac{x_{\mathrm{i}}}{f_{\mathrm{x}}} \\ \frac{y_{\mathrm{i}}}{f_{\mathrm{y}}} \\ 1 \end{pmatrix} z_{\mathrm{c}} + \mathbf{t}_{\mathrm{w/c}}. \tag{11.66}$$

We assume that the ground is flat, and the height of the UAV helicopter to the ground h is known. We also assume that $\mathrm{X_c}$- and $\mathrm{Y_c}$-axis translations from the camera frame to the world frame are smaller compared to the height and can be ignored, and $\mathrm{X_b}$- and $\mathrm{Y_b}$-axis rotations from the body frame to the world frame are smaller compared to the rotation of the pan/tilt servos and can be ignored too. We have

$$\mathbf{R}_{\mathrm{w/c}} = \mathbf{R}_{\mathrm{s/c}}(\mathbf{v}) = \begin{bmatrix} r_1 & r_2 & r_3 \\ r_4 & r_5 & r_6 \\ r_7 & r_8 & r_9 \end{bmatrix}, \qquad \mathbf{t}_{\mathrm{w/c}} = \begin{pmatrix} t_{\mathrm{x}} \\ t_{\mathrm{y}} \\ t_{\mathrm{z}} \end{pmatrix} = \begin{pmatrix} 0 \\ 0 \\ -h \end{pmatrix}, \tag{11.67}$$

which together with (11.66) yield

$$\begin{pmatrix} x_{\mathrm{w}} \\ y_{\mathrm{w}} \\ z_{\mathrm{w}} \end{pmatrix} = \begin{bmatrix} r_1 & r_2 & r_3 \\ r_4 & r_5 & r_6 \\ r_7 & r_8 & r_9 \end{bmatrix} \begin{pmatrix} \frac{x_{\mathrm{i}}}{f_{\mathrm{x}}} \\ \frac{y_{\mathrm{i}}}{f_{\mathrm{y}}} \\ 1 \end{pmatrix} z_{\mathrm{c}} + \begin{pmatrix} 0 \\ 0 \\ -h \end{pmatrix}. \tag{11.68}$$

Based on the assumption that the target is on the ground, z_{w} is equal to zero. We then can rewrite the last row in (11.68) and derive z_{c} as

$$z_{\mathrm{w}} = \left(r_7 \frac{x_{\mathrm{i}}}{f_{\mathrm{x}}} + r_8 \frac{y_{\mathrm{i}}}{f_{\mathrm{y}}} + r_9 \right) z_{\mathrm{c}} - h = 0, \tag{11.69}$$

$$z_{\mathrm{c}} = \frac{h}{r_7 \frac{x_{\mathrm{i}}}{f_{\mathrm{x}}} + r_8 \frac{y_{\mathrm{i}}}{f_{\mathrm{y}}} + r_9}, \tag{11.70}$$

which together with (11.68) yield

$$\begin{pmatrix} x_{\mathrm{tg}} \\ y_{\mathrm{tg}} \\ z_{\mathrm{tg}} - h \end{pmatrix}_{\mathrm{b}} = \begin{pmatrix} x_{\mathrm{w}} \\ y_{\mathrm{w}} \\ z_{\mathrm{w}} \end{pmatrix} = \begin{pmatrix} h \cdot \frac{r_1 x_{\mathrm{i}} f_{\mathrm{y}} + r_2 y_{\mathrm{i}} f_{\mathrm{x}} + r_3 f_{\mathrm{x}} f_{\mathrm{y}}}{r_7 x_{\mathrm{i}} f_{\mathrm{y}} + r_8 y_{\mathrm{i}} f_{\mathrm{x}} + r_9 f_{\mathrm{x}} f_{\mathrm{y}}} \\ h \cdot \frac{r_4 x_{\mathrm{i}} f_{\mathrm{y}} + r_5 y_{\mathrm{i}} f_{\mathrm{x}} + r_6 f_{\mathrm{x}} f_{\mathrm{y}}}{r_7 x_{\mathrm{i}} f_{\mathrm{y}} + r_8 y_{\mathrm{i}} f_{\mathrm{x}} + r_9 f_{\mathrm{x}} f_{\mathrm{y}}} \\ 0 \end{pmatrix}, \tag{11.71}$$

where $[x_{\mathrm{tg}}\ y_{\mathrm{tg}}\ z_{\mathrm{tg}}]_{\mathrm{b}}^{\mathrm{T}}$ is the coordinate of the target in the body frame. Due to the same reasons mentioned before, we replace $\mathbf{v}$ and $\mathbf{p}_i$ with $\hat{\mathbf{v}}$ and $\hat{\mathbf{p}}_{\mathrm{i}}$ and rewrite (11.71) as

$$\begin{pmatrix} x_{tg} \\ y_{tg} \\ z_{tg} - h \end{pmatrix}_b = \begin{pmatrix} x_w \\ y_w \\ z_w \end{pmatrix} = \begin{pmatrix} h \cdot \frac{\hat{r}_1 \hat{x}_i f_y + \hat{r}_2 \hat{y}_i f_x + \hat{r}_3 f_x f_y}{\hat{r}_7 \hat{x}_i f_y + \hat{r}_8 \hat{y}_i f_x + \hat{r}_9 f_x f_y} \\ h \cdot \frac{\hat{r}_4 \hat{x}_i f_y + \hat{r}_5 \hat{y}_i f_x + \hat{r}_6 f_x f_y}{\hat{r}_7 \hat{x}_i f_y + \hat{r}_8 \hat{y}_i f_x + \hat{r}_9 f_x f_y} \\ 0 \end{pmatrix},$$

where

$$\mathbf{R}_{s/c}(\mathbf{v}) \doteq \mathbf{R}_{s/c}(\hat{\mathbf{v}}) = \begin{bmatrix} \hat{r}_1 & \hat{r}_2 & \hat{r}_3 \\ \hat{r}_4 & \hat{r}_5 & \hat{r}_6 \\ \hat{r}_7 & \hat{r}_8 & \hat{r}_9 \end{bmatrix}.$$

As shown in Fig. 11.15, the relative distance between the target and the UAV is estimated, which is employed as the reference signal to guide the UAV to follow the motion of the target. The tracking reference for the UAV is defined as

$$\begin{aligned} \begin{pmatrix} x_{uav} \\ y_{uav} \\ z_{uav} \\ \psi_{uav} \end{pmatrix}_{ref} &= \begin{pmatrix} \begin{pmatrix} x_{tg} \\ y_{tg} \end{pmatrix}_n - \begin{bmatrix} 1 & 0 & 0 \\ 0 & 1 & 0 \end{bmatrix} \mathbf{R}_{n/b} \begin{pmatrix} c_x \\ c_y \\ 0 \end{pmatrix} \\ h_0 \\ \psi_0 \end{pmatrix} \\ &= \begin{pmatrix} \begin{bmatrix} 1 & 0 & 0 \\ 0 & 1 & 0 \end{bmatrix} \mathbf{R}_{n/b} \left[\begin{pmatrix} x_{tg} \\ y_{tg} \\ 0 \end{pmatrix}_b - \begin{pmatrix} c_x \\ c_y \\ 0 \end{pmatrix} \right] \\ h_0 \\ \psi_0 \end{pmatrix}, \end{aligned} \tag{11.72}$$

where c_x and c_y are the desired relative distance between the target and the UAV in the X_b- and Y_b-axes, respectively; h_0 is the predefined height of the UAV above the ground; ψ_0 is the predefined heading angle of the UAV; and $\mathbf{R}_{n/b}$ is the rotation matrix from the body frame to the local NED frame, which can be calculated in terms of the output of the onboard navigation sensors.

11.5 Experimental Results

The images of the target were captured by the onboard camera before the tests and analyzed off-line to estimate the feature model of the target using the proposed algorithm to realize the automatic target detection in flight. Shown in Fig. 11.16 is SheLion during an experimental flight test.

To verify the proposed vision system, multiple tests of the complete system were conducted. During these tests, SheLion was hovering autonomously at a certain position. If the moving target entered the view of the onboard camera, the target would be identified and tracked in the video sequence by the vision system automatically. Based on the vision information, the pan/tilt servo mechanism was controlled to keep the target in a certain position in the image as described in Sect. 11.4.3.1. The operator, then, can command the UAV to enter into the target following mode, in which the UAV followed the motion of the target autonomously based on the estimated relative distance, using the algorithm proposed in Sect. 11.4.3.2.

Fig. 11.16 Flight test of the vision-based target following

The experimental results of the vision-based target detection and tracking in flight are shown in Table 11.3, which indicate that the proposed vision algorithm could effectively identify and track the target in the video sequence in the presence of a disturbance of unknown motion between the UAV and the target. One example of the pan/tilt servo tracking control in flight is also shown in Fig. 11.17. The solid line in Fig. 11.17 indicates the expected position of the target in the image, and the dashed line indicates the actual location of the target in the image during the flight test. From Fig. 11.17, we can observe that in spite of the unknown motion between the UAV and the target, the pan/tilt servo mechanism can effectively control the target in a box-like neighborhood of the center point of the image by employing the vision-based pan/tilt servo control.

One example of the ground target following is described in Figs. 11.18 and 11.19, in which the target was manually controlled to move randomly on flat ground and the UAV followed the motion of the target automatically based on the scheme proposed in the previous sections. From Figs. 11.18 and 11.19, we observe that the UAV can follow the trajectory of the target and keep the constant relative distance between the UAV and the target. The results for the moving ground target

Table 11.3 Experimental results of target detection and tracking in flight

Test no.	Total time (s)	Total frames	Target frames detected	Accuracy
1	101.8	761	728	95.66%
2	77.2	591	518	87.65%
3	50.4	388	382	98.45%
4	75.3	572	501	87.59%
5	86.4	662	645	97.43%

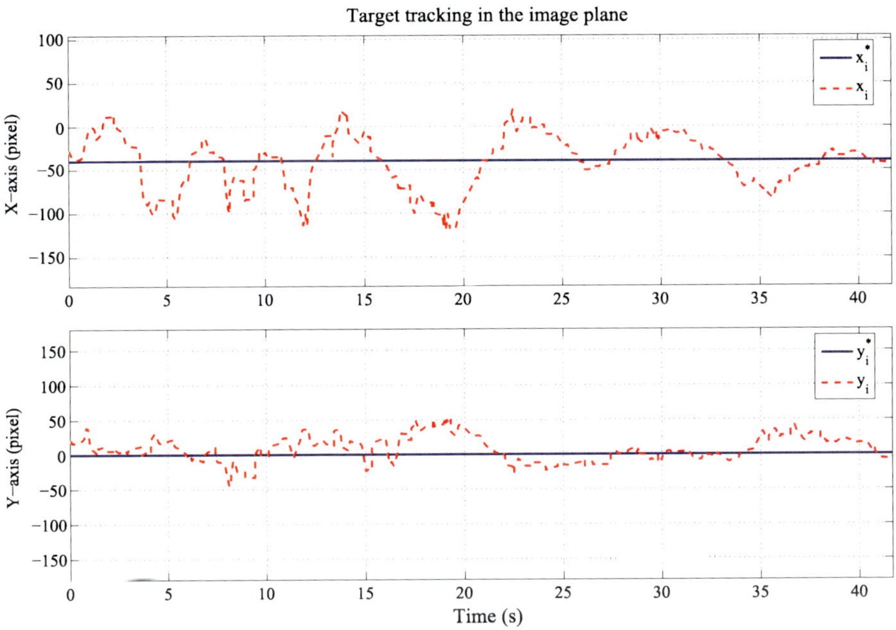

Fig. 11.17 The test results of the vision-based servo following

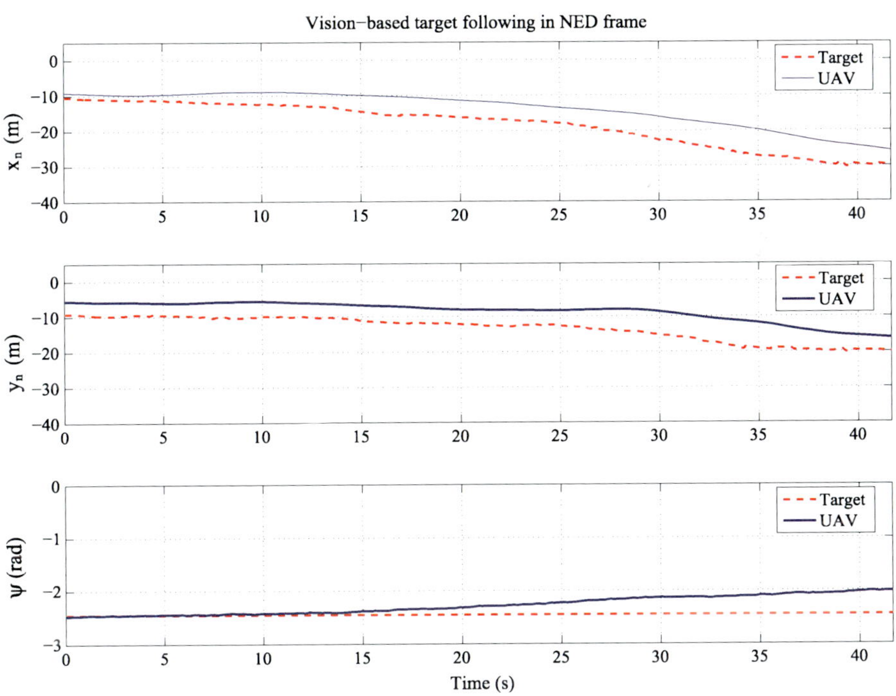

Fig. 11.18 The test results of the vision-based target following

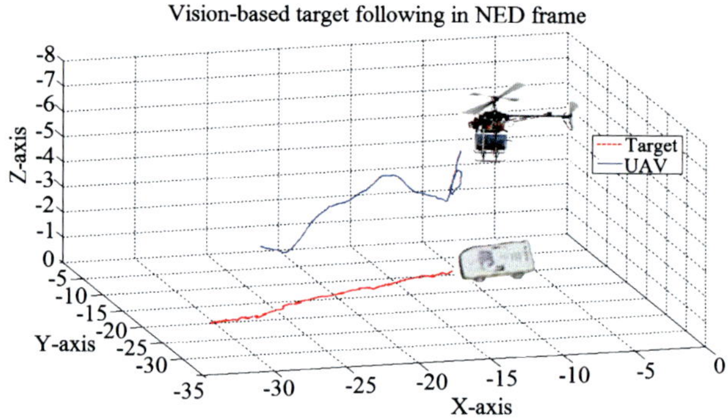

Fig. 11.19 The test results of the vision-based target following in 3D

following of the UAV indicate the efficiency and robustness of the proposed vision-based following scheme. The videos of the vision-based target following test can be downloaded from the official website of the NUS UAV Research Team [135] at http://uav.ece.nus.edu.sg/.

References

1. ADS-33D-PRF. Aeronautical design standard performance specification handling qualities requirements for military rotorcraft. U.S. Army Aviation and Troop Command; 1996.
2. AF25B helicopter. http://www.copterworks.com/. Cited Aug 2010.
3. Aladin UAV system. http://www.emt-penzberg.de/. Cited Aug 2010.
4. Amidi O, Kanade T, Miller R. Vision based autonomous helicopter research at Carnegie Mellon Robotics Institute 1991–1997. In: Proc American helicopter society int conf, Gifu, Japan; 1998. p. 1–12.
5. AutoCopter express UAV. http://www.autocopter.net/. Cited Aug 2010
6. Başar T, Bernhard P. H_∞ optimal control and related minimax design problems: a dynamic game approach. 2nd ed. Boston: Birkhäuser; 1995.
7. Berkeley aerobot team. http://robotics.eecs.berkeley.edu/bear/testbeds.html. Cited Aug 2010.
8. Bradski GR, Clara S. Computer vision face tracking for use in a perceptual user interface. Intel Technol J. 1998;Q2:1–15.
9. Bogdanov A, Wan E. SDRE control with nonlinear feed forward compensation for a small unmanned helicopter. In: Proc 2nd AIAA unmanned unlimited syst, technol, operations conf, San Diego, CA; 2003. AIAA-2003-6512.
10. Boykov Y, Huttenlocher DP. Adaptive Bayesian recognition in tracking rigid objects. In: Proc IEEE conf computer vision pattern recognition, Hilton Head, SC; 2000. p. 697–704.
11. Bramwell ARS. Bramwell's helicopter dynamics. Reston: AIAA; 2001.
12. Britting KR. Inertial navigation system analysis. New York: Wiley; 1971.
13. Cai G, Cai AK, Chen BM, Lee TH. Construction, modeling and control of a mini autonomous UAV helicopter. In: Proc IEEE int conf automat logistics, Qingdao, China; 2008. p. 449–54.
14. Cai G, Chen BM, Lee TH. Design and implementation of robust automatic flight control system for a small-scale UAV helicopter. In: Proc 7th Asian contr conf, Hong Kong, China; 2009. p. 691–7.
15. Cai G, Chen BM, Lee TH, Dong M. Design and implementation of a hardware-in-the-loop simulation system for small-scale UAV helicopters. Mechatronics. 2009;19:1057–66.
16. Cai G, Chen BM, Lee TH, Lum KY. Comprehensive nonlinear modeling of an unmanned aerial vehicle helicopter. In: Proc 2008 AIAA guidance, navigation and contr conf, Honolulu, HI; 2008. AIAA-2008-7414.
17. Cai G, Chen BM, Peng K, Dong M, Lee TH. Modeling and control system design for a UAV helicopter. In: Proc 14th mediterranean conf contr automat, Ancona, Italy; 2006. p. 600–6.
18. Cai G, Chen BM, Dong X, Lee TH. Design and implementation of a robust and nonlinear flight control system for an unmanned helicopter. Mechatron. 2011. doi:10.1016/j.mechatronics.2011.02.002.
19. Cai G, Chen BM, Lee TH. An overview on development of miniature unmanned rotorcraft systems. Front Electr Electron Eng China. 2010;5:1–14.

G. Cai et al., *Unmanned Rotorcraft Systems*, Advances in Industrial Control, DOI 10.1007/978-0-85729-635-1, © Springer-Verlag London Limited 2011

20. Cai G, Lin F, Chen BM, Lee TH. Systematic design methodology and construction of UAV helicopters. Mechatronics. 2008;18:545–58.
21. Cai G, Peng K, Chen BM, Lee TH. Design and assembling of a UAV helicopter system. In: Proc 5th int conf contr automat, Budapest, Hungary; 2005. p. 697–702.
22. Camcopter S-100 UAV system. http://www.schiebel.net/. Cited Aug 2010.
23. Camera calibration toolbox for MATLAB. http://www.vision.caltech.edu/bouguetj/calib_doc/. Cited Aug 2010.
24. Charles J. CMU's autonomous helicopter explores new territory. IEEE Intell Syst. 1998;13:85–7.
25. Chaumette F, Hutchinson S. Visual servo control part I: basic approaches. IEEE Robot Autom Mag. 2006;13:82–90.
26. Chaumette F, Hutchinson S. Visual servo control part II: advanced approaches. IEEE Robot Autom Mag. 2007;14:109–18.
27. Chen BM. Robust and H_∞ control. New York: Springer; 2000.
28. Chen BM, Lin Z, Liu K. Robust and perfect tracking of discrete-time systems. Automatica. 2002;38:293–9.
29. Chen Q. Hybrid system model and overlapping decomposition for vehicle flight formation control. In: Proc 42nd IEEE conf dec contr, Maui, HI; 2003. p. 475–88.
30. Cheng RP, Tischler MB, Schulein GJ. Rmax helicopter state-space model identification for hover and forward-flight. J Am Helicopter Soc. 2006;51:202–10.
31. Cheng Y. Mean shift, mode seeking, and clustering. IEEE Trans Pattern Anal Mach Intell. 1995;17:790–9.
32. Circular Error Probable (CEP). Technical paper 6 [report]. USA: Air Force Operational Test and Evaluation Center; 1987.
33. Civita ML, Messner W, Kanade T. Modeling of small-scale helicopters with integrated first-principles and integrated system identification techniques. Presented at 58th forum of American helicopter society, Montreal, Canada; 2002.
34. Comaniciu D, Ramesh V, Meer P. Kernel-based object tracking. IEEE Trans Pattern Anal Mach Intell. 2003;25:564–77.
35. Compact wire rope isolators. http://www.enidine.com/Industrial/CWRMain.html. Cited Aug 2010.
36. CONDUIT user's guide [report]. US NASA Ames Research Center, Moffett Field; 2009.
37. Conway A. Autonomous control of an unstable helicopter using carrier phase GPS only [PhD dissertation]. Stanford University; 1994.
38. Corban JE, Calise AJ, Prasad JVR. Implementation of adaptive nonlinear controller for flight test on an unmanned helicopter. In: Proc 37th IEEE conf dec contr, Tampa, FL; 1998. p. 3641–6.
39. Corfield SJ, Fraser RJC, Harris CJ. Architecture for real-time intelligent control of autonomous vehicles. Comput Control Eng J. 1991;2:254C–262C.
40. Curtiss HC Jr. Stability and control modeling. Vertica. 1988;12:381–94.
41. Delfly MAVs. http://www.delfly.nl/. Cited Aug 2010.
42. Dong M, Chen BM, Cheng C. Development of 3D monitoring for an unmanned aerial vehicle. In: Proc 1st int conf computer science edu, Xiamen, China; 2006. p. 135–40.
43. Dong M, Chen BM, Cai G, Peng K. Development of a real-time onboard and ground station software system for a UAV helicopter. J Aerosp Comput Inf Commun. 2007;4:933–55.
44. Dong M, Sun Z. A behavior-based architecture for unmanned aerial vehicles. In: Proc int symposium intell contr, Taipei, Taiwan; 2004. p. 149–55.
45. Dong X, Chen BM, Cai G, Lin H, Lee TH. A comprehensive real-time software system for flight coordination and cooperative control of multiple unmanned aerial vehicles. Int J Robot Automat. 2011;26:49–63.
46. Doyle JC. Lecture notes in advances in multivariable control. ONR-Honeywell Workshop; 1984.
47. Doyle J, Glover K, Khargonekar PP, Francis BA. State-space solutions to standard H_2 and H_∞ control problems. IEEE Trans Autom Control. 1989;34:831–47.

48. Draganflyer X4 quadrotor helicopter. http://www.draganfly.com/. Cited Aug 2010.
49. Enns R, Si J. Helicopter flight control design using a learning control approach. In: Proc 39th IEEE conf dec contr, Sydney, Australia; 2000. p. 1754–9.
50. Enns R, Si J. Helicopter trimming and tracking control using direct neural dynamic programming. IEEE Trans Neural Netw. 2003;14:929–39.
51. ETH unmanned helicopters. http://www.uav.ethz.ch/. Cited Aug 2010.
52. Fagg AH, Lewis MA, Montgomery JF, Bekey GA. The USC autonomous flying vehicle: an experiment in real-time behavior-based control. In: Proc 1993 IEEE/RSJ int conf intell robot syst, Yokohama, Japan; 1993. p. 1173–80.
53. Fancopter UAV systems. http://www.emt-penzberg.de/. Cited Aug 2010.
54. FH series miniature coaxial UAVs. http://www.ase.buaa.edu.cn/chenming/fh1.html. Cited Aug 2010.
55. Flight video gallery. http://www.model-helicopters.com/. Cited Aug 2010.
56. Flight video gallery. http://www.rcgroups.com/rc-video-gallery-271/. Cited Aug 2010.
57. Flight video gallery. http://www.runryder.com/helicopter/galleries/. Cited Aug 2010.
58. Flying-cam helicopter. http://www.flying-cam.com/. Cited Aug 2010.
59. Foley JD, Vandam A, Feiner SK, Hughes JF. Fundamentals of interactive computer graphics. Reading: Addison Wesley; 1990.
60. Fowers SG, Lee DJ, Tippetts BJ, Lillywhite KD, Dennis AW, Archibald JK. Vision aided stabilization and the development of a quad-rotor micro UAV. In: Proc 2007 IEEE int symposium computational intell robot automat, Jacksonville, FL; 2007. p. 143–8.
61. Francis BA. A course in H_∞ control theory. Berlin: Springer; 1987.
62. Franklin G, Powell FD, Naeini AE. Feedback control of dynamic systems. 4th ed. Upper Saddle River: Prentice Hall; 2002.
63. Frost W, Turner RE. A discrete gust model for use in the design of wind energy conversion systems. J Appl Meteorol. 1982;21:770C–776C.
64. Fujiwara D, Shin J, Hazawa K, Nonami K. H_∞ hovering and guidance control for autonomous small-scale unmanned helicopter. In: Proc IEEE/RSJ int conf intell robot syst, Sendai, Japan; 2004. p. 2463–8.
65. Futaba servo actuators. http://www.futaba-rc.com/servos/. Cited Aug 2010.
66. Gadewadikar J, Lewis FL, Subbarao K, Chen BM. Structured H_∞ command and control loop design for unmanned helicopters. J Guid Control Dyn. 2008;31:1093–102.
67. Gavrilets V, Mettler B, Feron E. Nonlinear model for a small-size acrobatic helicopter. Presented at AIAA guidance, navigation, and contr conf, Montreal, Canada; 2001.
68. Gavrilets V, Martinos I, Mettler B, Feron E. Aggressive manuevering of small autonomous helicopters: a human-centered approach. Int J Robot Res. 2001;20:795–807.
69. Glover K. All optimal Hankel-norm approximations of linear multivariable systems and their $\mathcal{L}_\infty$ error bounds. Int J Control. 1984;39:1115–93.
70. Gonzalez G, Woods RE. Digital image processing. Reading: Addison Wesley; 1992.
71. Guenard N, Hamel T, Mahony R. A practical visual servo control for an unmanned aerial vehicle. IEEE Trans Robot. 2008;24:331–40.
72. Hall JK. Three dimensional formation flight control [thesis]. Air Force Institute of Technology; 2000.
73. Hansen JL, Cobleigh BR. Induced moment effects of formation flight using two F/A-18 aircraft [report]. NASA TM-2002-210732; 2002.
74. Hanson CE, Ryan J, Allen MJ, Jacobson SR. An overview of flight test results for a formation flight autopilot. In: Proc AIAA guidance, navigation and contr conf, Monterey, CA; 2002. AIAA-2002-4755.
75. Harbick K, Montgomery JF, Sukhatme GS. Planar spline trajectory following for an autonomous helicopter. J Adv Comput Intell – Comput Intell Robot Autom. 2004;8:237C–242C.
76. Harris CM. Shock and vibration handbook. 4th ed. New York: McGraw-Hill; 1996.
77. He R, Bachrach A, Achtelik M, et al. On the design and use of a micro air vehicle to track and avoid adversaries. Int J Robot Res. 2010;29:529–46.

78. Heffley RK, Bourne SM, Curtiss HC Jr, et al. Study of helicopter roll control effectiveness criteria [report]. NASA CR 177404; 1986.
79. Heffley RK, Mnich MA. Minimum-complexity helicopter simulation math model [report]. Technical Report. NASA Contractor Report 177476, NASA; 1988.
80. Heikkila J, Silven O. A four-step camera calibration procedure with implicit image correction. In: Proc 1997 conf computer vision pattern recognition. Puerto Rico: San Juan; 1997. p. 1106–12.
81. HighEye UAV helicopter. http://www.higheye.nl/. Cited Aug 2010.
82. Hrabar S, Sukhatme GS, Corke P, Usher K, Roberts J. Combined optic-flow and stereo-based navigation of urban canyons for a UAV. In: Proc IEEE/RSJ int conf intell robot syst, Edmonton, Canada; 2005. p. 3309–16.
83. Hummel D. The use of aircraft wakes to achieve power reduction in formation flight. In: Proc fluid dynamics panel symposium, AGARD; 1996. p. 1777–94.
84. Isard M, Blake A. Condensation conditional density propagation for visual tracking. Int J Comput Vis. 1998;29:5–28.
85. Isidori A, Marconi L, Serrani A. Robust nonlinear motion control of a helicopter. IEEE Trans Autom Control. 2003;48:413–26.
86. Jacobs EN, Sherman A. Airfoil section characteristics as affected by variations of the Reynolds number [report]. NACA Report 586; 1937.
87. Jang JS, Liccardo D. Automation of small UAVs using a low cost MEMS sensor and embedded computing platform. In: Proc 25th IEEE/AIAA digital avionics syst conf, Portland, OR; 2006. p. 5C6.1–5C6.9.
88. Jang JS, Tomlin CJ. Design and implementation of a low cost, hierarchical and modular avionics architecture for the DragonFly UAVs. In: Proc 2002 AIAA guidance, navigation, contr conf, Monterey, CA; 2002. p. 446–77.
89. Johnson EN, Schrage DP. The Georgia Tech unmanned aerial research vehicle: GTmax. In: Proc 2003 AIAA guidance, navigation, contr conf, Austin, TX; 2003. AIAA-2003-5741.
90. Johnson EN, Calise AJ, Watanabe Y, Ha J, Neidhoefer JC. Real-time vision-based relative aircraft navigation. J Aerosp Comput Inf Commun. 2007;4:707–38.
91. Johnson W. Helicopter theory. Mineola: Dover Publications; 1994.
92. JR servo actuators. http://www.jrradios.com/Products/Servos-Air.aspx. Cited Aug 2010.
93. Kadmiry B. Fuzzy control for an autonomous helicopter [thesis]. Linkoping University, Sweden; 2002.
94. Kaminer I, Pascoal A, Hallberg E, Silvestre C. Trajectory tracking for autonomous vehicles: an integrated approach to guidance and control. J Guid Control Dyn. 1998;21:29–38.
95. KillerBee blended-wing UAV systems. http://www.raytheon.com. Cited Aug 2010.
96. Kim HJ, Shim DH, Sastry S. A flight control system for aerial robots: algorithms and experiments. Control Eng Pract. 2003;11:1389–400.
97. Kim J, Sukkarieh S. SLAM aided GPS/INS navigation in GPS denied and unknown environments. In: Proc int symposium GNSS/GPS, Sydney, Australia; 2004.
98. Kim SK, Tilbury DM. Mathematical modeling and experimental identification of an unmanned helicopter robot with flybar dynamics. J Robot Syst. 2004;21:95–116.
99. Kimura H. Chain-scattering approach to H_∞-control. Boston: Birkhäuser; 1997.
100. Koo TJ, Sastry S. Output tracking control design of a helicopter model based on approximate linearization. In: Proc 37th IEEE conf dec contr, Tampa, FL; 1998. p. 3635–40.
101. Kottmann M. Software for model helicopter flight control [report]. Eidgenossisches Technische Hochschule Zurich; 1999.
102. Krten R. Getting started with QNX neutrino 2: a guide for real time programmers. PARSE Software Devices, Kanata, Ontario; 1999.
103. Kruglinski DJ. Inside visual C++. 4th ed. Redmond: Microsoft Press; 1995.
104. Kwakernaak H. A polynomial approach to minimax frequency domain optimization of multivariable feedback systems. Int J Control. 1986;41:117–56.
105. Lama V4 coaxial helicopter. http://en.esky-sz.cn/home.html. Cited Aug 2010.
106. Larson G. Autonomous formation flight. Presentation to MIT 16.886 Class; 2004.

107. Li XR, Jilkov VP. Survey of maneuvering target tracking, part I: dynamic models. IEEE Trans Aerosp Electron Syst. 2003;39:1333–64.
108. Lin F. Development and applications of a vision-based unmanned helicopter [PhD dissertation]. Dept of Electrical and Computer Engineering, National University of Singapore; 2011.
109. Lin F, Chen BM, Lum KY, Lee TH. A robust vision system on an unmanned helicopter for ground target seeking and following. In: Proc 8th world congress intell contr automat, Jinan, China; 2010. p. 276–81.
110. Lin F, Lum KY, Chen BM, Lee TH. Development of a vision-based ground target detection and tracking system for a small unmanned helicopter. Sci China Ser F: Inf Sci. 2009;52:2201–15.
111. Limebeer DJN, Anderson BDO. An interpolation theory approach to H_∞ controller degree bounds. Linear Algebra Appl. 1988;98:347–86.
112. Liu K, Chen BM, Lin Z. On the problem of robust and perfect tracking for linear systems with external disturbances. Int J Control. 2001;74:158–74.
113. Ma Y, Soatto S, Kosecka J, Sastry SS. An invitation to 3-D vision: from images to geometric models. New York: Springer; 2004.
114. Malazgirt mini unmanned helicopter system. http://www.baykarmakina.com/heliuav. Cited Aug 2010.
115. Mali AD. On the behavior-based architectures of autonomous agency. IEEE Trans Syst Man Cybern, Part B, Cybern. 2002;32:231–42.
116. MARVIN Mark Series UAV helicopters. http://pdv.cs.tu-berlin.de/MARVIN/. Cited Aug 2010.
117. Mataric MJ, Sukhatme GS, Ostergaard E. Multi-robot task allocation in uncertain environments. Auton Robots. 2003;14:255–63.
118. MAVstar. http://www.robotics.unsw.edu.au/mavstar/. Cited Aug 2010.
119. Meingast M, Geyer C, Sastry S. Vision based terrain recovery for landing unmanned aerial vehicles. In: Proc IEEE conf dec contr, Atlantis, Bahamas; 2004. p. 1670–5.
120. Mejias LO, Saripalli S, Cervera P, Sukhatme GS. Visual servoing of an autonomous helicopter in urban areas using feature tracking. J Field Robot. 2006;23:185–99.
121. Mettler BM. Identification, modeling and characteristics of miniature rotorcraft. Boston: Kluwer Academic Publishers; 2002.
122. Mettler BM, Tischler MB, Kanade T. System identification modeling of a small-scale unmanned rotorcraft for control design. J Am Helicopter Soc. 2002;47:50–63.
123. Mettler BM, Tischler MB, Kanade T. System identification of small-size unmanned helicopter dynamics. Presented at 55th forum American helicopter society; 1999.
124. Miami police department. http://www.nwotruth.com/miami-police-plans-urban-test-of-honeywells-micro-uav/. Cited Aug 2010.
125. Micro flying robot. http://global.epson.com/. Cited Aug 2010.
126. Morris JC, Nieuwstadt M, Bendotti P. Identification and control of a model helicopter in hover. In: Proc American contr conf, Baltimore, MD; 1994. p. 1238–42.
127. MuFly autonomous micro helicopter. http://www.mufly.ethz.ch/index. Cited Aug 2010.
128. Musial M, Brandenburg UW, Hommel G. Inexpensive system design: the flying robot MARVIN. In: Proc 16th int UAVs conf unmanned air veh syst, Bristol, UK; 2001. p. 23.1–23.12.
129. Nassar S, Schwarz KP, El-Sheimy N. INS and INS-GPS accuracy improvement using autoregressive (AR) modeling of INS sensor errors. In: Proc ION 2004 national technical meeting, San Diego, CA; 2004. p. 936–44.
130. NAV420 series users manual. Crossbow Technology Inc.; 2004.
131. Naval rotary UAV. http://www.iai.co.il/. Cited Aug 2010.
132. Nickol C, Guynn M, Kohout L, Ozoroski T. High altitude long endurance air vehicle analysis of alternatives and technology requirements development. In: Proc 45th AIAA aerospace sciences meeting and exhibit, Reno, NV; 2007. AIAA-2007-1050.
133. Noureldin A, Sharaf R, Osman A, El-Sheimy N. INS/GPS data fusion technique utilizing radial basis functions neural networks. In: Proc IEEE position, location and navigation symposium, Monterey, CA; 2004. p. 280–4.

134. NRL Dragon Warrior UAV. http://www.designation-systems.net/dusrm/app4/vantage.html. Cited Aug 2010.
135. NUS UAV research. http://uav.ece.nus.edu.sg. Cited Aug 2010.
136. OAV iStar. http://defense-update.com/features/du-2-04/mav-oav.htm. Cited Aug 2010.
137. Observer-twin helicopter. http://www.bergenrc.com/. Cited Aug 2010.
138. Ollero A, Lacroix S, Merino L, et al. Multiple eyes in the skies: architecture and perception issues in the COMETS unmanned air vehicles project. IEEE Robot Autom Mag. 2005;12:46–57.
139. Padfield GD. Helicopter flight dynamics: the theory and application of flying qualities and simulation modeling. Reston: AIAA Press; 1996.
140. PC/104 embedded consortium. http://www.pc104.org. Cited Aug 2010.
141. Pelletier A, Mueller TJ. Low Reynolds number aerodynamics of low-aspect-ratio, thin/flat/cambered-plate wings. J Aircr. 2000;37:825–32.
142. Peng K, Cai G, Chen BM, Dong M, Lum KY, Lee TH. Design and implementation of an autonomous flight control law for a UAV helicopter. Automatica. 2009;45:2333–8.
143. Phang SK, Ong JJ, Yeo RTC, Chen BM, Lee TH. Autonomous mini-UAV for indoor flight with embedded on-board vision processing as navigation system. In: Proc IEEE R8 int conf computational technologies electrical electronics engineering, Irkutsk, Russia; 2010. p. 722–7.
144. Pixhawk. http://pixhawk.ethz.ch/. Cited Aug 2010.
145. Plataniotis KN, Venetsanopoulos AN. Color image processing and applications. New York: Springer; 2000.
146. Porikli F. Achieving real-time object detection and tracking under extreme conditions. J Real-Time Image Process. 2006;1:33–40.
147. Proud AW, Pachter M, DAzzo JJ. Close formation flight control. In: Proc 2002 AIAA guidance, navigation and contr conf, Portland, OR; 2002. p. 1231–46.
148. QNX momentics (v6.3 SP2) documentation. QNX Software Systems Corporation; 2006.
149. QNX neutrio RTOS. http://www.qnx.com/. Cited Aug 2010.
150. Raptor 90 SE helicopter. http://www.tiger.com.tw/. Cited Aug 2010.
151. Rodriguez PA, Geckle WJ, Barton JD, Samsundar J, Gao T, Brown MZ, Martin SR. An emergency response UAV surveillance system. In: Proc AMIA annual symposium, Washington, DC; 2006. p. 1078.
152. RT Linux RTOS. http://www.rtlinuxfree.com/. Cited Aug 2010.
153. SAE-AS94900. General specification for aerospace flight control systems design, installation and test of piloted military aircraft. Warrendale, SAE International; 2007.
154. Sathaye SS. Lift distributions on low aspect ratio wings at low Reynolds numbers [thesis]. Worcester Polytechnic Institute, Worcester, MA; 2004.
155. Schell FR, Dickmanns ED. Autonomous landing of airplanes by dynamic machine vision. Mach Vis Appl. 1994;7:127–34.
156. Schiebel camcopter. http://www.schiebel.net/. Cited Aug 2010.
157. Scheutz M, Andronache V. Architectural mechanisms for dynamic changes of behavior selection strategies in behavior-based systems. IEEE Trans Syst Man Cybern, Part B, Cybern. 2004;34:2377–95.
158. Seiler P, Pant A, Hedrick K. Analysis of bird formations. In: Proc 41st IEEE conf dec contr, Las Vegas, NV; 2002. p. 118–23.
159. Shakernia O, Ma Y, Koo TJ, Sastry S. Landing an unmanned air vehicle: vision based motion estimation and nonlinear control. Asian J Control. 1999;1:128–45.
160. Shakernia O, Vidal R, Sharp CS, Ma Y, Sastry S. Multiple view motion estimation and control for landing an unmanned aerial vehicle. In: Proc IEEE int conf robot automat, Washington, DC; 2002. p. 2793–8.
161. Sharp CS, Shakernia O, Sastry S. A vision system for landing an unmanned aerial vehicle. In: Proc IEEE int conf robot automat, Seoul, Korea; 2001. p. 1720–7.
162. Shim DH, Kim HJ, Sastry S. Control system design for rotorcraft-based unmanned aerial vehicle using time-domain system identification. In: Proc IEEE conf contr appl, Anchorage, AK; 2000. p. 808–13.

163. Shim DH, Kim HJ, Sastry S. Decentralized nonlinear model predictive control of multiple flying robots. In: Proc 42nd IEEE conf dec contr, Maui, HI; 2003. p. 3621–6.
164. Shin EH, El-Sheimy N. Optimizing smoothing computation for near real-time GPS measurement gap filling in INS/GPS systems. In: Proc US inst nav GPS, Portland, OR; 2002.
165. Shih YT. The reversibility of six geometric color spaces. Photogramm Eng Remote Sens. 1995;61:1223–32.
166. Sikorsky Cypher UAVs. http://en.wikipedia.org/wiki/Sikorsky_Cypher. Cited Aug 2010.
167. Skeldar V-150 VTOL UAV. http://www.saabgroup.com/en/Air/Airborne-Solutions/. Cited Aug 2010.
168. Sky Surveyor UAV helicopters. http://me2.tm.chiba-u.jp/uav/main/. Cited Aug 2010.
169. Smith AR. Color gamut transform pairs. In: Proc 5th annual conf computer graphics interactive techniques, New York; 1978. p. 12–9.
170. SR Series VTOL UAV helicopters. http://www.rotomotion.com/. Cited Aug 2010.
171. Stevens BL, Lewis FL. Aircraft control and simulation. 2nd ed. Hoboken: Wiley; 2003.
172. SolidWorks 3D CAD design software. http://www.solidworks.com/sw/products/cad-software-3d-design.htm. Cited Aug 2010.
173. Sugeno M, Hirano I, Nakamura S, Kotsu S. Development of an intelligent unmanned helicopter. In: Proc 1995 IEEE int conf fuzzy syst, Yokohama, Japan; 1995. p. 33–4.
174. Teo R, Jang SJ, Tomlin C. Automated multiple UAV flight: the Stanford DragonFly UAV program. In: Proc 43rd IEEE conf dec contr, Atlantis, Bahamas; 2004. p. 4268–73.
175. Tischler MB, Colbourne JD, Morel MR, Biezad DJ. A multidisciplinary flight control development environment and its application to a helicopter. IEEE Control Syst Mag. 1999;19:22–33.
176. Tischler MB, Colbourne JD, Morel MR, et al. CONDUIT—a new multidisciplinary integration environment for flight control development. Presented at AIAA guid, nav, contr conf, New Orleans, LA; 1997. AIAA-1997-3773.
177. Tischler MB, Remple RK. Aircraft and rotorcraft system identification: engineering methods with flight test examples. Reston: AIAA Press; 2006.
178. Tomlin C, Pappas GJ, Sastry S. Conflict resolution for air traffic management: a case study in multi-agent hybrid systems. IEEE Trans Autom Control. 1998; 43:509–21
179. TREX 450 helicopter. http://www.align.com.tw/alignhtml/EN/index.html. Cited Aug 2010.
180. Tsach S, Penn D, Levy A. Advanced technologies and approaches for next generation UAVs. In: Proc int council aeronautical sciences congress; 2002. p. 131.1–131.10.
181. Turbulence D3 helicopter. http://model.hirobo.co.jp/english/index.html. Cited Aug 2010.
182. UAV sWARM health management project. http://vertol.mit.edu/. Cited Aug 2010.
183. Understanding flybar mixing. http://rchelimag.com/pages/howto.php?howto=22. Cited Aug 2010.
184. UPK-2500 ultrasonic sensor. http://www.sntag.ch/englisch/distance_e.htm. Cited Aug 2010.
185. USC autonomous flying vehicle project. http://robotics.usc.edu/avatar/. Cited Aug 2010.
186. Valavanis KP, editor. Advances in unmanned aerial vehicles. New York: Springer; 2007.
187. van der Merwe R, Wan EA. Sigma-point Kalman filters for nonlinear estimation and sensor-fusion: applications to integrated navigation. In: Proc AIAA guid, nav, contr conf, Providence, RI; 2004.
188. Valavanis KP, Gracanin D, Matijasevic M, Kolluru R, Demetriou GA. Control architectures for autonomous underwater vehicles. IEEE Control Syst Mag. 1997;17:48–64.
189. Veeraraghavan H, Schrater P, Papanikolopoulos N. Robust target detection and tracking through integration of motion, color and geometry. Comput Vis Image Underst. 2006;103:121–38.
190. Vicacopter. http://www.vicacopter.com/. Cited Aug 2010.
191. Vine I. Risk of visual detection and pursuit by a predator and the selective advantage of flocking behaviour. J Theor Biol. 1971;30:405–22.
192. Visual C++ developer center. http://msdn.microsoft.com/en-us/visualc/default.aspx. Cited Aug 2010.
193. VxWorks RTOS. http://www.windriver.com/products/vxworks/. Cited Aug 2010.

194. Wan EA, Bogdanov AA. Model predictive neural control with applications to a 6 DoF helicopter model. In: Proc 2001 American contr conf, Arlington, VA; 2001. p. 488–93.
195. Wang B, Chen BM, Lee TH. An RPT approach to time-critical path following of an unmanned helicopter. To be presented at 8th Asian contr conf, Kaohsiung, Taiwan, 2011.
196. Wang T. Development of a micro unmanned vertical take-off and landing rotorcraft [thesis]. National University of Singapore; 2009.
197. WASP III UAV systems. http://www.aerovironment.com/uas_product. Cited Aug 2010.
198. Weilenmann MW, Christen U, Geering HP. Robust helicopter position control at hover. In: Proc American contr conf, Baltimore, MD; 1994; p. 2491–5.
199. Weilenmann MW, Geering HP. Test bench for rotorcraft hover control. J Guid Control Dyn. 1994;17:729–36.
200. Wikipedia. Compass. http://en.wikipedia.org/wiki/Compass. Cited Aug 2010.
201. Wikipedia. Color space. http://en.wikipedia.org/wiki/Color_space. Cited Aug 2010.
202. Wikipedia. Earth-centered, earth-fixed system. http://en.wikipedia.org/wiki/ECEF. Cited Aug 2010.
203. Wikipedia. Extended Kalman filter. http://en.wikipedia.org/wiki/Extended_Kalman_filter. Cited Aug 2010.
204. Wikipedia. Global positioning system. http://en.wikipedia.org/wiki/GPS. Cited Aug 2010.
205. Wikipedia. Inertial measurement unit. http://en.wikipedia.org/wiki/Inertial_measurement_unit. Cited Aug 2010.
206. Wikipedia. International aerial robotics competition. http://en.wikipedia.org/wiki/International_Aerial_Robotics_Competition. Cited Aug 2010.
207. Wikipedia. Miniature UAV. http://en.wikipedia.org/wiki/Miniature_UAVs. Cited Aug 2010.
208. Wikipedia. PC/104 standard. http://en.wikipedia.org/wiki/PC104. Cited Aug 2010.
209. Wikipedia. Proper acceleration. http://en.wikipedia.org/wiki/Proper_acceleration. Cited Aug 2010.
210. Wikipedia. Quaternion. http://en.wikipedia.org/wiki/Quaternion. Cited Aug 2010.
211. Wikipedia. Real-time operating system. http://en.wikipedia.org/wiki/Real-time_operating_system. Cited Aug 2010.
212. Wikipedia. World geodetic system. http://en.wikipedia.org/wiki/WGS_84. Cited Aug 2010.
213. Wills L, Kannan S, Sander S, et al. An open platform for reconfigurable control. IEEE Control Syst Mag. 2001;21:49–64.
214. WITAS UAV project. http://www.ida.liu.se/~patdo/auttek/introduction/. Cited Aug 2010.
215. Wood RJ. The first takeoff of a biologically-inspired at-scale robotic insect. IEEE Trans Robot. 2008;24:341–7.
216. Wren CR, Azarbayejani A, Darrell T, Pentland AP. Pfinder: real-time tracking of the human body. IEEE Trans Pattern Anal Mach Intell. 1997;19:780–5.
217. Yamaha Rmax unmanned helicopter. http://www.yamaha-motor.co.jp/global/. Cited Aug 2010.
218. Yamaha-R50-based UAV helicopters. http://www.cs.cmu.edu/afs/cs/project/chopper/www/. Cited Aug 2010.
219. Yun B, Chen BM, Lum KY, Lee TH. Design and implementation of a leader-follower formation flight control system for unmanned helicopters. J Control Theory Appl. 2010;8:61–8.
220. Yun B, Cai G, Chen BM, Peng K, Lum KY. GPS signal enhancement and attitude determination for a mini and low-cost unmanned aerial vehicle. Trans Inst Measurement Contr. 2011. doi:10.1177/0142331209342220.
221. ZALA 421-02 unmanned helicopter. http://zala.aero/en/uav/1205400770.htm. Cited Aug 2010.
222. Zames G. Feedback and optimal sensitivity: model reference transformations, multiplicative seminorms, and approximate inverses. IEEE Trans Autom Control. 1981;26:301–20.
223. Zhou K, Doyle J, Glover K. Robust and optimal control. Englewood Cliffs: Prentice Hall; 1996.
224. Zhu Q, Avidan S, Cheng KT. Flexible camera calibration by viewing a plane from unknown orientations. In: Proc 7th IEEE int conf computer vision, Kerkyra, Greece; 1999. p. 666–73.
225. Zhou QM, Aggarwalb JK. Object tracking in an outdoor environment using fusion of features and cameras. Image Vis Comput. 2006;24:1244–55.

Index

G. Cai et al., *Unmanned Rotorcraft Systems*, Advances in Industrial Control,
DOI 10.1007/978-0-85729-635-1, © Springer-Verlag London Limited 2011

W

Y

Other titles published in this series (continued):

Soft Sensors for Monitoring and Control of Industrial Processes
Luigi Fortuna, Salvatore Graziani, Alessandro Rizzo and Maria G. Xibilia

Adaptive Voltage Control in Power Systems
Giuseppe Fusco and Mario Russo

Advanced Control of Industrial Processes
Piotr Tatjewski

Process Control Performance Assessment
Andrzej W. Ordys, Damien Uduehi and Michael A. Johnson (Eds.)

Modelling and Analysis of Hybrid Supervisory Systems
Emilia Villani, Paulo E. Miyagi and Robert Valette

Process Control
Jie Bao and Peter L. Lee

Distributed Embedded Control Systems
Matjaž Colnarič, Domen Verber and Wolfgang A. Halang

Precision Motion Control (2nd Ed.)
Tan Kok Kiong, Lee Tong Heng and Huang Sunan

Optimal Control of Wind Energy Systems
Julian Munteanu, Antoneta luliana Bratcu, Nicolaos-Antonio Cutululis and Emil Ceangă

Identification of Continuous-time Models from Sampled Data
Hugues Gamier and Liuping Wang (Eds.)

Model-based Process Supervision
Arun K. Samantaray and Belkacem Bouamama

Diagnosis of Process Nonlinearities and Valve Stiction
M.A.A. Shoukat Choudhury, Sirish L. Shah and Nina F. Thornhill

Magnetic Control of Tokamak Plasmas
Marco Ariola and Alfredo Pironti

Real-time Iterative Learning Control
Jian-Xin Xu, Sanjib K. Panda and Tong H. Lee

Deadlock Resolution in Automated Manufacturing Systems
ZhiWu Li and MengChu Zhou

Model Predictive Control Design and Implementation Using MATLAB®
Liuping Wang

Predictive Functional Control
Jacques Richalet and Donal O'Donovan

Fault-tolerant Flight Control and Guidance Systems
Guillaume Ducard

Fault-tolerant Control Systems
Hassan Noura, Didier Theilliol, Jean-Christophe Ponsart and Abbas Chamseddine

Detection and Diagnosis of Stiction in Control Loops
Mohieddine Jelali and Biao Huans (Eds.)

Stochastic Distribution Control System Design
Lei Guo and Hong Wang

Dry Clutch Control for Automotive Applications
Pietro J. Dolcini, Carlos Canudas-de-Wit and Hubert Béchart

Advanced Control and Supervision of Mineral Processing Plants
Daniel Sbárbaro and René del Villar (Eds.)

Active Braking Control Design for Road Vehicles
Sergio M. Savaresi and Mara Tanelli

Active Control of Flexible Structures
Alberto Cavallo, Giuseppe de Maria, Ciro Natale and Salvatore Pirozzi

Induction Motor Control Design
Riccardo Marino, Patrizio Tomei and Cristiano M. Verrelli

Fractional-order Systems and Controls
Concepcion A. Monje, YangQuan Chen, Blas M. Vinagre, Dingyu Xue and Vincente Feliu

Model Predictive Control of Wastewater Systems
Carlos Ocampo-Martinez

Wastewater Systems
Carlos Ocampo-Martinez

Tandem Cold Metal Rolling Mill Control
John Pitter and Marwan A. Simaan